Induction, Probability, and Skepticism

Sri Garib Dass Oriental Series No. 156

Induction, Probability, and Skepticism

D. P. Chattopadhyaya

SRI SATGURU PUBLICATIONS
A DIVISION OF
INDIAN BOOKS CENTRE
DELHI-INDIA

Published by :
Sri Satguru Publications
Indological and Oriental Publishers
A Division of
Indian Books Centre
40/5, Shakti Nagar,
Delhi-110007
(INDIA)

First Indian Edition : Delhi, 1992

ISBN 81-7030-344-3

Printed at :
D.K. Fine Art Press, Delhi.

90-42945
CIP

PRINTED IN INDIA

To

Sankari Prasad Banerjee
Jitendra Nath Mohanty
and
Pranab Kumar Sen

for all their acts of kindness and friendship.

To

Sankari Prasad Banerjee
Jitendra Nath Mohanty
and
Pranab Kumar Sen

for all their acts of kindness and friendship

Contents

Preface

This study perhaps basically owes its origin to the influence of *history* on me and the *uncertainty* of theoretical life. To my mind, all forms of knowledge are basically historical; even formal sciences like mathematics are no exception. Equally strong is my conviction that uncertainty is the hallmark of the theoretical life. Whatever may be one's theoretical or philosophical commitment—essentialism, realism or absolutism—one can never be completely free from the native *human* sense of uncertainty. Over the years we all change, rather are obliged to change, our views about the various things of the world and their importance and relevance to our own lives. This comes about so silently and yet so compellingly that, on reflection, it can hardly be denied.

The sort of skepticism that underlies this study, though apparently akin to probabilism, historism, and fallibilism, in fact is differentiable from each one of them. It is the outcome of what I call *anthropological* rationalism as distinguished from the *sovereign* rationalism of thinkers like Śaṃkara and Leibniz and the *autonomous* rationalism of Kant. In a way this study may be said to be a part of larger view of the lived world, philosophical anthropology. Left to itself, skepticism, as I understand it, is neither destructive nor constructive. In its best available form it seems to be *self-critical*. And this *self*-criticism knows no finality. Yet, *practically* speaking, in spite of our being fallible, we have every right to be certain about many things of the world. What is interesting to note is that this right need not, perhaps cannot, be *fully* justified in terms of theory.

Preface

This study perhaps basically owes its origin to the influence of reality on me and the uncertainty of the theoretical life. To my mind, all forms of knowledge are basically hypothetical, even formal sciences like mathematics are no exception. Equally strong is my conviction that uncertainty is the hallmark of the theoretical life. Whatever may be one's theoretical or philosophical commitment—essentialism, scepticism or absolutism—one can never be completely free from the native human sense of uncertainty. Over the years we all change, rather are obliged to change, our views about the various things of the world and their importance and relevance to our own lives. This comes about so silently and yet so compellingly that on reflection it can hardly be denied.

The sort of scepticism that underlies this study, though apparently allied to probabilism, historism and fallibilism, is in fact distinguishable from each one of them. It is the outcome of what I call *anthropological rationalism* as distinguished from the so-called rationalism of thinkers like Samkara and Leibniz and the epistemic rationalism of Kant. In a way this study may be said to be a part of a larger view of the lived world, philosophical anthropology. Even in itself, scepticism, as I understand it, is neither destructive nor constructive. In its best available form it seems to be *self-critical*. And this self-criticism knows no finality, yet, intuitively speaking, in spite of our being fallible, we have every right to be certain about many things of the world. What is interesting to note is that this right need not, perhaps cannot, be *fully* justified in terms of theory.

Acknowledgments

Every author, in introspection, is deeply aware of his debt to so many persons. This indebtedness is of various kinds: direct and indirect, personal and institutional, conscious and even unconscious. Needless to say, my first debt is to the authors, classical and contemporary, referred to in the body of the book. But to the writings of some of them like Popper and Quine, my debt, as one could easily see, is indeed very deep. If to Popper it is more personal, to Quine it is perhaps more fundamental. I say *perhaps* because I strongly suspect that both of them will substantially disown my view. Like many others of my generation, I owe a lot to them.

It is with great pleasure that I confess my gratitude to John Watkins, J. N. Mohanty and Kali Krishna Banerjee. These three teachers and friends, with their characteristic kindness and warmth, have always stimulated my thinking and helped me in various ways over the years. It is a pity that Banerjee, who did so much for me and took so deep interest in my unconventional ways of philosophizing, is no longer alive to see this work in print. I am also grateful to my colleagues at Jadavpur, particularly Professors P. K. Sen, P. K. Mukhopadhyaya, H. Banerjee, S. R. Saha, and T. K. Sarkar. My gratitude to Dr. Krishna Roy, Dr. Chhanda Gupta, and Dr. Minakshi Roychoudhury are more personal than academic in nature. Over the years I have tested the ideas and arguments of this book on their heads. I appreciate their incredible patience and critical cooperation. I must thank Krishna and Chhanda also for carefully checking and correcting the notes and references of this book and also for preparing the index.

Gratitude of a different sort I owe to three institutions: the Indian Council of Philosophical Research, the University Grants Commission, and Jadavpur University. But for the material and moral help so generously extended by these institutions and their decision makers, I could not possibly have produced this book at all.

It is with pleasure that I would like to acknowledge here also my gratitude to Chandi Ghosh, Asit Datta, Renu Bala, Ajay Kumar, and D. P. Bhattacharya, for providing me excellent secretarial assistance.

Shefali, my wife, who has always supported me morally for more than three decades, hates to be thanked in print.

Introduction

I

The questions with which I deal in this book are indicated in its very title. First, I will try to show that induction, though totally denied at times, is somehow followed by us in our life, both theoretical and practical. Denial of induction is due to a misconception about it. Therefore, I propose to discuss different conceptions of induction. In the process I suggest that those who maintain that the only available scientific method is hypothetico-deductive fail to see the inductive background of the formation and use of hypothesis. In the contemporary period this methodological trend is strongly associated with the name of Popper and his followers, like Lakatos and Watkins. But I draw the reader's attention to the historical fact that several other thinkers, like Kant, Whewell, and Jevons, were quite aware of the plausibility of the approach highlighted by Popper.

Second, once the untenability of the distinction between induction and deduction is clearly understood, the whole question of justifying induction assumes a complex and interesting character. Its justification may be constructed both deductively and inductively. Such concepts as acceptance, extrapolation, and learning have their unmistakable inductive undertone. Even through deductive testing, repetition of tests, we add to our knowledge, strengthen the base of our accepted beliefs and learn more about what we have already learned. Our ways of learning are diverse and should not be brought under the exclusive heading of either induction or deduction.

It is perhaps instructive to recall that in Indian logic, for instance, we do not recognize two separate branches of logic, inductive and deductive. In fact in the five-membered structure of inference (*pancāvayavī Nyāya*) induction and deduction are used in combination. Let us look into the well-known structure of an inference:

1. This mountain has fire in it (*pratijñā* or hypothesis).

2. Because it is smoky (*hetu* or reason).

3. Whatever is smoky is fiery like a kitchen, unlike a lake (*udāharaṇa* or instantiation—both positive and negative—of a principle).

4. This mountain, because it is smoky, is on fire (*upanaya* or application).

5. This mountain is on fire (*nigamana* or conclusion).

Also it is of interest to note that in the Indian tradition, logic, epistemology, and ontology have not been sharply separated and discussed. On the contrary, thinkers in this tradition are of the view that their interrelated discussion adds to the clarity of our understanding of the concerned concepts. However, this is not to deny the possibility of formalization of Indian logic and even a part of its epistemology. In fact some interesting work has been done in this area in the recent past.

Third, probability, like induction, also is not to be taken in an omnibus way. There are different concepts of probability and different interpretations of each of these concepts. The original inspiration underlying the construction of probability theory was in many cases purely practical. Calculation of chances is a very pressing requirement of practical decision making. From betting to investment and planning, more or less in every sphere of life, we are always called upon to estimate the probable outcomes of our decisions and actions. What adds to the complexity of the situation is the persistent fact that in most cases we have to make a decision and perform actions under uncertain circumstances. Therefore, even the elements and extent of the concerned uncertainty demand some assessment, intuitive or, preferably, quantitative.

Finally, though we are destined to live amidst different sorts of uncertainty, practical as well as theoretical, we always try to minimize it. Uncertainty-minimizing and certainty-maximizing strategies are practically very pressing. For, unless we know our place in relation to our environment, both immediately and prospectively, it proves very difficult for us to plan our lives. Besides, without this understanding we cannot have that minimum *adjustment* with our physical situation, cultural environment, and fellow human beings necessary even for our biological survival. Reflection makes it clear that our interest in *truth value* (in theoretical discourses) and our concern with *survival value* (in practical affairs of life) are very closely interrelated.

These basic considerations run through the pages of this study.

II

Coming to the specifics of this book, perhaps it will not be out of place to recall my interest in the subjects delineated here. My interest in induction and probability dates back to my days of graduate studies at LSE with Popper, Watkins, and Lakatos. The most important book for us was, of course, *The Logic of Scientific Discovery* (1959), the English translation of *Logik der Forschung* (1934). The most fiercely attacked philosopher was Carnap for his continued defence of such "indefensible" views as inductivism and probabilism. Publication of *Conjectures and Refutations* (1963) kindled the spirit of attack against the psychological definitions of knowledge offered by Chisholm and Quine, on the one hand, and linguistic philosophers like Ryle and Austin, on the other. We used to hear and speak always in praise of "objective knowledge," "objective probability," and "falsificationism." But I had always an uneasy feeling within me: my critical rationalist colleagues were hypercritical of others and rather uncritical of their own positions. When external criticism is strongly discouraged, accumulated and unanswered questions make internal dissonance inevitable.

At the time I was working in a particular area of philosophy of history; namely, structure of historical explanation. Vico's basic principle that one cannot create or know anything that can be more certain than one is about oneself, appeared to me very insightful. The ruling paradigm of scientific knowledge at Vico's time (1675–1744) was Newtonian mechanics. Most of the working philosophers, barring one or two like Hume, accepted the Lockean underlaborer concept of philosophy. Kant's defense of Newton was yet to appear. In that historical situation, marked by unbounded confidence in scientific knowledge, Vico's sober hermeneutic principle, though extremely insightful, went almost unheeded.

Later on Hegel and Croce discovered the importance of Vico's hermeneutics and made critical use of it in the philosophical understanding of history and human knowledge. Although Kant's architectonic of knowledge is basically anti-Humean, antiskeptical and purported to justify the Newtonian paradigm of knowledge, Hegel tries to show that the process of justification goes well beyond natural sciences. Besides, the process of justification is gradual, dialectically sublative and leaves behind different forms of knowledge that, compared to metaphysics, are relatively uncertain and insecure. The Hegelian search for foundations of knowledge is historical and takes due note of the forms of knowledge like common sense and science that lack coherence, perfection, and the organic character of metaphysical knowledge.

III

At about the same time I was introduced to Husserl's *Cartesian Meditations* by Findlay. Findlay's seminar at King's College was scholarly, stimulating and thought provoking. While at LSE Popper was trying to show the futility of the logical positivists' attempts to lay the foundation of scientific knowledge in terms of direct reports of sense-experience, Findlay was painstakingly explaining the phenomenologist's way of defending foundationalism against psychological logic, naturalism, and skepticism. Apodictic certainty of rigorous science was the aim of the latter and truth seeking, growth orientation, and refutable knowledge were the main interests of the former.

These two approaches are not as antithetical as they may appear to start with. Both are antiskeptical in their inspiration and opposed to the enclosed image or finalistic character of knowledge. Husserl's concept of "horizonal expansion" substantially takes care of the Popperian ideal of epistemic growth. Both these views owe their origins to Descartes and Kant but have developed well beyond them. In brief I have presented this theory in one of my earlier works, *Individuals and Worlds: Essays in Anthropological Rationalism* (1976), and more recently in *Knowledge, Freedom and Language* (1989).

The fallibility of scientific theory can never be completely ruled out. Openness to evidence, pro and con, means openness to learn from experience. Fallibilism is to be distinguished from skepticism. It is neither negative in its inspiration nor universal in its scope. Like criticism, it is problem oriented, issue specific, and argument bound. Rationality of theory does not lie in the inductively available and accumulated evidence. The inductivists themselves admit that eliminative evidence is no less important than the enumerative evidence and that the method of difference is of stronger probative value than the method of agreement. In a roundabout way even Popper admits the significance of *additive* testing evidence. Scientific hypothesis, though required to be falsifiable in principle, must not in fact be easily falsified. It has to survive or withstand the falsifying evidential tests. Otherwise it would not be chosen as a good hypothesis at all. Besides, it would not be there to be tested any more. Survival is not mere survival. Through survival, theory *adds* to its content. When evidence marshalled to test theory fails to overthrow it, it turns out to be theory's ally, augmenting its empirical content. In his own admission, one finds a gap in Popper's main anti-inductive argument.

In this book one of my aims is to show that the Popperian approach to induction and the Carnapian one are not as opposed as they are at times claimed to be. In fact Carnap and Popper addressed themselves to different types of problems. Whereas Carnap is basically interested in developing formal Bayesian models of inductive logic, containing both the probability calculus and the rule of induction by simple enumeration, Popper's main interest is to use the propensity interpretation of the frequency theory of probability to explicate the concepts of empirical content of theory and parameters of theory choice. In a more comprehensive theory of probability developed by Hintikka and his followers like Niiniluoto they have tried to show that Carnap's practice-oriented concept of probability and Popper's theory-oriented one may be suitably reconciled. I have briefly referred to this theory together with its underlying liberalized ontology.

To the reader of the book it will be clear that my primary object is not to look into the *logic* of induction, still less to formalize it. I consider induction and probability in the larger context of theory of knowledge and its related ontology. The central concept of my epistemology is *finite and fallible human nature.* By implication I reject two other available and extensively used concepts of human nature; namely, (1) the knowing mind in its best possible, that is, transcendental, form is universal, absolute, and neutral among different systems of knowledge and (2) the knowing mind in its near-perfect form is Godlike and can construct a system of knowledge that has in it the regulative principles necessary for determining and removing from it all elements of skepticism. I am unable to accept the *sovereign rationalism* of Leibniz or that form of reason which refuses to recognize any trace of unreason, epistemic or ontic, in the world, nor am I reconciled to the Kantian view of *autonomous reason* or uncritical reason, according to which nothing in the empirical world can possibly assert itself as a definitive counterexample against what is constituted by our theoretical reason. My concept of *human reason* is essentially *human,* marked by its improvable finitude and irremovable fallibility. Anthropological rationalism, as I defend it, is committed to a sort of realism that, on the one hand. is substantially consistent both with the commonsense truisms and the scientific professional views. On the other hand, the human nature assumed in this theory is always open to learn from its experience of the world and of its own self. It is realism without essentialism; it is anthropological without being anthropomorphic.

It is not committed to metaphysical or external realism. Nor is it wedded to what is called *internal realism.* Positively speaking, I feel that the knowing mind is determined by its cultural milieu and historical situation. In a sense it is bound to be relativistic. But this relativism does not enclose its ability in a fixed or final manner, condemning it to a sort of solipsism. On the contrary, in spite of its situational determinants, it is more or less free, free to grasp the world given to it in different ways. From one and the same world it can carve out different worlds, commonsensical, scientific, and artistic. The sharp distinction often drawn between fact and fiction, finding and fabricating, seems to me arbitrary and untenable. This does not entail rejection of realism but only dilution of its pragmatic overtone.

The theory of knowledge partly defended and partly assumed in this book is intimately related to some particular concepts of induction and probability. We are inductive by nature. If we have clean canvaslike minds within us and an empty world around us, we cannot learn anything. To learn we must have minimal learning equipment within us and learnable materials around us. Neither our equipment nor our material, separately or jointly, can ensure certainty of what we learn or know. In a sense our inductive knowledge is destined to be uncertain. But exercising our abilities and, given a bit of luck, we can, and in fact do, enlarge the world of our knowledge and eliminate errors from it.

Our information-additive and error-eliminative capacities work simultaneously. We cannot indefinitely and uniformally go on enlarging the scope of our knowledge. Our openness to the world, other people, and their influences are sure to foster in us some wrong views and erroneous understanding. To put it from the other end, I am not the sole author of my knowledge. Knowledge is implicitly a community enterprise and community achievement with all their attending strength and weakness.

This noncumulative view of our knowledge, particularly of the history of knowledge, has some instructive implications. For example, history of science, rightly understood, is not linear. It is not necessarily progressive, but it has different sorts of shifts in it, progressive and regressive. Sometimes kings and dictators, sometimes scholars and learned societies impede the growth of knowledge instead of promoting it. History of philosophy and science abounds with examples of this kind. Therefore, an additive inductive history of science is neither possible nor desirable. Even it is artificially constructed, it remains dangerously vulnerable.

One of the fall outs of this critique of induction in epistemology is sociology of knowledge. Once we clearly realize the inadequacy of the additive view of induction and recognize the importance of eliminative induction, we begin to understand the growing importance of sociology of knowledge not so much as an example of relativism but as a justification of bringing in the social dimension or objectivity in the area of human knowledge.

For the relativistic view of knowledge Kuhn and Feyerabend are not alone to be held responsible. Wittgenstein and even Popper are also to be brought into the picture. The language-theoretic game cannot be played transculturally and universally, for the rules of the game by their very nature are culture specific or language specific. Similarly, it may be pointed out, human fallibilism and universally valid human knowledge cannot go together. No transcendental argument can lift any system of human knowledge from historical bounds or social limits. That is perhaps the minimal lesson one can derive from history.

IV

Let me be a little more specific about some of the general points that have been mentioned. Why do we accept certain beliefs and reject others? Availability of evidence, for and against, is perhaps the most simple answer. Evidence may be for acceptance. It may be against acceptance. Ordinarily, one single piece of evidence does not clinch the issue. But in exceptional cases a particular piece of evidence or experiment may assume crucial importance. In the normal course of acceptance or rejection of a belief, we depend on a number of pieces of evidence favorable as well as unfavorable. Both the quantity and quality of evidence prove relevant to the determination of actual acceptance or rejection of beliefs, theories, conjectures, and so forth.

In our day-to-day lives we are guided by so many beliefs and theories that we cannot possibly be aware of their number or individual truth values. I accept certain beliefs simply because most of my fellow social beings accept them. Many of them are rejected by me because experts are opposed to those beliefs. We readily fall in line. We do not raise the question whether their acceptance is ill-grounded or well-grounded. Often we subscribe to certain theories simply because they have been authored by eminent people in the field and on the basis of their institutionally recognized professional authority. In other words, the acceptance of many of our beliefs and theories does not depend on their own ascertained merits. In many cases we are not even competent to go into the question.

Ordinary beliefs, like "water flows downstream," "fire burns," "snow is white," "ravens are black," are readily accepted because of their long and uncontradicted inductive history. But at times we come across some such questions as, "Will the next raven I will see be black?" "Can there be no fire without burning capacity?" Against the well-known background of the concerned objects the likely answers to the questions can be easily anticipated. Unless I am in Australia, the next raven to be seen by me is not likely to be white. Unless I have on my body some protective chemical or device, it is going to be burned if it is set on fire. But that general truths of commonsense beliefs may be contradicted by experience cannot be ruled out a priori.

Knowledge has been defined as justified true belief—not that this psychological definition of knowledge in terms of belief has been universally accepted. But let us provisionally accept it and look into it. First, belief as belief is not good enough to be accepted as knowledge. It has to be true. What is truth is not an undisputed issue. Different views about the nature and test of truth are available. Coherentism, correspondence, and pragmatism are notable among them. Further, whatever theory of truth one accepts has to be justified. Arguments and examples in its favor are to be offered. Criticisms against it has to be met. Once the preferred view of truth is established, we have to be sure that its logical and epistemological requirements are satisfied by the belief that claims the status of knowledge. In brief, the truth claim of belief needs to be evidentially justified.

Can a factual belief be justified conclusively in terms of evidence? It is said that unless necessary and sufficient justifications for a belief are available, it cannot be regarded as knowledge in the strict sense.

Apart from this definitional problem, the substantive question is whether any empirical proposition can be absolutely certain. A proposition of universal scope can never be conclusively established for purely logical reason. Take the case of universally quantified proposition, "All bodies gravitate to the center of the universe." Enumeration of no finite number of observed cases of gravitating bodies can ensure the absolute certainty of the proposition. By experience we cannot hope to cover the entire scope of the proposition. Therefore, this sort of higher-level theoretical propositions remains always open to question and correction. The problem regarding the determination of truth value of singular existential proposition is also very disturbing. Unless it could be shown as subsumable under a group or circumscribed within a surveyable space-time region, its truth claim as a singular proposition can hardly be tested. When the singular predictions that we make on the basis of one or another inductively accepted general truth, we can never be sure about them in principle. It is true that in practice we speak of some such things as "the Second Hooghly Bridge will not collapse." There is no known law in the world preventing its possible collapse. But we have pretty good practical reasons to believe that it shall not collapse. Because one points out that all the necessary requirements of structural engineering have been satisfied by the concerned construction company. Even then one can feel skeptical about it. With almost equal plausibility another can feel certain about it.

V

Skepticism, like *dogmatism,* is a whipping boy. It is criticized before it is heard and understood. If it were so vulnerable to everybody's attack, no good thinker would have wasted time writing on the problem. Instructively enough, skepticism remains repeatedly "refuted" and "rejected," remains to be taken seriously and considered on merit. The instruction is unmistakable; viz., it symbolizes the questioning human spirit, critical awareness of the limits both of his body and mind, and our quest for more precise and more comprehensive knowledge.

Some philosophers and scientists maintain that chances cannot be eliminated either from human nature or physical nature. Imponderables and uncertainty are said to be *objective* in a sense.

The dubitable character of knowledge is at times ascribed to the real states of affairs of the world that includes the knowing minds within them. Negatively speaking, for doubt the knowing mind, or its alleged deformity, alone is not to be held responsible, dubitable knowledge may well be the outcome of sound mind and due to no deliberate fault of its own. In that case doubt turns out to be a "caused" epistemic phenomenon. Unless the mind is assumed to be more or less passive the causal status of doubt in it can hardly be accepted as logically tenable. Doubt then is neither entirely self-caused nor entirely caused by external states of affairs. This apparently very difficult situation has been tackled, though not solved, in different ways; viz., by assigning a priori probability, equal or unequal, to the concerned causal states and their clusters, by assuming that the knowing mind in its natural state is doubt free and that its vulnerability to skepticism is rooted in its will, biases, or prejudices: *saṁskāras*.

Dubitability of knowledge has at times been ascribed to insufficiency of evidence. Knowledge by its very nature is taken to be true, but not necessarily so. By *truth* some people mean "truth claim," some others mean "truthlike." Still other meanings of it—"useful," "successful," and so on—are also available. But in all these cases, barring perhaps the case of self-evident truth, what is claimed to be true is in need of evidence. In other words, *self-evident* knowledge alone is self-sufficient. In all other cases the truth or probability value of knowledge is dependent upon the quality and quantity of evidence.

If the truth claim of any proposition is found to be dependent on some evidence, problems of serious nature come up and deserve close consideration. First, if one admits that a proposition is in need of evidence to establish its truth claim, two other points are conceded by implication; viz., (1) the proposition in question suffers from some infirmity in respect to its truth claim, that is, it is yet to be confirmed, established, or proved, and (2) the propositions that as evidence may confirm, establish, and so on, the truth-claiming proposition must itself be deemed true or, at least, firmer than the latter. Second, serious doubt has been expressed about the *absolute* truth of every kind of empirical proposition, be it evidential or truth claiming. In the context of the arguments purported to do away with the customary distinction drawn between analytic and synthetic propositions the said doubt assumes added gravity. The absolute fabric of knowledge turns out to be subject to the laws of empirical or historical wear and tear. This is not to deny that the relatively firmer evidential propositions can, and do, lend support to or probabilify the truth-claiming proposition. In any case skepticism of some form or other seems inescapable.

In between the preceding two points the discerning eye notices another point. If no proposition is absolutely true or self-evident, addition of evidence, under either similar or even diverse conditions, cannot ensure truth of the truth-claiming proposition. In that case the distinction between "truth," "truthlikeness," "probability," and the like gets blurred. This means foundationalism fails or, one might say, fallibilism stands vindicated. One need not, at this stage, enter into the finer theoretical *distinction* between fallibilism and skepticism.

One finds that to refute skepticism two main forms of foundationalism, local or global (that is, holistic), are usually offered. Some empiricists maintain that basic statements or reports of direct experience can *not* be false and these form the firm foundation of scientific knowledge. But this is hardly convincing. For the higher-level scientific theoretical propositions cannot be deductively obtained from the empirical propositions of the *local* type. If these propositions speak or inform us of only what is there or happens in narrow, limited localities, in the small and separate regions of reality, how can we construct or infer from them a reliable global picture of reality as a whole, whatever might be its nature—mental or material. Most empiricists will readily use what are called *bridging* or *coordinative principles* or *rules of construction,* but they do not enter into, or remain elusive, about the question of their *truth value*. Only on the question of their *use value* are they eloquent; but that also in a roundabout way. Of the proclaimed use value of the rules of construction-inference, whether it is from particular to particular or from particulars to general-universal, one can never be certain, not even after the expected or predicted outcome is obtained, unless, of course, one takes a very restricted pragmatic or practical view of the whole issue. For what is inferable *here* today from some particular information may not be so *there* tomorrow. Space and time do make a difference both to our information and to what is inferable from it. To remove or minimize the possible skeptical miseffects it is "metaphysically" assumed that, amidst and in spite of all spatio-temporal difference of particular things of it, Nature has an underlying uniformity or limited independent variety of kinds. Metaphysical commitments, though thinned down, cannot be dispensed with totally even by the empiricist. Without minimal "transcendental" commitments neither inductive inference nor theory construction, especially construction of theories of wider scope and of probable character, is possible.

The inseparable relation between our experience and what is there, the content or ontological commitment of experience, opens up another horizon of knowledge wherein, intriguingly enough, both the "danger" and the possible "cure" of skepticism are found to converge. Whereas in the phenomenalist-empiricist idiom we are told of mind's *customary* tendency to move beyond particulars to the general, the phenomenologist speaks of *intentionality* or the object-ward tendency of every sort of experiential phenomena. From phenomenalism to phenomenology is neither a long nor a tortuous journey. Why, customarily speaking, do some particular experiences lead our *minds* to some other (more or less) definite and not *any* other indefinite particular or general truth? Why do psychological laws behave in this way? How do we account for the fact that *we* have in us this set of laws? Why are we all subject, though not uniformly, to this set of laws? The structure of our minds is such, the phenomenologist asks us to believe, that not only its ways of working but also the results thereof are bound to be intersubjectively sharable; that is, objective in character. The phenomenalist's "laws of association" and the phenomenologist's "modes of constitution" are theoretic-methodologic analogues and called upon to perform comparable double duties; viz., (1) how different experiences are brought together to *objectify* them so that they are identically or similarly presented to all persons, and (2) how to remove doubt about objects of knowledge.

VI

In the literature on the subject two broad approaches are discernible. First, our mind, aided by the laws of induction and probability and working on the stuff provided by sense organs, makes knowledge possible, detects its defects and also tries to correct them. The process seems to be endless, multiform, and primarily practical. No universally acceptable reason or experience is humanly available that could be shown to be conclusive and clinching to all alike. This approach, though aligned to empiricism, is not necessarily committed to skepticism or relativism. All skeptics are not equally radical. Some are skeptical only reluctantly because, they feel, unquestioned truths are *not* humanly available. Their skepticism is confined only to the theoretical domain. In the large domain of practice, they claim, their position hardly differs from that of their critics—the foundationalist, the absolutist, the self-evidentialist, the intuitionist, and their followers.

The second approach is evident in the works of those who maintain that skepticism and relativism of all shapes and shades are more or less untenable and that, positively speaking, one can achieve definitive knowledge; that is, has access to what is true. But when one raises the question, What is the paradigm of knowledge? the antiskeptics stand badly divided. Some say it is science. Those who want to be more specific say it is physics. But, historically speaking, the paradigms of physics, like those of other sciences, keep on changing. In the face of historical change of scientific paradigms the metaphysicians, broadly speaking, take one of the two possible stands. Metaphysics *supersedes* science. One could ascribe this position, though in different ways, to the Advaita Vedāntin, Hegel himself, and the Hegelians like Bradley. Another notable view is that metaphysics *sustains* physics. Descartes and Kant may be referred to as its defenders. Without the certification of the veracious God, the Cartesian says, the truth of scientific propositions cannot stand on its own ground. The Kantian maintains that the empirical propositions of science, unless backed up by the apperceptive synthetic unity of the transcendental self, the surrogate of God in all humans, cannot be established as objectively sharable, universally and necessarily true.

The said two modes of assimilation or subsumption of science under metaphysics, in spite of their differences, are in a way akin. If the Cartesian-Hegelian mode is called *substantive,* the Kantian one may be described as *formal.* Their kindred characters are to be found in their holistic foundationalism and antiskepticism. Unless reality is viewed, either substantively or formally, as a whole and science reviewed thereunder, the metaphysician fails to see how science as system of purely empirical knowledge can be vindicated, validated, or defended against the attack of the skeptic. Who, except God or his alter ego, the transcendental self, the metaphysician wonders, possibly can see and hold together the interlocked propositions of science as a unique system of truths?

The metaphysician's way of defending science against skepticism seems to imply, to use an expression of Merleau-Ponty, "distortion from above." The principles in terms of which scientific theories are sought to be justified are postulational or presuppositional and not propositional and therefore not even weakly testable. The presuppositions may be, as in Kant, strictly universal, or, as in Collingwood, historically delimited, that is, culture bound. In the latter position one notices implicit concession to historism and a sort of fallibilism. But, simultaneously, one must not forget the strong Hegelian thesis that the universals of *surface* thought or science and those of *deep* thought or metaphysics are identical "as bottom." Total rejection of this position seems to entail reductive naturalism or radical empiricism or both. But, the question is, Who proposes this sort of ridiculous position? Who defends "distortion from below" without reservation? Even in their heydays the logical positivists did not forget to take note of the inadequacy of a purely observational language. Even the skeptical Buddhist did not hesitate to accord an ontological status to *saṁsāra,* the empirical world. The latter may well be recognized without denying the transcendental (presupposition or reality) altogether. Rightly understood, the two go well together.

The yields of the preceding two approaches, "distortion from above" and "distortion from below," are numerous and often found to be radical in their stance. The *philosophical* extremity of their formulations clashes with one's common sense. This gives rise to a totally wrong, unfortunate, and avoidable impression that philosophy is remote from and unrelated to common sense and natural language. It explains, at least partially, why in philosophy, and not in common sense, we come across some such expressions as "eternal propositions," "truth-in-itself," "objective knowledge," "thought without a thinker," and "knowledge without a knowing subject." In common sense, science, and philosophy one keeps on hearing also the opposite ideas: "nothing is, everything flows," "infallible truths are not available to imperfect human beings," "only God is all-knowing," "even 'exact' sciences are historically changing," "nothing falls beyond the sweep of time." Recognition of the reality of time and of the validity of the historical mode of knowledge is closely associated with relativism and skepticism. If time is really real, it has been argued, nothing, not even the proposition, the bearer of truth value, is left unaffected (that is, unchanged) by it. This argument brushes aside, by implication, the logicist view that the proposition is a sort of entity that is timeless, that knows no history. In effect what is desired by the historist is the distinction between the empirical and the transcendental and that between the temporal and the eternal.

Historical change is not the only concept underlying and responsible for skepticism. Equally relevant to it are the concepts of chance, events, finitude of humankind, embodied character of the knowing mind, language-bound character of human knowledge, and the multiformity of nature.

VII

I draw an important distinction between skepticism and fallibilism. One's fallibilism does not commit one to skepticism. Nor does antifoundationalism lead to skepticism. We are not quarreling over words like *skepticism* and *fallibilism.* My main concern is the underlying concepts, their contents and arguments, for and against. Lack of tolerance, charity, patience, and a crusader's mentality are the grounds of refusal to listen to proskeptic arguments. In the name of realism different forms of antifoundationalism, weak and strong, have been arbitrarily bracketed with skepticism. Ignoring the different forms of realistic ontology, weak and strong, all sorts of fallibilism are branded as skepticism. I will try to show, among other things, that strong forms of foundationalism, transcendental or metaphysical, are bound to ignore the *growing* character of scientific knowledge and obliged to resort to untestable speculation and liberal postulation. Realism, rightly understood, does not prove inconsistent with fallibilism. The real world and the objects in it are not revealed to us all at once. We know them gradually, historically, and can never be *cognitively* sure of their absolute certainty. The knowing self and the known (and knowable) world are differently interlocked, interanimated, and interactive, both biologically and epistemologically. In our knowledge of the world we do not, rather cannot, stand totally apart from it. From the evolutionary point of view what is called *survival value* is in a way akin to truth value, truth of what our organism is informed of the world around us. Situated in the world, we have to know it. In spite of our ability to transcend our situation, in a limited way, we cannot totally get out of it. It is somewhat like our inability to jump out of our own skins and schemes.

These are some of the basic considerations that make me sympathetic to the skeptic's point of view. Self-critical rationalism, as I understand it, makes one patient toward the critic's point of view. In an intellectual democracy one must not be impatient with an opponent's argument and refuse to meet it. Antirefusalism and rationalism are inseparable. The basic thrust of the main arguments of this book is to show that the strong criticisms repeatedly leveled against skepticism are exaggerated, trivially true at best, and inconsistent with common sense at worst.

realism different forms of antifoundationalism—weak and strong—have been arbitrarily bracketed with skepticism. Paradigmatic of this [illegible] of realistic ontology, weak and [illegible] some of [illegible] are branded as skepticism. I will try to show, among other things, that strong forms of foundationalism, transcendental as well as empirical, are bound to ignore the growing character of human knowledge and obliged to resort to untestable speculation and kinds of postulation. Realism, rightly understood, does not prove inconsistent with fallibilism. The real world and the objects in it are not revealed to us all at once. We know them gradually, historically and can never be cognitively sure of their absolute certainty. The knowing self and the known (and knowable) world are inherently interlocked, interpenetrated, and interactive, both biologically and epistemologically. In our knowledge of the world we do not, cannot [illegible] cannot stand totally apart from it. From the evolutionary point of view what is called [illegible] is [illegible] truth value, truth of what our organism is informed of the world around us. Situated in the world, we have to know it. In spite of our ability to transcend our situation in a limited way, we cannot [illegible] it. It is somewhat like our inability to jump out of our own skins and schemes.

These are some of the basic considerations that make me [illegible] to the skeptic's goal of [illegible] critical rationality, as I understand it, makes one [illegible] toward [illegible] intellectual [illegible] must not be [illegible] an argument and [illegible] Antifoundationalism and [illegible] nationalism are inseparable. The basic thrust of the [illegible] of this book is to show that the strong criticisms [illegible] against skepticism are exaggerated [illegible] and consistent with common sense at work.

1

Induction, Probability, and Uncertainty

1. Cognitive Expansion and Uncertainty

In this chapter I will try to show that our cognitive experience, when it expands, encounters uncertainty. To the question, Can experiential expansion be certain? my response will be negative. This brief formulation of the issue needs elucidation.

Let me first distinguish the forms of experience that are cognitive and those that are not. As we all know, certain forms of experience are purely biological or psychological, and their initial cognitive import, except in a very trivial sense, is nil. We have aesthetic, ethical, and other modes of experience that form an important part of our psychobiological or even sociobiological history but are not of much informative value. About them we may raise many interesting structural or modular questions: How they are formed? How their forms are interrelated? and the like. Even the ontological and causal bases of these forms of experience may be of much help in understanding the workings of our mind. But as such these forms of experience are not cognitive. One may say that these processes of experience take place in our body-mind complex without adding anything to our understanding of ourselves and our circumstances.

There is another way of looking at these forms of experience. When we use some such words as *cognitive experience* and *noncognitive experience* we certainly have some concepts in view, and those concepts can be shared with others. Similarly, when I say, for example, "expansion of experience leads to uncertainty," I describe a situation that is intelligible to others, provided of course they know my language. In this way when we learn certain words denoting certain objects and concepts, we try, through those words, to convey the same idea to others. Ways of learning words and their uses, individual or sentential, of course are in a way silently informative. When, for example, children see pictures of animals and fruit and are told of their names, they certainly learn something not known to them before. Further, when they learn to use rightly some such sentences as "This tree is green" and "The dog is an animal," their horizon of knowledge undoubtedly expands. Through experience and successful use of language the so-called noncognitive experiences start disclosing their protocognitive implications.

To learn the uses of different types of nouns, pronouns, and abstract concepts certainly contributes to the growth of children's knowledge. In a way this is true also of the growth of knowledge of the adult human beings. When we learn how different types of words, or parts of speech, are related, that is, syntactically connected into meaningful sentences, we start moving toward complex and abstract forms of knowledge. More abstract and complex forms of knowledge are acquired when we learn to see the relations between sentences and the states of affairs they refer to. In other words, when we learn to argue and infer and to move beyond the mere use of words or isolated sentences, our world expands further. By using language and logic, logic in language, we expand the horizon of our experience.

The distinction needs to be drawn between how language acquisition adds to our expansion of knowledge and how we consciously plan the expansion of knowledge and design our experiences accordingly. For language acquisition may be studied as a part of linguistics, computation, biology, psychology, and sociology. That study will not form a part of what we call *epistemology.* Certainly they are related in a complex way; epistemology is grounded in those disciplines. But the description of that relation will not be of much help to us in ascertaining what is valid knowledge, *episteme*.

My main interest here is to analyze how consciously structured experiences, when expanded, run into uncertainty. The implicit structures of our knowledge, when laid bare or made articulate, exhibit, broadly speaking, two patterns, inductive and deductive. But strictly speaking, these two patterns or structures, except for the purpose of analysis, are not exclusive. Induction involves deduction. Deduction involves induction. It is not surprising that some logicians are of the view that, as *logic,* induction and deduction are not sharply separable, only their starting points and aims differ. This difference is primarily programmatic and heuristic and not intrinsic to their nature.

In a way logic, in all its forms, deductive, inductive, and probabilistic, constitutes the structuralization of our cognitive experience. At the stage of structuralization, the cognitive aspect may well be ignored. Given a set of premises, *P*, and set of rules of inference, *R*, we try to see what conclusion(s), *C*, can be derived. But even then the question of correctness, adequacy, or justification of how we structuralize our experience remains. If *P* is true and *R* followed, can *C* be false? The falsity of *C* may be a logical question or an epistemological one. If *C* turns out to be false, does it not infect *P* or at least a part of it? Therefore the question of justification is relevant not only in the cases of inductive logic or probability logic but also in the case of deductive logic. In brief, the question of justification may be rightly radicalized and taken to the basement or foundation of every form of logic. If logic itself is a human-made game, having no secure foundation of its own, must its limited use value (predetermined by definition and choice) be so glorified and taken so seriously?

2. Evidence, Acceptance, and Probability

Empirical realists, not committed to synthetic a priori truths, are of the view that neither in respect of empirical generalizations nor in respect of individual future events can one be absolutely certain. Available evidence can more or less confirm a universal generalization but cannot entail it. Our prediction about singular events may be more or less probable but cannot be certain, The question is, Though certainty seems to be elusive, when can (1) our *acceptance* of a hypothesis and (2) our *belief* that a particular event will take place be regarded as *rational* or *justified*?

Evidence *e* may at best make hypothesis *h* probable, but it does not make it true. For acceptance *AC* of *h* one may take *e* as necessary but not sufficient. To raise belief *b* to the level of rationality *R*, that is, above irrationality or absurdity, ~ *R*, *e* is necessary but not sufficient or adequate. Unless some quantitative value, *q*, can be assigned to *e*, it is difficult to define (the rule of) *AC* of *h* and *R* of *b*. Any value of *q* in between 1 and 0.5 may be regarded as high and in between 0.5 and 0 as low.

To say this is not to clarify the issue very satisfactorily. Unless the domain of the discourse is specified it may be misleading to state that hypothesis *h* is justified, supported, or confirmed by the evidence *e*. If the elements of a domain, the telephone system of Tokyo, for example, are *known* to be stably related and their behavior highly predictable, to state with reference to that domain that out of 100 local telephone calls made 75 go through is to assign low *probability* or *credibility* (*Cred*) to the system. But knowing as we do the poor functioning of the telephone system of Calcutta, if we say the same thing in the form, $P(h,e) = \frac{3}{4}$, informed people are likely to give high *P* or *Cred* to the system. For our experience is that out of 100 local telephone calls we make not more than 50 (on average) go through. This is because of bad maintenance or, during the rainy season, water-logging.

The parenthesized words "on average" brings to light another problematic aspect of the domain. It is possible that some lucky people of a particular area (say, exchange 75) find that out of 100 calls they make more than 70 go through. One can take the "subdomain" of the lucky Calcutta Telephone subscribers (of exchange 75) as an *isolated domain*. Against the known inductive (background) history of the Calcutta telephone system, one has to admit two things. First, with *specific* reference to the *new isolated system* the subscribers must not be treated as lucky. Because their short inductive background, unlike the long inductive background of the subscribers of the Calcutta telephone system as a whole, is different, the difference brought about by new and imported technology and improved functioning of the lucky system. If the lucky system is segregated from the not-so-lucky ("average") system, then the *expected* outcome changes considerably. Second, there may be some *very* unlucky groups living in some old (exchange) areas whose experience, quantitatively speaking, is extremely disappointing. Let us suppose that out of 100 calls they make not more than 25 go through. If the systems of those unlucky groups are also treated as isolated, then the probabilistic picture changes in the opposite direction. So, in calculating the weighted average one has to be clear about the types of the systems, isolated or overlapping (including the extent of overlap), the principles of differentiating the areas, exchanges, or groups, and also the principles of aggregating them statistically.

The whole domain of the discourse may be changed if the performance of the Calcutta telephone system and that of the Tokyo telephone system are taken together, ignoring, for the time being, the different groups or subsystems within them. It is not easy to determine the *Cred* of the conjoined or newly unified system.

The rating of the *Cred* of the systems becomes more uncertain if a new variable, *t* or time factor, is introduced. For example, during the monsoon season and as a result of frequent heavy downpours and the resulting waterlogging in Calcutta, many of its telephones go dead every year. Naturally the performance of the Calcutta telephone system as a whole and that of its subsystems, taken either exclusively or inclusively, goes down at a particular period of time during the year.

The problematic character of the matter may be put in an acute form in another way. Let us suppose (1) DPC is a telephone subscriber in the 44 exchange area and (2) 95 percent of the telephone lines in that area have been dead for the last six months, and, therefore, (3) almost certainly DPC's telephone line is dead. Now let us suppose (1a) DPC is a member of Parliament (MP) and (2a) less than 5 percent of the telephones lines of those who are MPs are dead, and, therefore, (3a), almost certainly, DPC's telephone line is not dead. Let us suppose, in addition, the premises of both the arguments are true. But even then, their conclusions turn out to be inconsistent.[1]

In this way it may be shown that the more the variables are introduced in the system, the more uncertain, if not downright inconsistent, becomes the outcome of its functioning. But it may be added here that if the values of the added variables and their mutual relationships are *definitely known,* then the elements of uncertainty can be *substantially* reduced. Because then it becomes a matter of correct calculation. But even then, if the range and magnitude of the calculable values of the variables are extraordinarily large or minute, calculation or computation itself turns out to be a contributory factor to uncertainty. Besides, "the definitely knownness" condition is arbitrary, unrealistic, and in a way begs the very issue under dispute.

The relativity of the domain of the discourse may be highlighted by introducing a new domain, say, of the survival rates of the pancreatic cancer patients. Because of the known high rate of mortality among the pancreatic cancer patients if one says at some future time, t_2, "out of 100 patients the lives of 40 have been saved," our reaction now (at time t_1) would be "it is very good; that is, the probability is pretty high."

The first point I am trying to make out by highlighting the domain-relative character of probability is the inadequacy of the Bayesian assumption of equal and a priori distribution of probability values over the different elements (or states of affairs) of the domain. Second, the probability rating of the hypothesis is not only domain relative and evidence relative but the weights of the evidence themselves are also domain sensitive in an indirect but significant way. I am aware that the concepts like "evidence relative" are intuitive and, therefore, cannot be given quantitative formulation. But this is precisely one of my arguments against the simple Bayesian approach and in favor of the epistemological one. In the wake of the latter approach comes up the relevance of the concept of *knower.*

In this inductive context the concept of *learner* seems to be more appropriate. In deductive logic or procedures to explain *new* concept formation, it is assumed that the decision maker (person) is not handicapped by imperfection or limitation (vis-à-vis information). But to make an inductive inference one must bear in mind the limited epistemic capacity of human beings. On the basis of a small number of samples it is not at all easy for one to solve complex and massive combinatorial problems of induction. It has been rightly pointed out that even the biggest conceivable computers cannot go through, say, $2^{2^{10}}$ possibilities and that, consequently, to make a *realistic and rational* decision, in respect to winning war, game, lottery, and so on, to phenomena of complex kind(s), is impossible for beings like ourselves.[2]

Third, my attention is directed primarily toward the epistemological problems arising out of finite powers of our memory and computational ability. Given this constraint, how do we rationally or justifiably come to expand what is available to us through experience? How do we systematize what we already have had in a presystematic manner? Rationality of inductive behavior and procedures have their descriptive as well as normative aspects. Our inferential ability is constantly being put to both cognitive and noncognitive uses.

Unreliability, incoherence, and inconsistency associated with induction make it very difficult for one to formulate suitably the program for justification of induction. At the same time, one finds it extremely difficult and counterintuitive to deny the necessity of what we regard as nondemonstrative inference for ampliation of our information and learning new things and new relations between already known things. It has been rightly observed that "in the philosophical study of induction, no task is of greater importance than that of giving a clear characterisation of inductive procedures: only when this has been done can the problem of justification [of induction] significantly be raised."[3]

3. Induction, Probability Calculation, and Rationality of Belief and Decision

The procedures known as inductive are numerous; viz., procedures pertaining to decision, acceptance, utility maximization, and learning. But, for the purpose of my study I propose to regard these different procedures as different forms of expansion. It is to be remembered that some writers like Carnap and Jeffrey are basically interested in explicating rules of inductive thinking within a static framework, whereas others like Popper and Lakatos basically are interested in explicating the parameters of decision making in the realm of competing hypotheses. The latter's framework is not static. And they are not willing to consider their logic as inductive.

Decision theory, as ordinarily understood, involves the concepts of utility and probability. Carnap understands *probability* as degree of belief. To start with, he takes it as a psychological concept in a descriptive sense, concerned with actual beliefs of actual human beings. Subsequently he addresses himself to the issue of normative decision theory by introducing some conditions of rationality. In its initial phase his conception of probability appears psychological or personalist. But, according to his own admission, he is interested mainly in a logical concept of probability.

In real life we have to make a choice among possible acts at a given time. We know the *possible* states of nature that are relevant to our acts. But ordinarily we do not know which of the possible states actually obtains. However, as the number of possible acts and possible states are finite, we can make a decision about a particular possible act in relation to a particular possible state of nature and also calculate the outcome of that decision. This outcome is determined by the concerned state of nature and the concerned act performed. It will also be reasonable to assume that we know the *expected utility* of the outcome of our act before we perform it.

Carnap's view is undoubtedly Bayesian in inspiration but that it has important differences from the Bayesian rule of decision making should not be lost sight of. According to Bayes, we are advised, under normal conditions, to make a decision or choose an act that could maximize the probability value of the outcome. This rule may be interpreted neutrally or psychologically and rationally or normatively. Corresponding to the interpretation we arrive at different theories, descriptive decision theory, and normative decision theory.

Once the ambiguity of the concept of probability is removed and the distinction between subjective or personal probability and the statistical or objective probability is made, the difference between the descriptive theory and the normative theory becomes clearer. The concept of personal probability confers on a proposition or event a particular value representing a person's state of belief. This concept of personal probability has its two versions: (1) expressing the *actual* degree of belief, and (2) expressing the *rational* degree of belief.

The moot point to be decided in this connection is the relative importance of the preceding two concepts of probability, the statistical and the personal, in determining the (probability) value of the decision principle. Understandably, mathematical statisticians prefer the statistical concept of probability, which makes no reference to the psychological belief or knowledge of this or that particular person. The numerical values of statistical probability are not cognitively available to the concerned person, the decision maker. On this ground some decision theorists are opposed to the concept. They favor the personal concept of probability. Carnap seems to endorse this view. However, this is not to deny altogether the importance of the statistical probability in some empirical disciplines and particularly in the context of decision principle.

At this stage we are required to draw an important line of distinction between *actual* decisions and *rational* decisions. The laws governing the actual degree of psychological belief are obviously empirical in character. To find out these laws one has to look into the actual behavior of belief under uncertain situations like betting and playing games. Sometimes to denote this psychological concept a technical term, *degree of credence*, or, in short, *Cred*, has been introduced. Different persons may entertain different degrees of credence. Others may have different kinds of credence. The betting behavior of a person shows his or her credence function. Measuring the credence function at the psychological level, it is needless to say, is extremely difficult. The same may be said of one's utility functions. It expresses the system of valuation and preference of the concerned person. It represents not only one's beliefs but also the strength with which the beliefs are entertained and likely to be acted upon. Even if utility and credence functions of one or two persons can be determined with reasonable certainty, it is not easy to expand the same to a larger group, especially when the composition of the group is heterogenous and the defining characteristics of the group are inexactly defined.[4]

Carnap's move from descriptive to normative decision theory, concerned with *rational* credence, is intended to link up descriptive decision theory with inductive logic. In different ways the works of Keynes, Ramsey, Savage, and de Finetti have clarified the concept of rational decision. But Carnap is opposed to the idea of characterizing *personal* probability as *subjective*. For subjectivity and rationality hardly go together in the context of logical probability. In this connection he rightly recalls de Finetti's observation that "probability theory is not an attempt to describe actual behaviour . . . [but] . . . coherent behaviour" and Keynes's explicit assertion that "probability is not subjective . . . once the facts are given which determine our knowledge, what is probable or improbable in these circumstances has been fixed objectively."[5]

The theory of *rational* credence or utility is established not by experiments but by requirements of rationality. On the question, What exactly are rational requirements? writers are not unanimous. Though at times the (Kantian) paradigm of complete rationality has been mooted, it is not generally favored. Whose credence function, except that of an all-knowing God, could possibly be perfectly rational? The realistic rationality requirements of credence function are (1) it must be *coherent* and (2) *strictly coherent*. Taking the time-factor into account (3) the third requirement of rationality may be indicated in terms of *information* or empirical data available to the decision maker at the interval between two moments, say, t_1 and t_2 of his or her life. *Information* may be taken exclusively (that is, as information only) or inclusively (that is, together with emotional and volitional factors). The third requirement (that is, information added between the concerned time interval) must satisfy the second requirement of *strict coherence*: it must be consistent with the information already available to the decision maker.

To lay the foundation of what he calls *inductive logic* Carnap introduces some simplifying concepts. Unlike other personal probabilists like Ramsey, de Finetti, and Savage, Carnap accepts initial credence and credibility for the purpose of constructing subjective probability and rejects the concept of "adult disposition" as a basic concept. Knowing fully well that his preferred concepts are "less realistic and remoter from overt behaviour" of the adult human beings he sticks to them apparently for two reasons. First, his credibility function satisfies the stronger requirements of rationality. Second, his concepts that are in consonance with this stronger requirements of rationality represent for him "permanent dispositions" of the normative type. This approach is analogous to that of the theoretical physicist who is concerned less with observable properties of objects and more with abstract properties of the same. Carnap's preference for "permanent dispositions" over temporary "mental inclinations" is obviously normatively motivated; that is, appropriate for the purpose of explicating utilities and acts of rational human beings. A person's belief system can hardly be judged fairly unless mental impulses and stray actions are not ignored. Positively speaking, to ascertain the rationality of someone's beliefs we have to concentrate on those beliefs, their contents, the evidential base of their confirmation, and how they are spread over different times of the concerned person's life.

In addition to *credence function* and *permanent disposition* the other two concepts crucial to Carnap's inductive logic are *measure function* and *confirmation function*. The credibility function of a person is deemed to be *perfectly* rational and backed up by an *unfailing* memory. For a logical function corresponding to credence Carnap uses (inductive) measure function; and for a logical function corresponding to credibility he uses (inductive) confirmation function. Given evidence *e*, a hypothesis *h* is said to be confirmed to a particular degree of belief, say, *r*. The measure function (*M*-function) or confirmation function (*C*-function) is defined in a purely logical or set-theoretic manner. The basic actions of inductive logic for *M*-functions are the usual axioms of the calculus of probability. This satisfies the requirements of coherence, regularity, and so forth. The other axioms mentioned by Carnap in his *inductive logic* are those of invariance. Also, he speaks of the axiom of symmetry. It falls under axioms of invariance. The latter remind one of the classical principle of indifference. All axioms of inductive logic are intended to state relations among the values of *M*-functions or *C*-functions. Carnap's inductive logic, as he keeps on reminding us, is *pure* and *not applied*. Yet his insistence on using the term *inductive* is noteworthy. He justifies it by saying that "this theory provides the foundation for inductive reasoning (in a wide sense)."[6]

4. Carnap's View of Inductive Logic: A Critical Review

Regarding the very nature of inductive logic or reasoning, there is a lot of controversy. In this respect Carnap rejects the view that inductive reasoning is an inference from some known proposition, premise, or evidence, to a new proposition or conclusion that may be a law or a singular prediction. This sort of reasoning is resorted to, according to the view in question, for *acceptance* or rejection of the new proposition. If we accept this view, Carnap feels we would not be able to refute Hume's position that induction has no rational grounds. To him, the paradigmatic piece of inductive reasoning concerning hypothesis *h* starts from some given evidence *e* and consists in assigning a probability value to *h*. Its schematic form is *c*(*h*, *e*). The term *c* stands for degree of confirmation of *h* in the light of evidence *e*. According to this formulation of inductive reasoning, the relation between *c*, *h*, and *e* is analytic. The task of the inductive logician, contrary to widespread belief, is not to infer the unknown on the basis of the known. His task is to calculate the probability value or degree of confirmation of *h* in terms of *e*, whatever the values of *h* and *e* may be.

Carnap's formulation of inductive logic seems to have lumped up three different things: (1) how *h*, given *e*, is confirmed to the extent of *p*; (2) how to draw the distinction between the choice procedure of *h*—whether a factual statement, a general hypothesis, or a particular prediction; and (3) how to decide the rationality or otherwise of a course of action *a*, given *e*, or a certain set of circumstances. Some writers are understandably opposed to this restrictive or analytic construal of inductive reasoning, which, they feel, ignores some very important distinctions between its different types. They are of the view that the question of acceptance, relatively ignored by Carnap, merits serious consideration.

The critics of the Carnapian system that I have in mind are Kyburg, Levi, Salmon, and Suppes. My main difficulty with Carnap centers around his *analytic* view of inductive reasoning. If one accepts this view, one cannot, I feel, find faults with the system of inductive logic that Carnap constructs systematically and step by step on the basis of that analytic *view* of inductive reasoning. Of course, by his own admission, it is not a very "powerful" system; nor has it been completed.

The main reasons for my uneasiness are briefly indicated later. In spelling out my reasons on some points I find myself in agreement or at least partial sympathy with the previously mentioned writers.

First, most of us, from the person in the street to the working scientist, are obliged to infer, explicitly or implicitly, something new on the basis of information already at our disposal. Even the radical anti-inductivists like Popper do not deny that for different needs, practical and theoretical, biological to epistemological, we try to transcend the bounds of the available information and discover new horizons of knowledge in terms of bold conjectures. Strictly speaking, Carnap himself does not deny this truism; but as a philosopher and logician the basic concept of inductive reasoning he has chosen for axiomatization is analytic, counterintuitive and departs both from common sense and science in some very important respects.

Second, if we are indifferent to rules of acceptance, it proves very difficult for us to explain how we learn from various sorts of experience and why we consult experts; for example, lawyers, medical practitioners, travel guides, in our day-to-day life. We draw upon others' experience, expertise, and skill, not only for expanding our cognitive universe but also to make our life more joyful and relatively pain free.

Third, without these rules of acceptance and rejection the higher reaches of our theoretic life prove elusive. To cite but a few cases. (1) The particular things we experience for the first time do not immediately appear reliable to us. Why, ordinarily, do we not rely upon strangers? We attach importance to repetition, familiarity, and confirmation or certification by others. (2) On the simple basis of direct reports of sense-experience we cannot accept or reject abstract and comprehensive views like Copernican hypothesis, the Harrod-Domar model of economic growth, EPR Paradox, or even "All emeralds are green." Nor on the *exclusive* basis, a basis not endorsed by others, of *my* falsifying (evidential) experience a well-established theory can be discarded. (3) If we do not accept rules we will not be able to draw a cut-off line between observational statements and theoretical statements.

Finally, acceptance rules have their nonepistemic utilities. In pursuit of a moral life one needs rules for accepting certain norms of values. Why should one do one's duty? Why should one be just? Certainly these questions and others of this sort, though in a sense nonepistemic, deserve answers based on some rules of acceptance. Unless one does one's duty one cannot rationally expect others to do the same. We may or may not accept Kant's Categorical Imperative but as a rule it is imperative that we have one or another such normative principle to guide us in our ethical decision making. Otherwise our civil society is bound to break down. To sustain and promote a civil society is more or less universally recognized as a normative requirement. It may be recalled here that, according to one interpretation, even the Categorical Imperative of Kant is a utilitarian principle of distribution; that is, it prescribes that all rational human beings must have equal access to primary goods like freedom.

Rawls's principles of difference may also be regarded as rules of acceptance in the context of alternative theories regarding just society. The point I am emphasizing is simple: the rules of acceptance or what Kyburg calls the rules of detachment are important not only in epistemic or theoretical affairs but also in nonepistemic, utilitarian, or practical affairs. Strictly speaking, one might even plausibly argue against the dichotomy between practical reason and theoretical reason, between epistemic utilities and nonepistemic utilities.

Some strong defenders of the Carnapian position like Bar-Hillel are of the view that rules of acceptance are irrelevant. According to their understanding, the rule of this type simply says that under such and such circumstances one can rationally accept *h*. Carnap clearly states that "rules of acceptance as ultimate rules for inductive reasoning are inadequate . . . [because they] give in some respect too much, in another respect too little."[7] Clarifying the point Carnap states that if the total evidence *e* available to a person *X* is adequate for the acceptance of *h*, it amounts to saying that for that person *h* is true. In other words, if *e* entails *h*, then by logic alone *X* could take the rational decision under a given set of circumstances. But that sounds counterintuitive. For, some other practical considerations enter into the picture; for example, the concerned person's varying nonepistemic utilities.

Carnap also tries to show how the traditional rules of acceptance are of no use to a person in making a decision under certain circumstances. If two possible courses of actions, under identical circumstances, are equiprobable or nearly equiprobable, the ordinary rules of acceptance will not be of much help to the concerned person in choosing his actual course of action. However, he concedes that "sometimes rules of acceptance may be useful." But the position taken by Bar-Hillel is apparently uncompromising. He says: "We do need further development of inductive logic, and further investigation into scientific methodology, in particular into the comparability of scientific theories, but nobody needs rules of acceptance of any kind."[8] But one wonders how Bar-Hillel reconciles the desirability of further development of inductive logic and further investigation into scientific methodology, on the one hand, and uselessness of rules of acceptance *of any kind,* on the other. In fact, Carnap himself shows in his own way how inductive logic is of help in taking practical decision under some specific uncertain situations. Even if the highly idealized concept of total evidence is weakened by disclosure or dispensed with altogether, it is difficult to deny the relation, positive or negative, justificatory or falsificatory, between *h* and *e*. The only logical requirement to be insisted on is that *e* must be relevant to *h*. Bar-Hillel's strongly negative position is clearly inconsistent with his own concession made to some of the rules of acceptance enumerated by Kyburg. He cannot deny that the estimation of probability of *h* has to be relativized to a corpus of evidential statements. Nor does he deny the importance of the rule expressed in terms of the information content of *h*, relativized to a corpus of evidential statements satisfying some principle of acceptability and also a condition of *consistent* totality. Equally understandable is his agreement with Kyburg that the *simplicity* and the fruitfulness of *h* is, as a matter of fact, recognized by working scientists as a rule of acceptance. It is baffling to see that having conceded this much Bar-Hillel should still underestimate the significance of these rules of acceptance.

I find his denigration of the concept of simplicity totally unwarranted. It is true that this concept has been differently interpreted, but nobody denies its importance in the context of theory choice.

It would be wrong to suggest that scientists or methodologists of science do not need any rule of acceptance. Both the person of common sense and the scientist do many things and follow many methodological rules without being aware of them. The working scientist, as we often find, is not necessarily his or her best philosopher. For example, many good scientists firmly believe that their method is inductive and that for the relevance of deduction one must turn to mathematics. The expressions like "the covering law model," "hypothetico-deductive model," and "deductive-nomological model" sound rather strange to their ears. Obviously this does not mean either that, in the structure of scientific explanation, there is no inductive component in the form of certain laws or theories, or that, from generalizations alone, without referring to some particular statements of initial conditions, one can explain or predict the explanandum.

The main difficulty with Carnap's system is it would not concede any degree of confirmation to any general statement other than zero. But this does not disturb Carnap or his followers. For, their basic concern, it seems to me, is not structure of scientific explanation and analysis of its different components but the degree of confirmation of h in the light of e and to systematize its rules in a deductive and axiomatic form. The p-value of r in the formula $p(h, e) = r$ is an analytic outcome of h and e. The term e being deductively closed total evidence. To determine the quantitative probability of h thus is a matter of calculation or computation and not conjecture. Given this view of inductive reasoning, it is not at all surprising that some writers like Bar-Hillel and even Hintikka, for different reasons, find nothing very philosophically important in the discussion on the rules of acceptance.[9]

Let us now forget about the weakness of Carnap's concept of confirmation function. Without rules of acceptance the problems of having a general hypothesis or law usable as a premise for deriving the explanandum (*ex*) remain. Without initial conditions (*ic*'s) from the general premise with a degree of confirmation, $c(h, e) = p$, the explanandum cannot be derived: *ex* cannot be derived either from (1) *ic*'s and $c(h, e) = p$; or from (2) *ic*'s, $c(h, e) = p$, and e; or even from (3) *ic*'s, $c(h, e) = p$, and e (as total evidence). For none of these forms of argument represents a valid deductive schema. Though (4) e (as total *true* evidence) entails *ex* is a valid deductive schema, it is hardly explanatory. If true premises *entail* a conclusion, that is deduction but not explanation. Most of our explanations are of the (deductive) form (5) *ic*'s ∴ *ex*. But without acceptance rules truth of *ic*'s (5) cannot be rationally claimed.

Comparable, if not more acute, difficulties are encountered in the inductive forms of explanation. The reason is obvious; viz., the structural similarity between the two forms of explanation. In both types of explanation one of the premises must be statistically general and the explanandum must be the conclusion of a logically valid inductive argument form. First, statistical generalizations, informative as they are, cannot be taken as certain. Naturally, the question of their acceptance is bound to come up. Second, because the explanandum is also informative and the form of inference is inductive, the acceptance of the questionable conclusion has to satisfy certain rules. Even if the rules of acceptance are followed, that does not ensure the certainty of the conclusion of inductive inference.

Frequentists like Reichenbach and Salmon turn to the acceptance rules because they are convinced that the Carnapian mode of inductive reasoning with the analytic degree of confirmation as its central concept has very little to offer us either in practical life or in theoretical pursuit. They readily concede that the Reichenbachian analytic estimate statements (estimate of values of limiting frequencies), too, like the Carnapian analytic degree of confirmation, fail to satisfactorily guide us in practical affairs. Because of this limitation they introduce factual truth claims and thus admit, indirectly, the need for acceptance rules. The need for the rules is obvious for those who interpret the frequency estimate statements as synthetic. Even the defenders of this interpretation are not dogmatically committed to the certainty or finality of their hypothesis. When one's accepted estimate statement or hypothesis is found inaccurate in the light of additionally available observation statements, that is, larger and more diverse samples, it is appropriately corrected. This process may be repeated in an ascending order, to move progressively toward accuracy and correctness. I will revert to this point later on in the context of probabilistic justification of induction.

Neither the empiricist's criticism of his analytic concept of inductive reasoning[10] nor the frequentist's stricture[11] seems to have deeply disturbed Carnap. Even in his latest publications he claims that his credence function takes care of both "factual basis" and "statistical probability"—rendering thus the notions superfluous for such practical affairs of life as rational decision making. He goes even further and expresses the hope that if induction is properly reformulated (presumably along the lines he has suggested) the common person's and the scientist's faith in induction can be reconciled with Hume's criticism of the same.

The old puzzle of induction consists in the following dilemma. On the one hand, we see that inductive reasoning is used by the

> scientist and the man in the street . . . and we have the feeling that it is valid and indispensable. On the other hand, once Hume awakens our intellectual conscience, we recognise that here is a serious difficulty. Who is right, the man of common sense or the critical philosopher? We see here, as so often, that both are partially right. Hume's criticism of the customary forms of induction was correct. But still the basic idea of common sense thinking is vindicated: induction, if properly reformulated, can be shown to be valid by rational criteria.[12]

Carnap's optimistic project of reformulating inductive logic, as we know, has been painstakingly followed up by different writers like Jeffrey, Hintikka, and Hesse in different ways. Jeffrey maintains that the degrees of our belief in various types of propositions are determined in complex ways by habit and experience, use of language, and modes of training. In a way confirmation theory is semantic and reflects the *meanings* of the expressions of conditional subjective probabilities. Jeffrey dispenses with the Carnapian notion of total evidence and, therefore, his way of determining one's credibility function differs from Carnap's.

Hintikka is conscious of the difficulties involved in making a conditionalization program truly realistic. He turns his main attention to developing a system in which two different tasks, formulation of singular inductive inference and that of inductive generalization, are first differentiated and then arranged together into a two-dimensional continuum. One parameter of the continuum shows how some interesting inductive procedures regarding singular predictions can be described. Another parameter describes the character of one's assumption about inductive generalization. The aim of Hintikka's approach is clear. Within his system he wants to show the compatibility of (1) determination of degree of confirmation of a particular prediction or decision in relation to a corpus of evidence and (2) determination of informative content of "inductive" generalization. This is clearly a departure from Carnap's λ-continuum dealing with the case *a* in which generalizations are accepted only on a priori grounds and, therefore, are informative only about singular predictions. This approach favors the Popperian view that, from among the competing hypotheses addressed to a particular problem, that one should be preferred which has high empirical content, more *prohibitive* power, and, therefore, in a sense is more falsifiable. Hintikka's a priori probability is Popper's absolute logical probability (in an atomistic universe). If the universe is infinite, the a priori probabilities all are zero. To have a posteriori probabilities above zero one has to assume some finite atomistic universe (which can well be large in size). Hintikka's *a* parameter is essentially Popperian in spirit. The small *a* is indicative of high falsifiability value simultaneously satisfying the requirement of acceptability.

Strictly speaking, the burden of Hesse's main argument in support of Carnap is to show how inductive support from evidence is available for universal generalization. A positive relevance theory of confirmation and high prior probabilities extended to the conclusions of analogical argument clearly suggest that convincing inductive support can be obtained for generalization in a finite domain. The sort of skepticism associated with the view that scientific generalizations all are of zero confirmation value is wholly unwarranted. Hesse's rescue operation is apparently confined to practical belief and finite domain. The sharp distinction that she draws between the problems of supporting universal generalizations and those of supporting individual predictions "derived" from them seems to be unrealistic. Besides, her relative indifference to the problems attending universal generalization is hardly in accord with the practice of the working scientist. In this respect she repeats the mistake of Carnap.

Personally speaking, I feel that Hintikka's reconstruction of Carnap's position is relatively more comprehensive and promising. But the point in which I find Carnap's position most puzzling is this. Although he and his followers repeatedly speak of their concern with the problem of determining rationality of belief and consequential action, they are apparently not very mindful of the problems obtained at the down-to-earth level. Certainly I do not deny that in constructing a system of inductive logic one is obliged to resort to some degree of idealization. But ultimately the question of referring the idealized concepts back to the practical issues of life can hardly be denied. This is important both for theory itself and also for testing its material adequacy. At the practical level we are bothered not only with the question of determining the degree of rationality of our beliefs but also how they undergo change in the light of new information, expected as well as unexpected. Besides, the problems of our finite memory and forming new concepts necessary to deal with new information are equally serious. Suppes has rightly drawn our attention to these aspects of a theory of *human* rationality. "A theory of rationality that does not take account of the specific human powers and limitations of attention, memory and conceptualisation may have interesting things to say but not about human rationality."[13]

By refusing to take into account the limitations of human

powers, cognitive and noncognitive, we try to avoid the problems raised by the skeptics and the relativists. This sophisticated maneuvre of avoidance impoverishes our theory of rationality and fails to reach the problems at all levels.

5. Noninductive Rationality of Science: Popper on Basic Statements

When we use some words, such as *rationality* and *induction,* we should be clear in our mind in what precise sense we are using them. Our brief survey of the concerned views shows that belief may be taken in its very ordinary psychological sense. If I see, as I am seeing now, a paperweight on my writing desk, I must believe in its existence. Ordinarily neither I nor anybody near me with normal vision is likely to ask me why I do believe in the existence of this paperweight. This sort of object of perception does not ordinarily raise the question of rationality of the belief in its existence. To say, for example, "I see it and therefore I believe in its existence" is likely to be accepted as a good enough ground or reason for my belief. Even the argumentative form of the sentence containing *therefore* may appear quite superfluous. Many people will feel that in such situations I have good enough reason to be *certain* about it, the existence of the paperweight before me, and the question of offering reason in support or to establish the belief does not arise at all.

But then why do we raise the issue of "reason," "ground," or "support" in this context? The answer to the question seems to lie in remembering the obvious distinction, too obvious to remember all the time, between the *psychological* sense of certainty and the *linguistic* mode of presenting it to others. As we do not believe in any private language, even in the case of the first person, in this case myself, the distinction between the psychological sense and the linguistic one needs to be drawn. Instead of drawing the distinction between the psychological sense and the linguistic sense we may perhaps reformulate it as the distinction between psychological *experience* and its *content.* Though experience is personal, its content, either through the explicit use of language or some other semiotic behavior, may be shared by, or communicated to, others.

The question of rationality of belief comes up when we move from psychology to cognitive psychology or epistemology. That some of our perceptual judgments, beliefs born out of perception, turn out to be more or less erroneous, at times even downright false, is hardly disputed. This partly explains why even *for* beliefs we are pressed to produce reason. If the concerned belief is sought to be defended in terms of content of psychological perception and its intersubjective sharability, another problem still remains. We can easily suppose a situation when we can well posit the content of perception but find ourselves unable to posit the existence of any percipient around us. What happens in that sort of case? Can one not then be skeptical about the content itself? In a way this is the argument the Berkeleyan raises against the acceptance of a principle like *esse est percipi* in an unqualified manner. If other human beings are not there to share the perceptual content, one could be easily skeptical about its very existence; or one might feel that what one sees before one is not a real thing but a delusion conjured up by a demon. The notion of a good God is introduced by Berkeley to stave off this sort of skepticism.

The philosophers who are opposed to the introduction of the idea of God to combat skepticism are obliged to follow some other strategy. If this transcendental strategy is discarded, some "firm empirical" basis has to be introduced. But then the question may be raised, If one's perceptual judgment is not self-certifying, how can it be certified by an other perceptual judgment that is logically or epistemologically on a par with it? One answer to this question takes pure pragmatic form; viz., we accept a perceptual judgment as self-justified, self-certified, or self-supporting simply on its *causally* satisfying ground. But those who are opposed to pragmatism and inclined to realism put a big question mark here.

In terms of what he calls *basic statement,* Popper tries to avoid the trilemma of dogmatism, infinite regress, and psychologism. For him, basic statements are acceptable as satisfactory because they are intersubjectively sharable; they are also sufficiently tested. And their dogmatic look is merely a semblance, an appearance. To Popper, this kind of dogmatism is innocuous, for they themselves are testable in principle. Popper also concedes that the charge of infinite regress may also be raised against them. But this sort of infinite regress also seems to him innocuous.

> And finally, as to *psychologism* . . . the decision to accept a basic statement, and to be satisfied with it, is causally connected with our experiences—especially with our *perceptual experiences*. But we do not attempt to *justify* basic statements by these experiences. Experiences can *motivate a decision,* and hence an acceptance or a rejection of a statement, but a basic statement cannot be *justified* by them—no more than by thumping the table.[14]

Popper's opposition to the word *justification* is understandable, because he is committed to falsificationism. But the *causal* connection between experiences and basic statements to which he alludes is not, strictly speaking, necessary. Undeniably there are cases when *we* do not agree about the question of what *I* see. Even if we agree, this agreement is purely decisional or conventional. In that case it is hardly distinguishable from the pragmatist's way of fixing belief. That basic statements are open to testing is admitted by Popper himself. If the test in fact is not undertaken, it is mainly because of *our* belief or faith in the correctness of our decision or convention. In Popper's scheme of thought, basic statements are primarily meant for testing or falsifying higher level statements, laws, and theories, for example. These statements themselves cannot be taken as *comparably* testable.

If the *acceptance* of basic statements is rooted in a *decision,* how do we determine the rationality of that decision? In Popper's own admission, basic statements, though testable, are not falsifiable. These and related considerations make one doubtful about the proclaimed firmness of the so-called empirical basis. Of course, it is true that Popper often speaks of baselessness or the unfounded character of scientific knowledge. If it is really meant, the difference between the Popperian antifoundationalism and skepticism becomes tenuous. In that case one of the basic theses of Popper's philosophy of science, the growth of scientific knowledge, seems to be very weak.

If to determine the question of rationality of lower-level statements like perceptual judgments raises so many problems, one can easily imagine that the problems attending the procedures of acceptance or rejection of higher-level statements are bound to prove even more serious. Can induction or inductive logic, rules of acceptance or rejection, or the analogous methodological principles show a reasonable way out? So far Popper and his followers are concerned, the answer is a clear No. For them, there is no thing like induction or inductive logic. Therefore, the question of using it for the purpose of determining rationality of belief, decision, action, and laws and theories does not arise at all. By *logic of science* Popperians mean some thing that has nothing to do with induction. Naturally, unlike the Carnapians, they do not take upon themselves the "nonexistent" task of reformulating inductive logic. To them the so-called logic of science is methodology, including epistemology, and this comprehensive discipline is partly positive and partly normative.

6. Induction without Justification: Ayer and Strawson

Some philosophers like Ayer and Strawson as a matter of fact recognize induction as a distinct type of reasoning, but they seem to be noncommmital about the necessity of justifying it. To them, especially Strawson, "justification" is a bracketed inductive concept. When *e* entails *h* or conclusion *c* is entailed by premises *pr*, it is a case of deduction. In that case the logically *necessary* character of *c* or *h* makes the issue of its justification pointless. Unless interested in the issues of foundation(s) of mathematics, one is unlikely to take it seriously in the specific context of induction. For ordinary people, who are neither deductive logicians nor mathematicians, induction is a matter of habit, habit formed by experience and that practically pays. The most appropriate characterization of it is that it is rational or reasonable. But there is nothing in or even beyond the universe, including the human beings in it, that can guarantee the necessary character of the outcomes of induction. Simultaneously it has to be admitted that the nature of the universe is not so chaotic as to make the formation of useful inductive habits impossible. Nor is the reasonableness of inductive procedures challenged by the *frequent* surprises produced by the universe. So, in spite of the presence and effectivity of imponderables in the universe, it is true that we have inductive habits and procedures, and that they are found to be reasonable and useful for the purpose of our learning from them and expanding our cognitive experience. In Strawson's own language the presupposition of induction is "synthetic apriori"; to be more explicit what in addition he could have added but did not add to it is "but not necessarily true."[15]

Strawson's is a very modest thesis. In effect what he says is this: we follow inductive procedures because they pay. To invoke usefulness as justification of induction is of very limited significance, particularly in view of the admission of the fact that its back-up principle or presupposition, though "synthetic apriori," is not necessarily true. The other and more serious objection that may be raised against this hybrid variety of "synthetic apriori" truth is that its claim of a priority is totally untenable. For it allows exceptions or, one may say, does not prohibit "surprises"; that is, counterexamples. The more disturbing point against this view is that it rests on circularity. That inductive procedures pay is a claim which itself rests on induction, remembered record of past and relevant experiences of success.[16] The consistent inductive defender of induction has an additional difficulty to meet: one may argue against the person that his or her inductive memory, besides being finite, is conveniently selective or oblivious of the cases of failure.

Most of these objections are traceable in some form or other to Hume's critique of induction. And that partly explains why it remains so influential till the day. But it is instructive to note that from the same critique different lessons are being drawn. The range of lessons, as we find, are amazingly wide. The Popperians tell us that there is no induction and therefore the question of "justifying" it does not make any sense. The followers of Carnap and Reichenbach feel that induction needs careful and systematic reformulation. Of course their proposed ways of tackling the task and also the construals of the task itself differ. Some writers maintain that induction needs no justification and that we may do with it as it is. There are others who are for defending it pragmatically.

From the said and related reactions against the Humean critique of induction one thing that comes out very clearly is this. If the critique stands unchallenged, skepticism wins the day and its fall out in the field of knowledge is bound to be total anarchy. But is this apprehension warranted, practically or even theoretically? To understand the related issues properly we perhaps should have a close relook at Hume's position itself.

2

Can Induction Be Justified?

1. Different Formulations of the Problem of Induction

The problem of induction received its classical formulation in the following words of Hume:

> It is evident that Adam, with all his science, would never have been able to *demonstrate* that the course of nature must continue uniformly the same. . . . Nay, I will go further, and assert that he could not so much as prove by any probable arguments that the future must be conformable to the past. . . .
>
> [T]here is nothing in any object, considered in itself, which can afford us a reason for drawing a conclusion beyond it; and . . . even after the observation of the frequent or constant conjunction of objects, we have no reason to draw any reason concerning any object those of which we have had no experience.[1]

The problem may be reformulated thus: on the *basis* of experience we cannot go *beyond* experience, and if we go beyond experience on the basis of some principle that is not derived from experience, then this going beyond (that is, inferring *universal* statements from *singular* ones) is not justified either inductively or even probabilistically. Attempts have been made to justify induction both inductively and deductively. The inductive justification of induction encounters all the difficulties that induction itself encounters. If it is said that induction has to be justified deductively, that amounts to admitting that induction as induction cannot be regarded as an independent and valid mode of inference.

1. Some empiricists of the Humean persuasion will admit that induction cannot be theoretically justified, and that its justification has to be sought elsewhere, perhaps in practice.

2. Some modern empiricists, notably Carnap and his followers, think that the principles of induction are analytic, and that systematically considered there is no fundamental distinction between deduction and induction; for both are concerned with logical relations between sentences—the former with the relation of L-implication and the latter with that of degree of confirmation as a numerical measure for a partial L-implication.

3. It has been contended by some empirically inclined analytic philosophers that the very problem of the justification of induction has been misconceived and that every attempt to formulate it clearly is doomed to fail; and they are trying to *dis*solve this pseudo-problem.

4. Conventionalists have also tried in vain to eliminate the problem of induction by *deciding,* perhaps rather arbitrarily, that the results of induction, however con*testable,* should be regarded as the norm for accepting the changes to be suggested by future observations and tests.

5. The philosophers of rationalistic persuasion justify induction either by directly assimilating it under deduction or by interpreting it as a sort of inverse deduction. The inverse deductive way of justifying induction has been characterized also as the inventionistic justification of induction, because its value is said to be dependent upon the invention or discovery of a hypothesis that not only can colligate the known facts but also serve as a premise from which other unobserved facts of the same kind may be deduced.

6. Influenced by rationalism, some contemporary empiricists prefer to draw a line of distinction between primary and secondary induction and believe that whereas the former, although really inductive, is not as such justifiable, when supplemented by the latter (which is deductive) it becomes so.

7. I hold that each of the above views breaks down at one or another very crucial point, and this is a pointer to the simple fact neglected (not uniformly) by the advocates of the preceding views: the problem of induction (as it is generally understood) does not really exist. My view must not be con-

fused with (3). I think that there is, as a matter of fact, no *pure* induction, just as there is no such thing as a *pure* given. So the question of justifying it in principle does not arise at all. We never argue from the singulars as such to a universal or universals.

2. Is Induction Not Inductively Justifiable? Donald Williams's Defense and Sample Induction

In this chapter on the *philosophy* of induction I shall not discuss each of the preceding views, nor shall I elaborate here my own view. I shall remain content by indicating only that justification of induction, if inductive, is not justification, and, if it be justification, it is not inductive, and that 4, 5, and 6—particularly 5—contain the elements of a constructive theory of induction.

Most of the inductive justifications of induction on analysis are found to be deductive. I do not know of many genuinely inductive attempts to justify induction. By a genuinely inductive attempt I mean an attempt that being pressed will not claim a priori either that nature is uniform, that there is a limit to the natural varieties, or that the human mind has some self-evident ability to determine the truth—or probability value—of isolated singular statements of experience. In other words, inductive justifications must eschew all a priori commitments, both ontological and epistemological. The question is, Starting from *pure* experience and relying on *simple* enumeration, can we justifiably generalize? If experience qua experience can be a reliable guide to general truths, we are then reasonably assured that purely inductive justification of induction is possible.

Without admitting anything extraempirical in experience (as it is ordinarily understood), it is difficult to imagine how we could possibly go beyond the given singulars. The relevance of the problem or particulars and universals to the problem of induction must not be overlooked, and I am inclined to believe that empiricists did overlook this relevance or that, for philosophical reasons, in effect did underestimate its significance. Their special difficulties with the problem, can, I think, be explained, at least partially, on the basis of this lapse on their part.

Logicians, in fact, never have been hopeful of the possibility of inferring, by enumeration, general truths from experience. Bacon and Mill hardly cared to conceal their distrust of induction through simple enumeration. Bacon says: "The induction which proceeds by simple enumeration is puerile, leads to uncertain conclusions and is exposed to danger from one contradictory instance" (*Novum*

Organum, vol. 1, p. 19). Mill counts this method as induction improperly so-called and points out that it cannot be justified by assuming the uniformity of nature (*System of Logic,* vol. 3, p. iii). Aristotle makes it clear that induction (by complete enumeration) *leads* us *to* no universal propositions if the particulars that are enumerated are not restricted to species or kind (*Prior Analytics,* 68 8–37). This implies that induction presupposes specific identification of the particulars that are enumerated (whether completely or otherwise).

The difficulties of enumerative induction did not make all logicians go straight to eliminative induction, which is more or less motivated by faith in deduction—at least some tried to take an intermediary stand. It is in this way that I understand the transition from simple enumeration to sample(d) enumeration.

Among those who have argued recently in defence of sample induction, Donald Williams may be taken to be a very competent representative.

> Given a fair-sized sample . . . from any population, with no further material information, we know logically that it very probably is one of those which match the population, and hence that very probably the population has a composition similar to that which we discern in the sample. This is the logical justification of induction.[2]

According to this view, the probability value of sample induction increases, provided of course most of the possible samples approximate the composition of the population; that is, the induction must be based on those samples that resemble the population more closely than others. If the chosen samples nearly match the population, then with the increase in the number of that *type* of samples the *probability* of their being representative of the population-composition *logically* (that is, statistically) increases. This view may be put in a statistical (syllogistic) mode of argument of the following form:

Of all things that are P, m/n are Q

x is a P

x (with a probability of m/n) is Q

Here m and n are integers, x is an individual thing and P and Q are empirical predicates. Although presented syllogistically, its inductive import may be easily shown. Let us suppose we have seen n

Indians and found *m* of them brown, the rest not brown. If *n* is a fair-sized sample, and the Indians are finite in number, then it can be proved algebraically that the great majority of the *n* membered sub-classes of the class of Indians differ relatively little from the whole class in regard to the fraction of their members who are brown.

It is to be noted that the significance of enumeration in the preceding form of induction is dependent upon two assumptions (one ontological and the other mathematical) that are open to question. First, the nature of the things enumerated (and to be enumerated) is, and will continue to remain, uniform. Second, the chosen samples are fair. But both assumptions are a priori and necessary. Except induction from past experience, there is no empirical ground for assuming the uniformity of nature. To appeal to it is simply to beg the issue. Besides, to appeal to the fairness of the samples is also empirically unwarranted. The defenders of this theory may point out that the plausibility of this view has never been claimed to be *basically* empirical, and that its justification is logical and lies in a truth of mathematics. From my point of view, this contention is very important, for it indirectly concedes what I have been insisting upon; viz., the *unaided* principle of experience can hardly be regarded as a justification of induction.

Even for the purpose of Williams's own (logical) justification of induction, his view is inadequate. How can he be justifiably assured of the fairness or representative character of his samples? Fairness of the samples depends upon their truly *random* character, and Williams thinks that samples can be assumed to be random if they are not known as biased (or not random). About the necessity of the rule of randomness some philosophers, Peirce and Kneale, for example, are completely convinced, but they are not quite sure how to make proper use of it in practice. Peirce observes: "The needfulness of this rule is obvious: the difficulty is to know how we are to carry it out."[3] It is not difficult to define randomness in terms of equal probability and according to the Principle of Indifference (or of *In*sufficient Reason) that may be traced back to Laplace and Bernoulli. But how to decide rationally that the alternatives are equally probable? The negative answer, "they are equiprobable, if they are not known to be not equiprobable," would seem to suggest that total ignorance is the test of the equiprobability of alternatives. But in practice those who collect samples (of population, for example) are not generally found to be ignorant of the homogeneity or otherwise of their field of inquiry; on the contrary, in selecting sample collectors due consideration is given to their "personal acquaintances" with the field besides technical qualification. The Principle of Indifference on which

Williams's thesis rests has been irreparably damaged by Von Kries and Kneale among others; and I do not intend to reproduce their arguments here. I remain content with referring to one point made by Kneale and another by Keynes that are relevant to the course of my own argument. Kneale observes:

> That principle [of indifference], which purports to provide a rule for the determination of probabilities *a priori* from the consideration of our own ignorance, must be rejected entirely. Probability statements may be modest assertions, but even so they cannot be justified by mere ignorance.[4]
>
> Before accepting a sample as fair the investigator must naturally use all the knowledge he already possesses about the composition of the population and the mutual relevance of various characters, but here again the general theory of chances will not provide detailed guidance.[5]

The fate of sample induction depends upon the defensibility of the Principle of Indifference, but serious defense of it turns out to be a priori. So even if it is assumed for argument's sake that its a priori defense is valid, the sort of validity induction that would derive from it would be purely deductive; and such validity is quite irrelevant to (the principle of) experience. Nothing in experience forbids us to believe that fair samples, assuming that they are fair, are not ultimate or simple; that is, they include an indefinite number of subsamples that are structurally different from the fair samples themselves. To justify sample induction, some methods have been suggested to randomize the samples and subsamples, but the significance of these methods depends upon another a priori assumption; viz., that the variety of the kinds of objects is limited or exhaustibly enumerable (at least in principle). Keynes rejects the Principle of Indifference in its crude form, for it leads to a rule by which judgments of preference and indifference can be based on direct judgments of relevance and irrelevance. He writes (*A Treatise on Probability*):

> The principle [of indifference] remains as a *negative* criterion, two propositions cannot be equally probable, so long as there *is* any ground for discriminating between them. The principle is a necessary, but not, as it seems, a sufficient condition.[6]
>
> There must be no relevant evidence relating to one alternative, unless there is corresponding evidence relating to the other;

our relevant evidence, that is to say, must be symmetrical with regard to the alternatives, and must be applicable to each in the same manner. This is the rule at which the Principles of Indifference somewhat obscurely aims.[7]

The theory of sample induction, although it looks like enumerative, is deductive in structure and not inductive. Purported to be a probabilistic justification of induction, Williams's theory falls back first on the Fair Sample Postulate and then on the Principle of Indifference, and this shows its drift toward deductivism. But it must be noted here that, unlike Keynes, Broad, and Jeffreys, Williams does not use the principle of inverse probability and succeeds in avoiding the associated difficulties. The main difficulty of his theory, however, arises, I think, from his belief that the fairness of his samples cannot be disturbed by any disorderliness or heterogeneity in the field of inquiry; and unless he can prove that the apparent disorders are objectively ordered, his theory even as a piece of a priori reasoning remains unsatisfactory for the purpose of justification of induction.

One might ask me why I have decided to examine sample induction under (1) and not in connection with probabilistic (frequentist) justification of induction (4), as on examination it is found to be much closer to the latter than to the former. My only excuse for doing so, as I have already indicated, is to show the untenability of the view that only by enumerating fair samples, without any other material information, can we justify our inductive inference. The value of sample induction becomes significant if it is combined with either some principle of elimination (as recommended by Keynes) or with a much improved frequency theory of probability, particularly its axiom of randomness (as required by Von Mises and Reichenbach). But even then, I anticipate, satistical extrapolation on the assumption of symmetry can never be more than a conjecture.

3. Harrod's Defense of Induction

The basic requirement of an inductive justification of induction is to get rid of all a priori assumptions that reduce all so-called inductive and probabilistic justifications either directly or indirectly to the deductive form. Not many thinkers would defend induction without any a priori assumption whatsoever: and it is in this connection that the view of Sir Roy Harrod is worth examining. He claims:

> In the past it has been thought that the validity of induction depended on prior acceptance of such principles as the Law of Universal Causation and the Uniformity of Nature. More recently it has been claimed that we have to assume initial prior probabilities for certain beliefs. All such prior postulates have been totally dispensed with here. If induction is to be vindicated, it must be vindicated without any prior assumptions about the nature of the universe whatever. In this I subject myself to Hume, the great master. Starting from his basic principles I have, I would claim, rebutted his sceptical conclusions.[8]

His claim sounds all the more challenging because he asserts that "the essential principles of induction are . . . co-equal in validity and certainty with the principles of deduction." to show the parity between induction and deduction in respect of validity and certainty there are two opposite approaches—one upgrades induction by incorporating it in a deductive framework; and the other, which is rare, downgrades deduction by questioning the self-evident character of its basic principles. Harrod is firmly committed to the latter approach. He rejects both the Carnapian sort of justification of induction because of its underlying deductive character and the Millian sort because of its irretrievable circularity. Because of his persistent refusal to use any a priori principle he also does not endorse the attempts to justify induction by Laplacean inversion of Bernoulli's theorem using Bayes's formula.[9] Very roughly speaking, Bernoulli's theorem is an argument from probabilities to probable values of relative frequencies; but what is needed is derivation of probable values of relative frequencies from probabilities, and hence the necessity of inversion. Bernoulli wanted to show that the smaller segments of chancelike sequence often exhibit wide disorder, and the larger ones order and stability. To invert Bernoulli's theorem, Laplace was obliged to use the Principle of Indifference and the notion of a priori probability; for, without using the former, probability values cannot be assigned at all. Assuming the stability of statistical frequency (or the Law of Great Numbers) that states that "in the long run" it becomes infinitely probable that the relative frequency with which an event occurs offers us the correct value of its future probability from an event's actual relation frequency, the most likely value of its probability may be indicated (according to the inversion theorem). Sometimes the theory of inverse probability is characterized as a theory of inductive probability or of the probability of "hypotheses" or "causes." If the "causes" of an event e are mutually exclusive and jointly exhaustive, then in relation to each of these 'causes' *e* possesses a certain likelihood, and each 'cause' a certain a priori probability of being true. Among the a priori equiprobable 'causes' that one is accepted (according to the Inverse Principle) as the most probable, which confers the highest probability upon *e*. A priori assigning of initial equiprobability to possible "causal" candidates (or "hypotheses") is "based" upon ignorance and 'justified' by the notorious Principle of Indifference. It has also been shown that Laplacian inversion does not logically allow him to justify universal induction; that is, the induction of laws.[10]

Harrod condemns the Principle of Indifference "not only as unwarrantable in itself and liable to raise false hopes, but as capable of obscuring the true principles of induction."[11] Together with this Principle he is ready to give up the postulates and theorems dependent upon it. This readiness puts Harrod in a really difficult situation; that is, he has to vindicate induction on purely empirical grounds. Fortunately, he believes, such a vindication is possible without any a priori chance calculus or assumption about uniformity of (or limited uniformities in) nature or causation.

Harrod has the highest admiration for Williams and finds in the latter's work a right step toward the solution of the problem of induction. Remember that Williams does not use the Principle of Inversion and thus show his eagerness, at least partly, to avoid the related difficulties. Accordingly, to Harrod, the validity of sample induction can be sustained without the fair sample postulate (as it is ordinarily understood). The only postulate claimed to be sufficient is that "our experience as a whole is a fair sample"[12] and there is nothing a priori—no deductive presupposition in this postulate. Simple induction, holds Harrod, underlies sample induction. His notion of simple induction deserves some attention; for it has been said that the failure to justify induction is largely due to the total neglect of this most *primitive* form of induction. The problem of determining the fairness of samples does not disturb the primitive or the child, and the theorists of induction have a lot to learn from his experience.

The principle embodied in simple induction is that the less we go beyond our experience, the more likely is our extrapolation to be correct; and by sufficiently reducing the extrapolation, the probability number of the neighborhood forecast may be brought very close to 1. This implies *continuity* (or continuities) in nature but this implication may be said to be present in our experience and not a matter of postulation; nor does it entail any sort of necessity in the structure of nature. Harrod's analysis starts, as (he insists) every analysis of induction must start, *"from a condition of total nes-*

cience,"[13] and assumes nothing more than what is empirically permissible; viz., the continuity-informing character of *memory.* Relying on the Principle of Indifference, one could say without any investigation that the probability of the die's falling on to one of its six sides is 1/6; but dispensing with the principle, one could legitimately define one's total area of nescience as $(I\text{-}p\text{-}q)$, where p stands for the postinvestigation probability that the die will show a six is *at least* p; and q stands for the postinvestigation probability that the die will show a face other than six is *at least* q. To establish that neither p nor q is biased, that is, not fair, is to raise p and q so that $(p + q)$ tends to unity. In any case, the distance between 1 and $(p + q)$ remains always finite. Empirical evidence never justifies the assignment of any particular number to the hypothesis (H) that the die will show a six, unless it is qualified by *at least.* Harrod attaches special importance to what may be called the *at least principle,* which is purported to show the reason why we are not empirically justified in assigning a probability number to not-H. From the simple fact that H and not-H add to 1 and that the probability of H is 1/6, one is not inductively justified to assign 5/6 to not-H. Those who are interested in non-a priori but frequency justification of induction must not give up the *at least principle,* for otherwise probability and inductive theory cannot be welded together. Inductive (evidential) determination of the probability value of H, for example, tells us nothing about the probability of not-H, nor about the *improbability* (except in a popular sense) of H.

This shows Harrod's refusal to make an a priori use of the Law of Large Numbers, but he admits that our experience provide us the fairest possible sample of samples—the best (possible) dependable continuity of continuities. Our experience does not warrant us to say that we frequently come to the edge of continuities or that we frequently encounter irregularities even if our extrapolation is very modest. Let us assume someone in a state of total nescience regards one-tenth of a given spatial distance as natural unit of measurement and has proceeded (continuously) 10 yards. Then by applying his fundamental axiom of probability (*If it is the case that B is true if, and only if, A is true, whatever probability pertains to A pertains to B also*) it can be asserted that the continuity will proceed for at least one yard more. For whatever is probable in relation to one-tenth of 10 yards (A) is probable in relation to 1 yard (B). The resulting probability of the hypothesis that continuity will proceed at least 1 yard more comes up to $100/121 = 10^2/(10 + 1)^2$. If the person has proceeded (continuously) for 20 yards, then the probability of proceeding 1 yard more is $20^2/(20 + 1)^2 = 400/441$. Relatively to one-tenth the probability of one-twentieth has risen substantively; that is, (109.747/121). Harrod gives this law of increasing probability continuance the slightly misleading name the Principle of Experience, which may be defended by a restricted frequency calculus without reference to any experience whatsoever.

To extend the application of this principle beyond a particular class of continuity it is necessary to postulate that if enquiries concerning the continuance of continuities are made in a large number of cases, then these will be equiproportionally disturbed among the various classes of continuities. This postulate is purported to negate the disturbing effect of sporadic answers to the enquiries concerning continuities within a particular class. There is nothing to prevent *abnormal* crowding or accumulation of sporadic answers in a particular field of enquiry. It is not easy to define *normal* in terms of *experience*; so inductive significance may be attributed to the use of the word *abnormal* only under an a priori commitment to the Law of Large Numbers. Harrod's claim that his way of attributing equiproportionate dispersion of continuities (and discontinuities) among all classes of continuities (and discontinuities) needs no fair sample assumption seems to be acceptable. To say that the class of continuity of continuities, the fairest sample, is warranted by our experience as a whole is to lend an empirical look to an otherwise a priori piece of reasoning. *Experience* emphatically tells us some samples are biased and some are not; but this "telling," on analysis, is found to carry little weight unless there is an assumption about the composition of the totality of the field from which the samples are chosen (randomly *per* assumption). Harrod refuses to assume uniformity of (or limited variety in) nature and thus puts a tremendous load on *his* Principle of Experience, more than it could possibly bear and carry. The law of increasing probability of continuities is not based on "free" experience, but on experience under some a priori conditions, without which measurement of increase (or decrease) in probability is impossible. I doubt if Hume would have agreed to accept Harrod's Principle of Experience as a principle of *experience*.

His theory embodies an attempt to justify induction by a frequency theory of probability without using any a priori assumptions. This anti-a prioristic claim is not as serious as it is intended to be. Note that the sort of experience on which his theory is established is highly selective and preconditioned by certain necessary formal requirements. The conditions of his crucial example of the journey of continuity from a state of total nescience are too artificial to be counted as genuinely inductive. The condition of total nescience is useful for Harrod's purpose, but clashes with common experience: a clean sweep of background knowledge at any stage of knowledge is unrealistic. When the background knowledge has nothing to do with the choice of the direction of our journey, our actual journey is *indifferent* to what we are experiencing immediately. When the journey starts, the importance of memory becomes obvious. The Principle of Experience depends upon the reliability of memory, and because this reliability is to be shown inductively from previous memory performances, the question of justification reopens once more at this level. That apart, the question of the reliability of memory cannot be decided in a general way; for it varies from person to person; and, given this fact, it is not easy to develop a *logic* of probability entailment on the principle of experience. Unless experience is defined and used as something impersonal, it cannot be a basis of a neat frequency theory of probability. To say that our experience *as a whole* is a fair sample is to say something like this: the unbiased alternatives are equiprobable *in the long run*. Neither the phrase *as a whole* nor the phrase *in the long run* can be given any clear and pure empirical significance: one can always ask, When is the *whole* fair? or How *long* is the run required to be? Clearly, there is no *general* answer to such queries.

For the purpose of common sense I am ready to admit, on purely intuitive grounds, that when I extrapolate my experience very modestly, I feel more or less sure I am not at the "extreme edge"; but for higher theoretical purposes, for framing law hypotheses, for example, modest extrapolation will not do. Conservative empiricism may be saved only by seriously crippling bold scientific investigations. One who is not ready to save empiricism but rather to test whether it will work with the idea that continuity is sometimes threatened by (the discovery of) discontinuity. To justify induction Harrod needs to assume some sort of uniformity of nature, but intending to be thoroughly empiricist he overtly eschews any such assumption. His law of increasing probability of continuities is a substitute for the uniformity assumption. It is to be noted that the substitute law fails to do what it is called upon to do. Natural laws are said to be known by induction, but the conservative Principle of Experience fails to justify them; for its scope is extremely limited.

The attempts to justify induction by sample induction and simple induction fail basically for the same sort of reasons. When the field of induction is infinite, the problem of selecting fair samples cannot be solved by experience and within experience. The largeness of the number and the variety of the samples do not touch in *principle* on the problem of universality (of laws). By bringing the samples under the rules of chance calculus the sort of justification that is extended to induction is structurally deductive in character; and by counting the frequency of "chances" (without any uniformity *assumption*) the "necessity" of universal laws cannot be justified.

Many problems of sample and simple inductions reappear in the frequency theories of probability; but, as I said before, I decided to discuss Harrod's view separately only to examine the claim that by setting out from Hume's basic (inductive) principles, one could break through his impasse; that is, show the rationality—justification—of induction. Irrespective of whether Harrod did succeed in his attempt (I think he did not), his bold and ingenious attempt has some intrinsic value. It brings once more to the fore the old paradox involved in inductive justification of induction: when the Principle of Experience is most reliable (psychologically), that is, when extrapolation is extremely modest, its worth is negligible, and when it is called upon to prove its mettle, that is, to justify induction of laws and theories, it retreats to the protective (deductive) custody of a frequency theory of probability.

4. Induction and Discovery: Mill, Whewell, and Jevons

We do not always argue from *explicit* premises to a conclusion. Sometimes it happens that having reached the conclusion of an inductive argument, we argue back to (or deduce) the premises. This sort of inductive argument is used particularly in connection with the discovery of *general* laws and theories. In the case where we induce (or predict) a *single* occurrence, and arrive at the conclusion stating it, the question of ascertaining its premises or ground does not arise at all. For then the necessary arguing back from a singular statement cannot be claimed to be deductive, and therefore, the proposed justification remains problematic. Even when the conclusion (law or theory) is *general* and the arguing back assumes a deductive character, it is not easy to ascertain whether the original operation of hitting the target (that is, discovering a law or theory) was inductive. The difficulties involved in ascertaining the character of the *original* process of discovering laws have led different thinkers to quite different views. Some are of opinion that the process is undoubtedly inductive, however inarticulate that might be and that otherwise the great scientific discoverer could hardly be distinguished from a clairvoyant.[14] Some others think that these difficulties indicate nothing less than the very impossibility of inductive logic.[15] There are still others who hold that although inductive logic is possible, yet that does not solve the difficulties; for a *philosophical* problem underlies those difficulties (first indicated by Hume), which cannot be satisfactorily tackled within the framework of any inductive *logic*.[16]

Mill thinks that induction is "the operation [both] of discovering and proving general propositions." This view fails to bring out the distinction between generalizing and discovering. But Mill's supporters may rightly point out that this view unmistakably brings to focus what is Mill's basic philosophy in the matter; viz., that induction is the only genuine scientific method which enables us not only to generalize but also to justify what we generalize. Thus Mill is led to the doubtful conclusion that inductive justification of induction is not only possible but the only *logical* justification available.

Mill's conclusion that generalizing and discovering are the same (or a parallel) process has been rejected by such logicians as Whewell and Jevons who think that the operation of discovering is inverse to that of generalizing. Those who hold that discovery requires a literally *incalculable* genius are totally against assimilating discovery to induction and argue to the effect that discovery is due to intuition or lucky guess or "love of reality" and *not* induction in Mill's sense. The idea of *inductive* probability has been rejected by Whewell:

> The doctrine [of] which *hypothesis* of the deductive reasoning . . . is the *inference* of the inductive process. The special facts which are the basis of the inductive inference are the conclusion of the train of deduction. And in this manner the deduction establishes the induction. The principle which we gather from the facts is true, because the facts can be derived from it by rigorous demonstration.[17]

The sort of probability acceptable to Whewell is hypothetic; that is, it pertains to some hypothesis. And the probability value of a hypothesis is said to be determined by the facts deducible from it. As per this view, induction is justified by deduction. It is not clear if Whewell would discount the very possibility of induction. In forming a *probable* hypothesis what we do *psychologically* as a matter of fact is *perhaps* inductive, but this cannot be logically demonstrated. So the logical claim of induction, if any, seems to be too weak to be *logically justified* by induction. The psychology of discovering or framing hypothesis cannot be assimilated to any articulate logical method. Whewell operated with two different but intimately connected concepts of induction—hypothetic and colligating. Inductive inference may be regarded as colligating in this sense that it connects, or is superimposed on, some facts of the same type. The facts that are deducible from a hypothesis, or may be colligated by an induction, are not to be necessarily interpreted as *observed* facts, for the unobserved facts also may be connected by a hypothesis provided their type-affiliation is available and understandable. Hypotheses are intended to colligate facts, both observed and unobserved: if they are restricted only to the observed ones, then knowledge of the future turns out to be a vain quest.

Whewell's two concepts of induction depends upon their interconnectibility. Inductive hypotheses not only interpolate the facts but also extrapolate their patterns (in so far as some patterns are discernible in them). A scientist who observes certain particular features of an object seeks a hypothesis not only to colligate them but also to connect other *possible* features of the *same* object. Kepler's discovery of the planetary path of Mars and Snell's law for the refraction of light are often cited as instances supporting Whewell's theory regarding the justification of induction.

Whewell's theory has been criticized mainly on the ground that it fails to distinguish between "primitive" (or inventive) and derivative (or generalizing) induction. It concentrates attention only on the justification of the former and overlooks the fact that this justification does not cover the latter; that is, generalization. Kepler's law, for example, explains not only the *observed* positions of Mars at various points in its elliptical path but also *all* its (future) positions in its elliptical path. Mill criticizes Whewell for the latter's alleged failure to realize that it is *not* Kepler's *discovery* of the law of ellipses *but* rather the *generalization* regarding it that is induction. Whewell's position and Mill's criticism of it represent two fundamentally different theories of induction: whereas the former is rationalist in its inspiration and transcendental in its implication, the latter is pronouncedly empirical.

The rudiments of Whewell's theory may be traced to the none-too-systematic ideas of Leibniz and Huygens. Huygens thought that laws may be regarded highly probable (that is, just a little short of complete certainty), if their consequences agree with known facts, particularly if such agreements (or verifications) are numerous, and above all if the predictions made according to them are found to be correct. On the basis of formal (logical) similarity between deduction and prediction, the hypothetist sometimes tries to strengthen his or her theory by elaborating the notion of consilience of induction; that is, by pointing out how different generalizations may be deduced from, or subsumed under, a more general law. It has been said that the probability value of Newton's law of gravitation, for example, increases when (together with his laws of motion) it explains and entails both Galileo's law of falling bodies and Kepler's laws of planetary motion. Whewell laid great stress on the notion of the consilience of inductions, for it accords well with his rationalist leanings. The inductions found to be consilient to one another are all primary inductions and their consilience is shown always *via* another secondary generalization of a relatively higher order. The consilience of inductions increases the probability not only of the higher generalization that entails the primary inductions but also of the latter. If a hypothesis may assume and augment its probability values by showing the consilience of different inductions under it, then, it is often argued, even a nonempirical hypothesis may be said to have probability value. Second, it has been said that the hypothetist fails to explain the increase in the probability value of a *universal* empirical hypothesis, for the existential statements (the verification of which might inversely and indirectly probabilify it) are not deducible from it *alone*.[18]

I do not think that these defects in the formulation of the method of hypothesis are irremediable. I shall try to show later on that these defects in fact have been remedied by some contemporary logicians. Those who believe that the "essence" of induction is generalizing and not hypothesis framing (to connect the facts) seem to be laboring under an inductive misconception of science and are doing too much with the distinction and too little with the affinity between primary and derivative (or secondary) induction.[19] The relation between primary and secondary induction may be studied both negatively and positively: negatively speaking, neither primary nor secondary induction starts from (observ*ed*) particular facts as such; positively speaking, both primary and secondary inductions are generality oriented or concerned with observ*able* facts.

It should be remembered that *facts* are *not* bare ontological posits or entities, but transcendentally oriented in the literal sense. Both *generalizing* and *hypothesizing* (or framing hypotheses) are modes of *transcendentally orienting* ontological posits (facts). The positions of the posits are oriented always toward some (existential) or other (universal) form of generality. To *induce* is to *relate*: induction involves relation, relation of facts or properties of facts. Relating means grouping ontological posits from a certain point of view. I am inclined to believe that Jevons is right in holding that grouping or "classification is not really distinct from the process of perfect induction."[20] Both classification and induction are based upon identification of properties *of* facts; so it has been observed that there is no point in taking the bare ontological posits as the basis of induction. Inductive inference is said to be dependent upon the invention of hypotheses that accord with experience. Jevons writes:

> Such accordance renders the chosen hypothesis more or less probable, and we may then deduce, with some degree of likelihood, the nature of our future experience, on the assumption that no arbitrary change takes place. We can only argue from the past to the future . . . particulars as particulars can no more make an inference than grains of sand can make a rope. . . . [T]here always is the possibility that some unknown change may take place between past and future cases. . . . Our inferences, therefore, always retain more or less of a hypothetical character, and are so far open to doubt.[21]

All knowledge is inductive in Jevons' sense. But one must note in what sense Jevons uses the term *induction*. According to him, "all inductive reasoning is but the inverse application of deductive reasoning." And, following Laplace, he tries to justify induction by the principle of inverse probability. The theory of probability is purported to show "how far we go beyond our data in assuming that new specimens will resemble the old ones." The possibility of going beyond the given data, it will be recalled, depends on the *number* and *variety* of the *samples* or specimens of the data. Because Jevons believes that our inductive hypotheses can never avoid the risk of being falsified by future contingencies, he seems to be free from ontological commitments regarding the uniformity (or uniformities) of nature, and, therefore, his theory, like that of Williams, depends on and is open to the criticism of the postulate of sampling.

Discovery, according to Jevons, is nothing but the discovery of hypotheses purported to colligate or connect known facts and foresee the unknown "facts" of the variety of the "known" specimens; and as the hypotheses can never rise above doubt, the projected justification of induction by discovery is never beyond doubt. If the facts (of a kind) to which the inductive hypothesis is intended to apply are infinite, its justification by a finite number of (deducible and observed) cases is inadequate for strict logical reasons. Except in cases of perfect induction, this sort of justification can never be adequate. If the scope of inductive generalization is unrestricted, then its accordance with experience can lend it only a "partial justification"; but this justification is *not* the result of the verification of that generalization; for a probable and unrestricted generalization cannot be conclusively verified.

Inventionistic justification shows in effect that although inference from particulars as such to particulars is absolutely impossible, still it is possible via some hypothesis. Prior to finding that the empirical consequences of the suggested hypothesis accord with experience it cannot be claimed that the suggested hypothesis is justified merely by virtue of its power to connect some phenomena. The necessary justification has two aspects: in the first place, it must be able to connect the phenomena of a *kind,* and in the second, to anticipate other verifiable phenomena of the same kind. Only when the second part is fulfilled, at least partially, are the initial facts (to connect which the hypothesis is suggested) admitted as the premises of the conclusions entailed by the hypotheses. Two points need to be distinguished here: (1) verification of anticipated facts may justify the generic feature (indicated or covered by the hypothesis) of the facts but *not* the hypothesis itself and (2) verifications of the consequences of hypothesis may test the hypothesis itself. In the latter case the *logical* status of the initial facts (awaiting connection) and the deduced (or anticipated) facts is hardly different. The property or properties in terms of which a set of facts is defined are indifferent to the time reference of the facts. Hypotheses may be infirmed or modified not only by *future* but also by *past* facts. This is understandable if we remember that facts as such neither confirm nor infirm any hypothesis, and interpreted facts alone may be said to have modifying influence.

It is not clear how invention or discovery may be logically regarded as lending justification to induction. First, the hypothetist's use of the term *induction* is so fundamentally different from the inductivist's that one can never be sure whether by *justification of induction* they mean the same thing. It seems that, although the hypothetist intending to justify induction is in fact trying to legitimize his or her hypothesis, the inductivist is trying to show that the inference of general truths from particular ones is rational. Analyses reveal that neither the framing of a hypothesis nor the way of legitimizing it is inductive in the inductivist's sense, and that inferring without any a priori presupposition regarding the structure of nature or some analytic principles is not possible. Second, inductivists themselves are not definite whether induction is *the* (or even *a*) method of discovery. Without totally discarding the vocabulary of modern logical literature, it is not wise to claim that one can construct an inductive machine that will be able to discover. Third, because the distinction between primary and secondary induction cannot be defended on strict theoretical grounds, it is idle to assert that even if secondary induction (used to frame higher-level hypotheses) cannot be justified inductively, the primary one can be. It has been said (by Kneale, for example) that primary and secondary induction are acts of policy necessitated by our wish to go beyond the limits of our actual experience, and that, whereas primary induction is concerned with laws or probability rules, secondary induction is concerned with theories. The plausibility of this argument too depends upon the alleged distinction between laws (or probability rules) and theories. To this question I shall return later on, but at the moment let me say straightaway that both laws and theories are hypothetical in character and *probable* only in this sense that they can never be conclusively established and may always be con*tested.* Besides, a mere *wish* to go beyond experience cannot be regarded as justification of induction (as a matter of policy). It is not quite clear what precisely makes a policy inductive: if it is nothing other than the observed cases, then it does not explain the *direction* in which we wish to extrapolate (and it is to be remembered that *un*directed extrapolation takes us nowhere); and if it is our wish to connect the observed cases *and* extrapolate their *kind,* then the policy (even at the primary level) is already hypothesis oriented, that is, not inductive in any ordinary sense.

It is to *decide* the direction of inductive extrapolation that some justifiers of induction favor the adoption of conventionalist stratagem. Instead of seeking the principle of induction in reason or experience, says the conventionalist, we may accept a convention (as it were, as a matter of policy) to guide and regulate our experience. When we say, for example, that the melting point of phosphorus is 44°C we may initially believe that this is an inductive generalization based on numerous observations, and that in each case the identification of phosphorus depends on the observation of some of its properties such as color, smell, and taste; but *practice* shows that something "having" those properties is not regarded as phosphorus, *if* it does not melt at 44°C. This refusal to regard something ("having" the properties of phosphorus) as phosphorus is claimed to be *rational* on the grounds of an accepted convention that *defines* what *is* phosphorus. Rationality or justification of induction then seems to result from a convention (defined a priori), which cannot be falsified (but unilaterally verified) by any experience (regulated by that convention). The eternal immunity of conventions from error has often been interpreted as the vacuity of *analyticity* of the general statements of science (subsumable under, and justifiable by, some infallible conventions).

The conventionalist's indifference to the verdict of experience has been rightly criticized by the logical empiricist on the grounds that it is intended primarily to eliminate the problem of induction and not to justify induction. In one very important respect my sympathy lies with the conventionalist, who clearly realizes that inductive justification of induction is *logically* impossible, and that any justification of induction worth considering must be a priori. But this does not mean that the a priori principle(s) must be *analytic*. Nor does it imply that the *general truths* are mere definitions. According to some radical conventionalists (Le Roy, for example) scientific truths are matters of *choice*; but this radical view has been rejected by, among others, Poincare, who takes the question of truth very seriously. But one doubts if the question of *general* truths may be satisfactorily tackled within the framework of conventionalism (or instrumentalism).

It may be incidentally remarked that even those who like Carnap think that the principle of induction is analytic fail to satisfactorily answer the question, How may experience be shown *relevant* to the establishment of (general) truths of scientific laws and theories?

3

Probabilistic Justifications of Induction

1. From Inductive to Probabilistic Justifications

I have tried to show in the previous chapter, though rather briefly, that the so-called inductive justification of induction is *logically* destined to fail; and that a tenable view of justification demands, among other things, reformulation of the problem so that the justification of the problem may not be *logically* impossible, the recognition of the necessity of accepting a norm or rule for identification and weighing of *relevant* evidence (both pro and con), and admission of the open texture or nonconclusive character of justification. My purpose in this chapter is mainly critical and very limited. I intend to show the instructive errors of the probabilistic theories of justification with special reference to Keynes, Reichenbach, and Carnap. Later in this book, I shall try to justify induction from a prophenomenological point of view, bringing the concept of hypothetic probability to the center of discussion and taking a nonatomic model of the world.

2. From the Basis of Experience to Beyond It

"On the *basis* of experience we cannot go *beyond* experience." Thus did I reformulate Hume's problem. This reformulation appears to be too simple and, therefore, a bit ambiguous unless, of course, it is further spelled out. The initial ambiguity centers around the concepts of *basis, beyond,* and *experience.*

Basis may be construed, broadly speaking, in two different ways. First, some observation reports, unconnected by any law or unoriented by any point of view, may be regarded as a *basis* of such generalization or hypothesis as is not logically entailed by the said observation reports. Second, the observation reports *relevant* to a hypothesis are counted and weighed as the *basis* for the acceptance or rejection of that hypothesis. The basis-*of* view of *basis,* if constructable at all, is of little significance in any scientific or humanistic inquiry. The basis-*for* view of *basis* is a necessary part of any critical inquiry and should not be rejected together with the former view on the alleged ground that it too leads to genetic fallacy.

The narrow inductivist view of scientific inquiry, as Hempel calls it, has now been largely abandoned, and in this matter Popper's anti-inductivist ideas proved very influential. In an obvious sense every generalization may be said to go *beyond* the observation reports on which it is based but that do not logically entail it. Even this obvious sense of *beyond* was once contested by logical atomists. In a less obvious sense we might assert that laws or lawlike generalizations are destined to remain more or less *beyond* the bounds of experience, that is, observation reports, on the basis of which these are accepted or rejected.

The concept of experience is notoriously vague. The phenomenalist takes *experience* as intrinsically formless, and all of its forms are said to be adventitious imposition. Some rationalists are committed a priori to the views that experience is intrinsically form laden and that humans are blessed with the intuitive power to *discover* those forms. Critical rationalists admit the structured or form-laden character of experience but deny that we enjoy any intuitive and infallible access to the forms already implicit in our experience. Both the macro and micro forms of experience, they hold, are too complex to be apprehended a priori and *necessarily* aright. The phenomenologists' program to avoid the devil of forms without matter or experience and the deep sea of formless matter is, although very well-intentioned, difficult to execute. I do not know how to justify the claim that knowing is a sort of widening the horizon wherein we are already living, unless, of course, I believe that knowing and generalization are all self-validating or self-authenticating processes. One must bear in mind the difference between "self-correcting" pragmatic rules and "self-validating" phenomenological or horizonal expansions.

However, I am inclined to hold that different concepts of *basis, beyond,* and particularly *experience* have, as I will try to show, important bearings upon the problem of justification of induction. The relatively insignificant position of epistemology in the current discussion on induction and probability is very distressing indeed.

Whoever raises epistemological issues in this area runs the risk of being accused of psychologism or subjectivism. I am not as yet quite clear how one could possibly construct a purely formal and interesting theory of justification of induction that would have nothing to do with epistemology. True, separable issues should be separated and then discussed, but we must not lose sight of their connection either. To look at the issue from different sides might help us grasp it better.

Before we start discussing the issue of justification of induction, we should, I think, be clear about the type of inductive inference we are going to discuss. This question seems to me very important for two reasons. The first reason is rather obvious: there are different, although not strictly exclusive, types of inductive inference. The second reason is not so obvious: different logicians, while speaking of justifying induction, have different types of inductive inference in their mind. In other words, they address themselves to different problems, but their similar terminology creates confusion, which, given caution, I think, is avoidable.

Carnap speaks of five types of inductive inference: (1) *direct* (that is, Population-to-Sample); (2) *predictive* (that is, Sample-to-Sample); (3) *analogical*; (4) *inverse* (that is, Sample-to-Population); and (5) *universal* (that is, Sample-to-Universal Hypothesis).[1] Whereas some logicians like Carnap are mainly concerned with (2), others like Popper are with (5).

3. From Demonstration to Probabilification: Keynes's Position Examined

The premises of inductive inference can never *demonstrate* the certainty of the conclusion but can only *probabilify* it. The problem of probabilistic justification assumes special significance when the conclusion transcends the boundary of factual information embodied in the premises, and it is with this type of probabilistic justification that we will be mainly interested in this chapter. Inductive inference is often bracketed with probable inference to contrast it with deductive inference.

The probabilistic approach to the problem of justification assumes two important forms that must be distinguished. One purports to show that inductive inferences (or predictions) are pragmatically justified, if not in short run, at least "in the long run." How long "the long run" will be, one wonders. This views of justification is based on the frequency of the concept of probability (henceforward FIP), and has been defended, among others, by Von Mises and Reichenbach. The other view, represented by Keynes, Broad, and Jeffreys, discounts the pragmatic-predictionist approach and tries to show that the question of justification is purely logical and not empirical. In this section I propose to examine the second view as it has been developed by Keynes. He writes:

> the validity of induction . . . depends, not on a matter of fact, but on the existence of a relation of probability. An inductive argument affirms, not that a certain matter of fact *is* so, but that *relative to certain evidence* there is probability in its favour. The validity of the induction, relative to the original evidence, is not upset, therefore, if, as a fact, the truth turns out to be otherwise.
>
> The clear apprehension of this truth profoundly modifies our attitude towards the solution of the inductive problem. The validity of the inductive method does *not* depend on the success of its predictions. Its repeated failure in the past, may, of course, supply us with new evidence, the inclusion of which will modify the force of subsequent inductions. But the force of the old induction *relative to the old evidence* is untouched. . . . The validity and reasonable nature of inductive generalisation is, therefore, a question of logic and not of experience, of formal and not of material laws. The actual constitution of the phenomenal universe determines the character of our evidence; but it cannot determine what conclusions *given* evidence *rationally* supports.[2]

This is a very clear statement of how "inductive generalisation" can possibly be validated, and, I think, this has important bearing upon our problem. The points of special importance to be noted in this connection are (1) to what extent given evidence rationally supports or probabilifies a generalization is a purely logical issue and has nothing to do with the experience of the said evidence; (2) failure (but not success) of past predictions influence future induction; and (3) the constitution of nature determines the character of our evidence. In what follows I shall try to clarify each of these points in some details.

First, (1) *probability relation,* according to Keynes, obtains between two sets of propositions. Propositions are either true or false. Conclusions of probabilistic *arguments* being propositions must be either true or false and cannot be doubtful, uncertain, or probable. What is probable is our *rational belief* in the conclusion and not the conclusion itself. The rationality of our belief in a proposition "depends on our circumstances . . . a *corpus* of knowledge, actual or hypothetical."[3] Because our circumstances differ and background knowledge is not uniform, different individuals might entertain "different rational beliefs" in one and the same proposition. This might lead one to think that Keynes is developing a subjective theory of probability. Both Carnap[4] and Popper[5] admit that Keynes's view is subjective in no ordinary sense. Those who have accused Keynes of subjectivism did not take due note of his such statements as "probability is not subjective . . . not . . . subject to human caprice" or "there is some real objective relation between [for example] Darwin's evidence and his conclusions, which is independent of the mere fact of our belief." Hanson seems to be justified in holding that "despite Von Mises' contention to the contrary . . . one can characterise [Keynes's concept of probability relation] in a time-independent way."[6]

According to Keynes, *q* confers probability *i* on *p* if *p* is *derivable* from *q*, and *q* confers probability zero upon *p* if *p* and *q* *contradict* each other. Between the extremes of *derivability* and *contradiction* lie other *probability relations* that may be put in this way: given *q*, *p* is probabilified to the extent of *r* (symbolically $p/q = r$). The numerical probability value *r* of *p* depends upon *q;* that is, *r* is the measure of the *rationality* of our belief in *p* in the light of our knowledge of *q*.

"The light of our knowledge," of *p* or *q* has nothing to do with the truth value of *p* or *q*. It is either true or false, but not both. One and the same proposition may be more probable in the light of *my* knowledge *and* less probable in that of *your* knowledge. Even then we will be less than fair in attributing a subjective theory of probability to Keynes. For he is clearly against interpreting variation in rational belief in terms of psychological causes and defining probability as measuring the degree of *subjective* belief. Keynes is interested in measuring the rationality and not subjective or psychological intensity of our belief in a proposition. "Probability is the study of the grounds which lead us to entertain a rational preference for one belief over other."[7] Very rightly has he pointed out that even Venn, an early advocate of the statistical (frequency) theory of probability, could not deny the "real value of the (non-statistical) 'process' or 'causes' of variation in our belief in a proposition." Although Venn rejects the consideration of these "processes" or "causes" as purely "preliminary" on the ground of (a) subjectivity and (b) immeasurability, Keynes holds (a) that "Objective probability" presupposes "subjective probability" and (b) that the presupposition of *accurately* measurable probability is itself untenable. In fact, the explicanda of Venn and Keynes are different, and it is no wonder that their explicata too should differ.

The sort of knowledge that we obtain by *argument* is said to be derived from and justified by direct knowledge. Keynes thinks that the logical relations (of probability) between the premises (supplied by direct knowledge) and the conclusions derived from them measures the *rationality* of our belief in a proposition or set of propositions. The object of his theory of probability is to systematize the processes of probability inference on the basis of given evidence or drawn from a *limited* body of premises. Keynes was the first to emphasize the relative character of probability.

He is concerned with propositions and not with "events" or "happenings," for he thinks these terms are "vague and ambiguous." But he does not tell us why he thinks so. Probability for Keynes is a property of *argument* and not of event or happening. Arguments define relations between propositions that are either true or false in respect of events or happenings. Although some philosophers like Popper have used such terms as *events* and *occurrence* ("the material mode of expression"), most of them are persuaded of the superiority of "the formal mode of expression."[8]

Keynes's idea of proposition as object of knowledge and belief seems to have been abandoned by modern philosophers. He thinks that we know propositions directly by contemplating the objects of acquaintance and indirectly by argument, that is, by perceiving the probability relation obtained between propositions. Propositions in this view are like quasi-Platonic entities, true or false forever *in themselves,* and their probability values, which are always *relative,* have nothing to do with their truth values. Consequently, Keynes is obliged to draw a very sharp distinction between "logic of truth and logic of probability." And Reichenbach is strongly opposed to drawing this line of distinction without indicating the difference adequately.[9] As for himself he thinks that truth logic is "approximative" and not absolute, and that it is continuous with probability logic. He counts Keynes among the a priorist logicians who treat logic as "a science with its own authority, whether it is founded in the apriori nature of reason, or in the psychological nature of thought, or in intellectual intuition or evidence," and characterizes his own conception of logic as formalistic. "Probability logic," according to him, "determines the methods of scientific investigation," and it can be interpreted in a formalistic way. This view clashes with the view that probability logic is a system of rules of rational belief. The difference becomes obvious when we remember that, although Reichenbach says "that the theory of probability needs the definition of probability as the limit of the frequency,"[10] Keynes holds that "a *definition* of probability is not possible."[11] The latter is emphatic on the point that there is no direct logical passage from truth to probability or the converse. "Probability begins and ends with probability."[12]

Second, (2) the defensibility or otherwise of the Keynesian justification of induction depends upon his ability to meet Reichenbach's claims (a) that the frequency concept of probability supersedes the concept of truth and (b) that a probability statement sustains and is sustained by prediction.

It seems that many contemporary philosophers like Carnap[13] and Popper[14] are against the supersession of the concept of truth by that of probability. Carnap thinks that Reichenbach's belief in a multivalued probability logic superseding the two-valued logic is "based on a lack of distinction between 'true', on the one hand, and 'known to be true', 'absolutely certain', 'completely verified', on the other." Popper is also against replacing *true* by *probable* and in favor of maintaining the distinction between the two, and he does not think that this distinction is a legacy of classical logic. Truth seems to be an absolute ideal concept. We do not know what is absolutely true; rather our knowledge is regulated by this ideal. The distinction between "true" and "known to be true" is very important and should be observed.

We will have occasion later to discuss in detail Keynes's arguments against Reichenbach's claims (a) and (b). Let us now proceed taking Keynes's conclusions on these two points for granted. (a) Every proposition, according to him, is *necessarily* true or false, but its probability is *contingent* upon "the limitations of our knowledge" of it.[15] (b) Predictions as such are of little weight for the probabilification of a hypothesis unless they are backed by a uniform principle limiting the *variety* of the predictions to be counted (and weighed) as relevant.[16]

In terms of the second (b) point Keynes's emphasis on "negative analogy," unsuccessful predictions rather than successful ones, is to be understood. He draws an important distinction between *Analogy* and *Pure Induction* and a further distinction between *Pure Induction* and *Inductive*. Pure induction is concerned with "the mere repetitions of instances," but the number of additional instances as such is of little importance unless from previous knowledge we could gather the *various* type-affiliations of the instances. Keynes uses the word *analogy* in the sense of similarity or resemblance. *Positive analogy* measures the resemblances and *negative analogy* measures the difference between the characteristics of the instances of a generalization. Keynes draws another useful distinction between *total negative analogy* and *known negative analogy*.

If the negative analogies are already known to be uniform, there is little use of counting further instances. But if this uniformity is not known, each additional instance may be expected to add to the negative analogy. This expectation, of course, presupposes that the instances are variously limited in nature (objectively).

The existence of various characteristics in nature is definitely somewhat more than a matter of *presupposition* with Keynes, and this seems to underlie his suggestion that "inductive methods are necessarily confined to the objects of phenomenal experience and empirical question" or that he is not ready "to preclude from the outset the possibility of their use in abstract and metaphysical inquiries."[17]

Following Hume, Keynes maintains that analogy is the basis of every inductive argument. A general proposition is arrived at on the basis of partial analogy of some of its instances. The generalization $g(\phi, f)$ is true if ($f(x)$ is true for all those values of x for which $\phi(x)$ is true. To assert "All swans are white" is equivalent to asserting "'x is white' is true for all those values of x for which 'x is a swan' is true." The proposition $\phi(a)\cdot f(a)$ is an instance of $g(\phi, f)$. A positive analogy ϕ between set of instance $a_1 \ldots a_n$ may be written as

$$A\ (\phi)^1$$

$$a_1 \ldots a_n$$

and a negative analogy ϕ as

$$\bar{A}\ (\phi')^1$$

$$a_1 \ldots a_n$$

The value of the generalization $g(\phi, f)$ depends partly upon the comprehensiveness of the conditions ϕ and f. The more comprehensive ϕ and the less comprehensive f, the greater a priori probability value is assumed by g. The significance of additional instances

> arises not out of their number as such, but out of their tendency to limit and reduce the comprehensiveness of ϕ_1, or, in other words, out of their tendency to increase the negative analogy ϕ', since $\phi_1\phi'$ comprise between them whatever is not covered by $\phi\ f$. The more numerous the instances, the less comprehensive are their superfluous resemblance likely to be. But a single additional instance which greatly reduced ϕ_1 would increase

the probability of the argument more than a large number of instance which affected ϕ_1 less.[18]

Addition of positive instances help us to determine the proportion between the *total* negative analogy to the *known* negative analogy. But the probability of a generalization is influenced by negative instances as well. When the instances of ϕ or f or part of them are *known to be false,* the generalization $g(\phi, f)$ is certainly influenced. Instances of the form $\phi(a) \,.\!\sim f(a)$ certainly ruin $g(\phi, f)$. But instances of the form $\sim \phi(a) \,.\!\sim (a)$ are favorably relevant to the generalization. Our object in the case of scientific generalization is to increase the negative analogy. Predictive verifications of the generalization depends not so much on their *number* as upon the variety of the circumstances in which they are fulfilled. In advanced scientific inquiry, when our background knowledge is very rich and analogy good, the role of pure induction is of little importance.

Keynes shows clearly, I think, that all evidences, positive and negative, are not of equal weight. But this raises a problem that affects his very notion of probability. Probability, according to him, is indefinable, and it is made intelligible in terms of degree of rational belief. This view seems to clash with his assertion that probability in some cases is, and in other cases is not, measurable. I find nothing wrong in the view itself that "there are some pairs of probabilities between the members of which *no* comparison of magnitude is possible." What perplexes one is Keynes's claims (a) that the distinction between probabilities that are measurable and that are not "is not fundamental," and (b) that both measurable and immeasurable types of probability could be satisfactorily tackled within the framework of *one* unified notion of *indefinable* probability.

Keynes is a continuer of the Bacon-Mill tradition in the field of induction with a particular emphasis on elimination and *not* enumeration. Bacon thinks that laws of nature cannot be directly established through induction. It is rather in the light of negative instances that the generalizations *about* the laws of nature are to be precisified and validated. The importance Bacon attaches to negative evidence is betrayed by his remark, *major est vis instantiae negativae*. That additional instances, particularly the negative ones, can help us to determine the range and character of laws with varying degrees of probability, according to him, is due to the limitation of what he calls *generating natures* in relation to *generated natures*. As the number of generating "phenomena" or the properties is relatively less than that of the generated ones and as there are coordinating laws or forms between them, it is *logically possible* to establish "causal" relation between them by counting and weighing their copresence, coabsence and covariance. Bacon seems to have been influenced by a sort of Aristotelian belief that *natura naturans* and *natura naturata* are ordered in hierarchy of forms. A somewhat similar line of argument is followed by Keynes when he writes:

> the almost innumerable apparent properties of any given object all arise out of a finite number of generator properties, which we may call $\phi_1\phi_1\phi_1$. . . . Some arise out of ϕ_1 alone, some out of ϕ_1 in conjunction with ϕ_2, and so on. . . . Since the number of generator properties is finite, the number of groups also is finite. . . . Since the total number of apparent properties is assumed to be greater than that of generator properties, and since the number of groups is finite, it follows that, if two sets of apparent properties are taken, there is, in the absence of evidence to the contrary, a finite probability that the second set will belong to the group specified by the first set.[19]

And this brings me to the third and perhaps the most important point in Keynes's argument in justification of induction.

There are two different ways of justifying induction (3) by subsuming it under some deductive system. First, it may be shown that every induction, some particular minor premises apart, presupposes a suppressed major premises. But the main difficulty involved in this approach is this: the inductive conclusion cannot be shown to be a logical consequence of the minor premises in spite of the backing of the suppressed major. This difficulty led some logicians like Carnap and Kneale to try the second way of deductive justification. The conclusion of an induction, they hold, is not itself a scientific hypothesis "but is instead a proposition assigning a number to the hypothesis in relation to the evidence, this number measuring the 'probability' in the sense of Kneale's acceptability or Carnap's 'degree of confirmation'."[20] It must be said, as it has been by Braithwaite, for example, to Keynes's credit that he has made it abundantly clear that the second way of deductive justification cannot be logically defended unless it is combined with the first. In order that an inductive hypothesis may assume a definite numerical value (that is, "posterior probability") in relation to certain "confirming" instances its "initial (that is, prior) probability" should be greater than zero. To be assured of this we need a supreme major premise to back up the minor premises, the confirming instances. The Keynesean way of justifying induction requires both a theory of (degrees of) confirmation *and* a supreme major premise. We have already made some brief remarks about the first requirement. It will not, perhaps, be unfair to remark here that subsequent works in this field have largely superseded the conclusions of Keynes.

About the second requirement Keynes seems to have said something of lasting interest. He realized clearly that without a supreme major premise inductive generalization is not possible. He also felt that the Law of Uniformity of Nature as propounded by Mill would not do. "The kind of fundamental assumption about the character of material laws, on which scientists appear commonly to act," says Keynes, "seems to me to be much less simple than the bare principle of Uniformity."[21] The Uniformity Principle assumes the *whole* of Nature to be *uniform*. This implies that the generalizations that hold good in one part of Nature also do so in all other parts in spite of difference in space and time. The assumption has been characterized also as a "generalized judgment of irrelevance"—irrelevance of space and time to generalizations that are not otherwise circumscribed by space and time. This assumption does not appear to be satisfactory to Keynes for several reasons. First, it is too simple to explain the complex phenomena of Nature. Second, this assumption is committed to an atomic (as opposed to the organic) view of the world; viz., that the world consists of atoms "such that each of them exercises its own separate, independent, and invariable effect." It has not been always realized, alleges Keynes, that if every configuration of Nature were unique and subject to a separate and independent law, prediction would be impossible and induction useless. Finally, Keynes thinks that the principles of uniformity and causation merely assert the inferability of some posterior events from *some* given data and "do not seem to yield as much assistance in solving the inductive problem proper, or in determining how we can infer with probability from *partial* data."

The sort of supreme major premise that Keynes thinks we need to justify induction must be indicative of some such fact as that any two individuals in Nature differ only in a limited number of respects. This fact may be expressed in various ways. First, the number of the *ultimate* constituents and the necessary laws of Nature is less than that of the members belonging to *various* types or groups (of constituents and laws). One may put the thing in a negative way: an inductive generalization is not possible at all if the objects or individuals over which it ranges have an infinite number of qualities. Keynes writes: "we need some such assumption as that the amount of variety in the universe is limited in such a way that there is no one object so complex that its qualities fall into an infinite number of independent groups."[22] Or, second, it may be said that to justify the inference of the conclusion *C* from the given evidence *E* we need a hypothesis *H* such that *C/H*, the a priori probability of *C*, has a finite value. This value is increased with the increase of *E*; that is, *C/HE*, *C/H*. Or, third, an exhaustive description of an individual or a set of individuals is in principle possible. Or, finally, "we believe," says Keynes, "that all properties of numbers can be derived from a *limited* number of laws, and that the same set of laws governs all numbers."

The most damaging criticism that has been raised against Keynesean way of justifying induction is that it is no less viciously circular than the predictionist justification of induction. If the predictionist is accused, as he has been by Keynes, for example, of circularity on the ground that predictions presuppose a hypothesis, which in fact can never be established without being backed by a similar presupposition, the critic may point out that probabilistic justification (of the Keynesean type) itself presupposes that *empirical* (that is, nonpostulational) character of the Principle of Limited Independent Variety (PLIV), which is again logically vulnerable. The empirical character of the PLIV cannot be vindicated without induction, but then the PLIV is ruined. Or, as Braithwaite argues, the postulational character of it has to be admitted.[23] Keynes foresaw the difficulty and also realized that every attempt to get rid of it gives rise to another or several other difficulties. From this one might conclude that the problem of induction is not genuine and that there is no point in trying to solve it. Keynes does not entertain any such skeptical attitude. He admits that larger issues of epistemology and metaphysics are involved in this problem and does not pretend to have given any perfectly adequate *reason* for accepting the PLIV. "It seems to be neither a self-evident logical axiom nor an object of direct acquaintance."[24] Ultimately Keynes is obliged to conclude that the PLIV is itself made probable by induction. This reminds one of Mill's justification of the Principle of Universal Causation.[25]

Nicod alleges that the reasoning underlying the PLIV rests on an unacceptable hypothesis and clashes with the second condition (for the probability of a law or generalization to approach certainty by the multiplication of its instances to infinity). Nicod's second necessary and sufficient condition has been put in this way: to increase the initial (nonnull) probability of a law beyond any limit by an infinite number of cases, "*it is necessary . . . that on the hypothesis that the law is false, its successive verification in an infinite number of cases is infinitely improbable.*"[26] Keynes assumes that the number of the individual cases in Nature is finite. From this he feels entitled to determine a finite lower limit of the probability that the next case (provided it is *randomly* chosen) is one of the exceptions to the law. Nicod argues that what is finite is not the number of individual cases but only the number of their species (that is, the number of distinct cases). Keynes appears to be committed to the "strange assumption" that distinct cases, however large their number may be, do not constitute more than a single instance. Given Keynes's assumption, a law, say *P*, of the form "*x* entails *A*" is always infirmable as there is at least one species of characters where *x* is but *A* is not. Under the condition of equiprobability, the probability of a *species* of *x* being not *A* is always higher than a finite value. Nicod's point is that "*we do not by any manner or means actually draw a species, but always an individual.*" Moreover the *expectation* that an unexamined individual of the genus *x* will also be a member of the species *A* rests upon our *ignorance*.

Incidentally it may be remarked here that, although Keynes is in favor of justifying induction by infirmation, Nicod intends to vindicate induction by confirmation. The defects of Nicod's theory of confirmation have been clearly pointed out, among others, by Hempel.[27] Some philosophers go so far as to assert that Nicod's concept of confirmation is not inductive at all.[28]

It has been pointed out that the Keynesean rule of assigning numerical values to different infirming instances derives its *rationale,* if any, from the inadequately modified Principle of Indifference (PI) or from the consideration of our subjective ignorance. That the issue of equiprobability (rather the rule of its determination) may be freed from subjectivism was not clearly realized by Keynes. In this respect he is a follower of Laplace and Bernoulli. Ellis writes: "Mere ignorance is no ground for any inference whatsoever: *ex nihilo nihil.*"

In fairness to Keynes we must admit that for measuring probabilities he does not rely *merely* on the assumption of subjective ignorance. One might say that the sort of ignorance Keynes speaks of is relative—relative to the field of total (possible) evidence regulated by PLIV. Our actually available evidence being always partial we are obliged to assume a pattern of the distribution of the rest of our *relevant* evidence. A somewhat similar argument has been raised by Barker against Carnap's notion of total evidence.[29] The PI depends upon the *intuitive* judgments of relevance and irrelevance, which are said to be inexplicable in terms of frequency theory. Ultimately relevance or irrelevance presupposes some such principle as the UN (Uniformity of Nature) or the PLIV. This is another way of saying that the relation between (and weighting of) different infirming (or eliminatively confirming) instances is determinable because of a definite structure of the universe. The range of the possibility of success of an anti-inductive demon is defined a priori by the structure of the universe, whatever that might be. But most important question perhaps is, How is that structure to be identified?

If it is held that the structure is identified intuitively and that the intuition is self-evident, it amounts to foreclosing any rational discussion. And then the sort of justification that the PLIV offers is dogmatic. But Keynes says that the PLIV itself is probable. *Self-evident* and *self-probable* are kindred concepts—indicative of theoretical necessity without adequate rational backing. The problem centers partly around the concept of rationality. One might think that induction is always rational whatever might be the structure of the universe. For there are pragmatic or phenomenalistic ways of determining what is or is not rational. Even assuming that there is a very successful anti-inductive demon, this does not prove induction to be irrational. It might suggest nothing more than the necessity of restricting the range of ampliation or inductive extrapolation. The assertion about the success of the fictitious demon itself is inductive. The question of rationality (of a belief in the light of some evidence) has no necessary connection with the structure of the world.[30]

Keynes was right in keeping the issues of truth and probability separate but wrong in conflating those of probability and rationality. Assuming certain methods of induction are rational he raised the question, What must the structure of the universe be to make them so? He did not raise the more fundamental questions, Is induction possible at all? and, assuming it is, Is is rational?

Keynes's discussion of the problem of justification of induction highlights several points.

1. To justify induction by assimilating it to deduction *both* a theory of probability (that is, degree of confirmation) and a supreme major premise like the PLIV are necessary.

2. The plausibility of the predictionist justification rests on a confusion between truth and probability, resulting in forgetting that "a mere guess, the lucky fact of [an a priori hypothesis] preceding some or all of the cases which verify it adds nothing whatever [of truth-value] to its [probability] value." There is no logical passage from probability to truth.

3. Eliminative induction is superior to enumerative induction.

4. Probability is in a sense subjective and circumstantial and "not subject to human caprice."
5. Some probabilities are measurable, some are not, but the distinction is not fundamental.

4. Examination of Reichenbach's Frequency Interpretation of Probability and the Predictionist Justification of Induction

I now propose to discuss a theory of probabilistic justification of induction that is inconsistent (more or less) with each of the preceding five points. This is known as the frequency interpretation of probability (FIP), and I think Reichenbach's presentation of the view, though a little outdated, is the one most comprehensive and careful.

Reichenbach offers a predictionist justification of induction without presupposing any PLIV-like principle. But before doing so he suitably redefines the Humean problem of induction. Hume's objections against the justifiability of induction are said to be the consequences of his wrong assumption that "inductive inference must lead to success" or an inductive conclusion must be true.[31] It is to be noted here that Reichenbach is mixing up the issues of truth and success. The question to which he addresses himself primarily is, "Is it necessary, for the justification of inductive inference, to show that its conclusion is true?" And he returns a negative answer to the question, with the argument that the proof of the *truth* of the inductive conclusion is only a sufficient and not a necessary condition for the justification of induction. To justify induction, he thinks, we must furnish "the best assumption concerning the future," that is, "a necessary condition of [future] success." The justification of induction is somewhat like that of an emergency operation performed on a gravely ill person. If asked, the surgeon tells us: "I'm not quite sure whether the operation will save the man, but if there *is* any remedy, I think, it is the operation." But an important distinction must be made between the justification of induction and that of an operation of this sort: whereas the latter does, the former does not, presuppose, that is, inductive generalizations of empirical character. Reichenbach claims that this can be *proved* without being caught in a vicious circle and as a preliminary to that proof he again suitably reformulates the aim of induction as *"to find series of events whose frequency of occurrence converges towards a limit.*"[32] This reformulation of the Humean problem suggests that Reichenbach was convinced that, given Hume's own formulation. the possibility of a validation of induction is ruled out, although its vindication is possible.[33] Even if induction could be justified, it falls short of Hume's demand (did Hume really demand it?) that it will imply a proof of the truth of the inductive conclusion. Out to find a rule of convergence (RC) to determine the limit of frequency, Reichenbach proposes to offer a probabilistic justification of induction—induction that at best can only probabilify the conclusion. To show the comprehensive character of his explicatum he tries to make out two further points. The usual distinction drawn between (1) truth and probability, on the one hand, and (2) laws of nature and statistical laws, on the other, is not tenable.

Reichenbach tries to justify induction within the framework of the FIP and he thinks it is consistent not only with our day-to-day practice but also with scientific practice. Moreover, it enables one to formulate the problem in a precise and general way. He says:

> Inductive inference cannot be dispensed with because we need it for the purpose of action. . . . It is no justification of inductive belief to show that it is a habit. It is a habit, but the question is whether it is a good habit, where "good" is to mean "useful for the purpose of actions directed to future events."[34]

> Scientific method pursues the aim of predicting the future, in order to construct a precise formulation for this aim we interpret it as meaning that scientific method is intended to find the limits of the frequency.[35]

This formulation of the problem of induction invites an immediate objection: assuming induction can find a *limit* somewhere in the *infinite* series, has it any *practical* significance for the layperson or the scientist? One might even think that the world is so disorderly that it is impossible for humans to construct series with a limit, or that it is unpredictable. Reichenbach anticipated the objections and has tried to meet them in this way. First he introduces the concept of practical limit.[36] Reichenbach's emphasis on practical limit or predictive success is at least partly due to this pragmatic orientation. Second, to avoid the charge of vicious circularity Reichenbach proposes to construct the RC without making use of any previous induction. Truth of inductive inference cannot be guaranteed, for, he argues, that would commit us to one or another a priori regularity assumption. He is in favor of "determination *a posteriori*" of the RC; that is, on the basis of *statistically* observed data. The relative frequencies observed in a segment of an infinite series are assumed (because the opposite assumption leads to absurdity) to hold good more or less for any future prolongation of the series. For an exact formulation Reichenbach says:

> We assume a series of events A and A (non-A), let n be the number of events, m the number of events of the type A among them. We have then the relative frequency:
>
> $$h^n = m/n$$
>
> The assumption of the determination *a posteriori* may now be expressed:
>
> *For any further prolongation of the series as far as s events (S > n), the relative frequency will remain within a small interval around h: i.e., we assume the relation*
>
> $$h^n = E < h^s < h^n = E$$
>
> *where E is a small number.*[37]

He calls this assumption the *principle of induction* and tries to save it from the charge of retrogression and show its nontautological character. Those who are open to the charge of retrogression misconceive the aim of science and think mistakenly that to foresee the future one has only to report the past. And they do this under the questionable assumption that the future is the prolongation of the past or that time is irrelevant to the *kinds* of natural events.

Reichenbach concedes that his assumption may also be questioned, for it is not logically necessary that h^n remains always within the interval $h + E$. If this concession is deemed to be a victory for Hume's critique of induction he points out that "any logic of science remains a failure so long as we have no theory of induction which is not Hume's criticism."[38]

Reichenbach expresses his surprise particularly at the empiricists' failure in taking the right lesson from Hume's critique of induction, and this he attributes to their untenable logical apriorism. To vindicate his RC he makes use of a *reductive* argument. It is true, he confesses, that we cannot prove the existence of a limit of the frequency. "But the absence of proof does not mean that *we know* that *there is no limit;* it means only that *we do not know whether there is a limit.*"[39] In the absence of a proof to the contrary Reichenbach feels provisionally justified in trying the *posit h* in the hope that in *repeated applications* it would lead to the greatest ratio of successes. For us h^n is a *blind posit,* for we do not know its appraised weight, but as it is the value of the limit we take it as our best wager.

At this stage Reichenbach's argument seems to be open to a very serious criticism. To repeat this argument: if a sequence of events possesses a limit and if his rule is consistently used in iterative application, ultimately we will be able to estimate that limit to any preassigned degree of approximation. He claims that the validity of his argument rests entirely on logico-mathematical principles. But in fact it assumes the existence of a limit. He defends this minimum assumption on the simple ground that if we choose some end we must choose some means to realize (that is, to try to realize) it as well, and that it has no cognitive import. The question remains, Why should we think that the rule of induction alone can enable us to find the limit or lay the best wager? We can well imagine noninductive rules for this purpose. To say we use inductive rules rather than noninductive ones in pursuit of our ends is not of much consequence. For either doing so may itself be questioned or, in case it is not questioned on the plea that doing so is vindicated by pragmatic success, it may be shown to be viciously circular. Reichenbach's measure of pragmatic success is itself inductive. He shows that inductive success is a necessary condition for the success of any rule, inductive or not. What is more, he claims that induction is sufficient condition for finding a limit. But this presupposes that the statistical sequence must observably converge. It is not necessary to think that his presupposition has to be vindicated only by the inductive method. There is no earthly reason why noninductive rules of convergence should be disallowed. *Per contra,* it is not enough to show that inductive posits converge asymptotically toward the limit. For noninductive rules may, and perhaps do, have this character as well. The claim of induction must in some respect be superior to that of noninductive rules of convergence. Reichenbach's argument for the rejection of the clairvoyant (that is, noninductive) method of determining the value of the limit makes sense only in his prior commitment to the principle of induction, which, he thinks, ultimately decides whether a clairvoyant is a good clairvoyant or not.[40] Even to say, as he does, that this principle is the principle of undertaking the smallest risk is also not of much value, for he is already committed to the principle for measuring the risk itself. In other words, Reichenbach makes the very *applicability* of the principle a necessary condition of the existence of a limit of the frequency. In spite of all its ingenuity, Reichenbach's argument seems to have failed to break the circle.

Reichenbach anticipated the problem posed by "indefinitely delayed convergence" and "divergence within limits." He thinks that this problem may be successfully tackled within the general framework of the FIP by pointing out the *simplicity* and *self-correcting* character of induction. To take the point of self-correcting first, induction has been claimed to be the best self-correcting method. Long before Reichenbach, Peirce spoke of this marvelous character of induction. The essential point of self-correction is rather simple. Divergence from the point in the first order has to be corrected on the basis of its appraisal in the second order. But remember that the correcting posits of the second order are themselves generalizations about proportion and, therefore, subject to correction. It is true that we may further correct the values of the posits in the second order and use them for correcting our past calculations of the appraisals for the posits of the first order. Thus we may get an infinite hierarchy of posits and appraisals and corrections. That this accelerated approach of the points of convergence in sequences of relative frequencies is not of much epistemological value has been well argued, among others, by von Wright.[41] It is not at all clear in what respect this method marks an improvement on the inductive principle itself. Besides, it leaves us always undecided as regards the number or orders of appraisals and corrections we have to pass through to get to the limit. Second, Reichenbach's defense of the rule of induction formulated in terms of *descriptive* simplicity and *inductive* simplicity appears unconvincing. Descriptive simplicity is said to be a matter of psychological taste, convenience, or economy and has nothing to do with facts described (or "co-ordinated to the description"). For example, the metric system of measurement is preferred to the system of yards and inches purely on the grounds of convenience: although the former does, the latter does not, permit the application of the rules of decimal fractions. In the cases of inductive simplicity what governs our choice is predictive power and not economy. The criterion of inductive simplicity applies only to those constructions that are logically equivalent. The assumption that "cannot be justified by convenience"; for "it has a truth-character and demands a justification within the theory of probability and induction."[42] Now it is clear that in the matter of choice of the RCs (Rules of Convergence), which are not all logically equivalent, the criterion of descriptive simplicity has no relevance. The application of the other criterion (that is, of inductive simplicity) is open to the charge of vicious circularity, for it is applicable to the constructions having different predictive powers, and, according to Reichenbach himself, the difference is to be measured and justified within the theory of probability and induction.

Finally, it may be observed that Reichenbach's formulation of the problem of justifying inductive generalization is open to the general criticism leveled against the FIP. If we are to regard, as Reichenbach insists we must, every inductive generalization as a statement about relative frequencies in infinite series, we immediately face certain highly questionable consequences.[43] Whenever we assert an inductive generalization such as "All ravens are black," we are ontologically committed to the untestable assumption that there exists an infinite number of ravens. Given Reichenbach's analysis, what is strange is this: such a generalization may be held to be true even if innumberable ravens were not black. Incidentally this also exposes the hollowness of (simple) enumerative induction.

5. Carnap's Concept of Confirmation: A Review

Besides Keynes and Reichenbach, those few who have made significant contribution to the *philosophy* of induction in the recent past include Rudolf Carnap and Karl Popper. Their views on the problem of justification of induction deserve careful study. Historically speaking, although the comprehensive statement of Popper's view[44] was made available earlier than Carnap's, for the purpose of our study I propose to take up the latter's view first. This I propose to do for two reasons. First, although Carnap does, Popper does not, believe in the possibility of the probabilistic justification of induction. Second, an idea of Carnap's view and its limitations, I hope, will help us to review critically the current controversy between Carnap and Popper (and for that purpose between an improved version of inductive probability and that of FIP) in proper perspective. Popper set before himself two tasks: to develop the FIP without the axiom of convergence and with a somewhat weakened axiom of randomness; and to restate the relations between probability and experience for solving the problem of decidability of probability statements within the system of his *propensity interpretation of probability* (PIP).

Carnap has made it abundantly clear that the controversy between Keynes, Jeffreys, and Broad ("subjectivists"), on the one hand, and Von Mises, Reichenbach, and Popper ("objectivists"), on the other, is "futile," for laboring as they were under "the erroneous belief that they had the same explicandum" ultimately arrived at different and mutually incompatible explicata.[45] It has been rightly pointed out, among others, by Hintikka that even the current controversy between Carnapians and Popperians over the problem of the criterion of acceptability of inductive generalization is confusing, because the controvertists, often unduly influenced by the word *probability*, tend to overlook the distinction between their explicanda.[46] Even Hempel shares our impression "that Popperians and Logical Empiricists have often talked past each other, especially on the subject of induction."[47] Carnap holds that there are two concepts of probability—probability1 (degree of confirmation) and probability2 (relative frequency)—and that the acceptability of an inductive generalization (for example, "all swans are white") depends upon its qualified instance confirmation (C^*) of the form that the next instance of the swan (or one after that) to be observed will be white.[48] Moreover, he does not think that the problem of justifying the choice of *generalization* is of central importance in inductive logic. The most important kind of inductive inference, according to him, is *singular predictive inference*, that is, the inference from model to model.

We will have occasion later to see how Popper criticizes this view. But let us note immediately some of his basic points. First, in addition to Carnap's probability1 and probability2, Popper claims, his degree of corroboration has to be recognized as an independent theory of degree of confirmation. Second, the acceptability of a generalization is determined neither in terms of probability1 nor in terms of probability2 but in terms of its *corroborability* (that is, *im*probability). And, third, "Carnap had inadvertently two different and incompatible "explicanda" in mind with his probability1, one of them [Popper's] *C*[orroboration], the other [his propensity] *P*[robability]."[49] Finally, what is most important to bear in mind, Popper does not regard what is ordinarily named as induction as induction at all. He emphatically denies the very possibility of induction and, therefore, is not concerned at all with the Humean problem of induction as it is ordinarily understood. Underlying every generalizing hypothesis, he admits, is a psychological process, but he refuses to identify it as inductive and proposes to "justify" (that is, corroborate) it by deductive testing.

Carnap, on his part, is convinced that there is induction, that it cannot be justified by PLIV-like synthetic principles, and that the principles of inductive logic are all analytic. Intending to justify induction probabilistically he finds probability "not suitable as the basic concept of inductive logic." "The concept of probability on which inductive logic is to be based is," according to him, "a logical relation between two statements or propositions; it is the degree of confirmation of a hypothesis (or conclusion) on the basis of some given evidence (or premises)."[50] He concedes that it is not possible to lay down the exact rules of induction or to construct an automatic and effective inductive machine capable of solving the problems of decision procedures involved in reasoning from particular information to general hypothesis. But he refuses to regard this concession as an endorsement of the Popper-Einstein thesis that there is no (inductive) logical passage to generalizations or that generalizations are due to intuition. Inductive generalizations have little or no augmentative significance for him. Both inductive and deductive logic are said to be concerned with logical relations between sentences; the latter with the relation of L-implications, the former with that of degree of confirmation as numerical measure for a partial L-implication. The much-debated problem of decision procedure is not peculiar to induction. As soon as we leave behind the most elementary part of propositional calculus we fail to devise a satisfactory method of decision for all sentences. The rules necessary to prove the inductive theorem of the form '$c(h, e) = r$' are said to be not less rigorous than those necessary to prove the deductive theorem of the form "*e* L-implies *h*." Carnap achieves the parity between inductive and deductive logic by assimilating the former under the latter. Inductive logic, to quote him, "presupposes deductive logic . . . [and is] constructed out of deductive logic by the introduction of the definition for *c*," "which represents the rules of induction." The sort of justification that Carnap provides for induction in terms of his C^* turns out to be purely deductive.

The introduction of Carnap's concept of degree of confirmation involves three steps: (1) the definition of regular *c*-function; (2) the definition of a symmetrical *c*-function; and (3) the definition of degree of confirmation, C^*.

To follow the fuller import of Carnap's inductive logic one has to understand his language systems of semantic rules, L, to which the former is applied. There is an infinite sequence of finite language LN (and an infinite language L, which we will ignore here). These languages are forms of lower functional logic with identity. In LN systems there is a finite number, *N*, of individuals. Rules are laid down for determining the ways of combining the signs of L systems into sentences. An atomic sentence consists of a predicate of degree *n* with individual constants. In LN *i* is a *state description* (henceforth SD) if and only if *i* is a conjunction that contains as components exactly one sentence from every basic pair in LN. In order that the state descriptions describe possible states, the interpretation of the individual constants and the primitive predicates must satisfy the *condition of logical independence*.

We need semantic rules for each LN that would determine, for each sentence *j* and each SD *i* formulable in LN, whether *j* hold in it. From the rules, then, we can gather whether *j* would be true if *i* describes the actual state in the domain of individuals. The states with which Carnap's L(-inductive logic) is concerned are discrete, isolated and very unlike the real world (that is, space-time continuum) described in physics. The set of primitive predicates in an L-(inductive) system must be sufficient to express every qualitative attribute of the individuals in the universe of L. This Carnapian *condition of completeness*[51] reminds one of Keynes's PLIV,[52] for it too assumes that an exhaustive description of any physical entity is possible. The class of SDs in LN where *j* holds constitute the *range* of *j* in LN. A sentence whose range is universal (that is, holds in every SD) is L-*true* (logically, necessarily, and analytically). A sentence whose range is null is L-*false* (impossible and self-contradictory).

A sentence that holds in some but not in every SD (that is, neither L-true nor L-false) is factual. Two sentences with the identical ranges are L-*equivalent*. If, for example, there are two sentences, *e* and *h*, such that the range of *e* is included in the range of *h*, that is, R*e* is a subclass of R*h*, then *e* L-implies *h*.

To be able to express in numerical terms how much of the range of *e* is a part of *h*, Carnap constructs a metric for ranges, a *regular measure function* or *m*-function. For every SD in LN its *n*-function is a positive real number, and the sum of the values of *n* for all SDs in LN is *i*. For any L-false sentence *j* in LN $n(f) = 0$. For any non-L-false sentence *j* in LN $n(j)$ is the sum of the values of *m* for the SD in R(ange)*j*. Given any regular *n*-function, Carnap now offers the definition of its corresponding *c* function. For any pair of sentences *e* and *h* in LN, where $m(e) = 0$, $c(h, e) = m(\mathrm{e, h})/m(e)$; for any *e*, where $m(e) = 0$, $c(h, e)$ has no value.

Carnap's next step consists of narrowing down the class of *c*-functions to a class containing only *symmetrical c-functions,* which treat all individuals on a par with respect to the determination of $c(h, e)$. A regular *n*-function is called a *symmetrical n-function* if it assigns equal values to any two isomorphic SDs (in LN). And a *c*-function that rests on such an *n*-function is called a *symmetrical c-function.*

The final step of Carnap's argument is designed to fix the distribution of *n*-values for the *structure-descriptions* in LN. If *i* is an SD in LN and there are *ni* SDs isomorphic to *i*, for example, *i*, *i*′, i″, and so on, then all these SDs exhibit the same structure with respect to all the properties and relations designated by the predicates in LN. This common structure of *ni* may be expressed by their disjunction t v t′ v t″ v . . . , and this disjunction, say j, is a structure description; and $m(j)$ is the sum of the m-values for these SDs. Given that m is symmetrical, these values are equal. Carnap assigns equal n-values to each of the structure-descriptions. And he chooses a particular m-function m^* and a particular c-function c^* based on m^*. If a sentence is not L-false, its m^* value is the sum of those SDs contained in its range; if it is L-false, its n^* value is 0. For any pair of sentences e and h in LN, where e is not L-false, $c^*(h, e)$ is taken as the quotient $n^*(e, h)$ over $n^*(e)$.

Of the infinite possible symmetrical c-functions Carnap has chosen one, c^*. What is the justification, if any, of *this* choice rather than any other? The question is analogous to the one raised against Reichenbach's choice of the RC. Carnap rejects Wittgenstein's *Cw*,[53] for example, and accepts c^* on the basis of induction. When induction itself is under question, to cite its authority in support of the decision to accept c^* or reject *Cw* is of no consequence. But Carnap asserts that "a practical decision is reasonable if it is made according to the probabilities with respect to the available evidence, even if it turns out to be not successful."[54] This echoes Keynes's antipredictionist argument and is sought to be justified "with the help of inductive method, because the uniformity of the [atomic] world is probable." To show that his principle of uniformity is analytic Carnap formulates it in terms of probability and thus intends to avoid the charge of circularity. But then, it leads to other difficulties. The choice of c^* is made to depend upon the general interdependence or uniformity of the atomic world, which turns out to be a matter of mere decision. Moreover, this implies that any two sentences having the same predicate are interdependent or positively correlated (probabilistically). This is not only counterintuitive but also unsubstantiated. Any attempt to substantiate or justify it presupposes (or demands the acceptance of) c^*. Popper observes: "It is equivalent to the acceptance of as many natural laws as there are predicates, each asserting the same degree of dependence of any two events with like predicates in the world."[55] This aspect of Carnap's inductive logic has been criticized, among others, by Hintikka, Hilpinen, Pietarinen, and Toumela.[56] Kemeny tried to minimize some of these difficulties, but he seems to have achieved very little.

The most perplexing result of Carnap's scheme, even as modified by Kemeny, is "that the c^* of a law is very small in a large system and 0 in the infinite system."[57] Carnap does not regard it as a serious problem. According to him we do not need laws with high c^* but with high *qualified instance* c^*. He goes as far as to hold that laws are "not indispensable for making predications" and that these are nothing more than "efficient instruments for finding . . . singular predications." But it may be shown that, given Carnap's assumptions, (1) that singular terms must appear both in *h* and *e* and (2) that each singular term is unique, which are clearly incompatible, instance confirmation itself is not possible at all. This also indicates the difficulties involved in his concepts of relevance and irrelevance. Under assumptions (1) and (2), it is difficult, if not impossible, to show, for example, how an additional evidence *i* brings about change in *c* of *h* by *e* (prior evidence). Under the restriction imposed by his system, Carnap first admits the irrelevance of laws in science and, then, finds the impossibility even of instance confirmation.

His insistence on the point that any inductive statement or statement on probability (re: *e* and *h*), if true, is analytic, that is, purely logical, makes it obvious that the sort of justification he offers for induction is inductive or probabilistic in a very extraordinary sense and that it is deductive and, therefore, has nothing to do with empiricism. And it seems that his probabilistic justification has clearly failed. The system of his inductive logic is not only inadequate but also basically defective.

4

Probable Knowledge: Confirmation and Correction

1. Reason and Experience in Induction

Unless we become clear in our minds regarding the nature of induction, the question of its possible justification can hardly be answered in a satisfactory manner. That induction is used as a method by scientists and even by common people is rarely disputed. The dispute basically centers around the *philosophical* explication of the relative primacy of *reason* or that of *experience* in induction. When I say that the issue is mainly philosophical and not technical, what I mean is this. Whatever theory of probability one defends, subjective or frequency, one cannot but admit, directly or indirectly, that we learn from experience. The ability to learn from experience may be construed broadly or restrictively. According to one construal, the ability to learn from experience is not peculiar to human beings and that at least some other species of animal also show this ability, at least to some extent. According to another construal, only we, human beings, *rational* animals, have this ability to learn from experience or, more specifically speaking, from mistakes. In self-praise we often claim that we alone are able to identify those experiences that are inconsistent with our *aim* of knowledge or action or both.

In other words, it is being implicitly claimed that we alone are able to entertain aiming at knowledge and action that are more or less remote from us, that is, mediate in nature, and to be realized through a somewhat complex set of related steps. These steps may be theoretical or practical or, what we ordinarily come across, composite; that is, partly theoretical and partly practical.

The preceding explication of what *experience* is and what role *reason* performs in our *lives* are sure to remind us of the old and unending debate between rationalism and empiricism in theory of knowledge. To segregate theory of knowledge, epistemology, from theory of action, praxiology, is to misconstrue the relation between theory and practice, between knowing and doing. In the name of *analysis* we often tend to forget that knowing involves doing and doing involves knowing. Our theories both presuppose and anticipate practices. Our practices prove fruitful in the light of theories. The latter are corrected by the former. There is a sort of circularity between knowing and doing, between theory and practice, but it is not vicious. Positively speaking, they are interpenetrative in character and mutually supportive and corrective. Hegel and Marx would have liked to characterize their relations as dialectical.

Somewhat similar observations are in order in the context of the reason-experience relationship. Reason is nothing mysterious, "essential," and nonempirical. The rationalist often finds in human reason a superhuman, if not divine, and infallible capacity, enabling us to know whatever is real, get into the very complex structure of reality. At times, to strengthen this thesis another collateral thesis is offered; viz., the structure of reality and that of human reason are "essentially same," "identical at bottom." The empiricist, on the other hand, modestly affirms that experience is reliable only when it is age-old, that is, traditional, or immediate. Most of the empiricists are unwilling to say anything about "the deep and distant structures" of reality. They prefer to remain noncommittal, or minimally committal, on ontological issues. On scrutiny it is found that most of the epistemologists unduly honor the unfounded divide between "the epistemic" and "the ontic," between experience and reason.

It is interesting to note the division between experience and reason that is highlighted in most of the Euro-American philosophical systems is nonexistent in the Indian philosophical tradition. Reason and experience, realism and pragmatism, sail here in the same boat. The Western association of *reason* with *form* is not to be found here. The problems regarding form are taken care of in terms of experience. And evidently the latter concept, experience, is interpreted very comprehensively. This synthetic or integral outlook is not peculiar to Indian philosophical thinkers. In different ways, Descartes, Kant, Husserl, Popper, and Quine have tried to bring about a rapprochement or reunion between empiricism and rationalism (in epistemology), between idealism and realism (in ontology). But it seems that the "scandal" of the old (and, to my mind, unfounded) divide still remains influential in many quarters. The demand for "the justification of induction" is an upshot of the said fancied divide and has been drawing its sustenance from an unnatural craze for epistemic certainty that is not humanly available.

2. Biology and Psychology of Learning and Quine on the Justification of Induction

The problem of induction in many respects is identical with the problem of justifying it. Different facets of the problem are a fallout of the principle of empiricism. It is not at all surprising to note that it is primarily the empiricists of different hues who felt bothered by this problem, not knowing how to solve or tackle it. It is equally interesting to note that the rationalists of strong persuasion either straightaway deny the very existence of induction, dismissing with it the question of its justification as a mere pseudo-problem, or interpret it as a camouflaged or inverse deduction.

There is another view according to which induction is self-justifying or self-corrective and therefore hardly needs any external justification. Even to ask for a justification is criticized as misplaced. For, it is argued, science and its method cannot be justified by philosophy, at least not that sort of philosophy which is not rooted in and accountable to science. "Asking for a justification of induction is like asking for a first philosophy in support of science."[1] Induction may be taken to be justified if as a method of science it is found to be paying or working well.

Exposed to the heterogeneous environment around us, we, our bodies, are unevenly informed and educated, enthused and inhibited. This process is continuous but our awareness of it is not. Sometimes it is silent, sometimes it is loud. Sometimes it is pleasing and, therefore, attractive. Sometimes it is painful and, therefore, understandably repulsive. This account of our organism's familiarity with and learning of the environment is somewhat oversimplified, if not deceptive. Pleasure-and-pain principles, the forces of attraction and repulsion, though influential, are not compulsive. The third option of the body not to be pleased or pained by its environment, its refusal to learn what comes from outside, may be ascribed to the instructions from past experience, remembered past, or our willpower, or a combination of the two. And this partly accounts for the gap between "receptual similarity" and "behavioral similarity," on the one hand, and that between "perceptual similarity" and "behavioral similarity," on the other.[2] Unless will is accorded its due importance these gaps cannot be satisfactorily explained in terms of the traces of experience alone. Even the very existence and lack of traces are partly due to our volitional, partly valuational, disposition toward them. Pleasure and pain entirely on their own cannot account for the unevenness or unpredictability of the volitional-valuational attitude. One might ascribe this fact to what we call freedom of will. We all know, given the "same" sensory input, our perceptual similarity cannot be ensured. Similarly, given the "same" perceptual input, our behavioral similarity cannot be ensured. At every stage we are obliged to recognize the role of volition, the value exercise of the body.

Our learning process is interfered with and deterred by many imponderables, none of which can be generalized beyond a point. Neither somatic nor epistemic utilities, strictly speaking, can be compared interpersonally. Still less are we in a position to define the standard said to be necessary to make the comparison possible. However, interestingly enough, this uncertainty, the incommensurable utilities, do not prevent us from undertaking interpersonal commerce. In spite of these difficulties we carry on our communication, making due allowance for the margin of inaccuracy or even error. Money or gold (as "standard") makes our commerce possible. In some countries in the past and in some even now, "incommensurable" barter makes commerce possible. Words are our main instrument of communication. But, strangely enough, words are also responsible for creating confusion and breaking down communication. By gesture and even silence we may succeed in communicating our meaning where words fail us. Notwithstanding these qualifications and imponderables, the main two forces that promote and retard our learning process are pleasure and pain, hedonic reward and punishment.[3] The sensory input that is pleasantly informative to the body is generally welcomed by it. To the sources and reception of painful information the body is antipathic. In memory storage pleasant experiences are not only frequently revived but also elaborately ornated. The behavior that paves the way to success helps us in getting to the chosen goal and is not only accepted but also well received. Consequently, the paying behavior is recommended by the agent to himself as well as to others. Over a period of time and in a given community this becomes gradually more and more known and accepted. It is in this way that rules of behavior, norms of conduct, are learned, formed, taught, and gain currency.

The silent features of experience that teach us or from which we learn our position in the world and the world itself and those that corroborate and correct this view are basically same. Although we arrange them differently, as of "now" and "then" empirically and historically and so on, their originary elements are same. Which parts of our self-view and world-view will be preserved and which ones have to be changed depends primarily on us, our attitudes, interests, and interpretations of received information. If we *want* to preserve our existing views and refuse to entertain questions about them, we will resist attempts by others to change and correct our concerned views. However, this does not *totally* rule out the possibility of correction of our favored views by experience and views induced or imposed by others. In the face of persistent adverse verdicts of experience against our fancied views it is not easy to maintain them indefinitely.

At one stage or other our own views and those of others are bound to get mixed up. The reason of the *community* of views, despite difference and conflict, are not far to seek. After all views are not news from nowhere nor can they survive in a vacuum. Positively speaking, views grow out of experience, are rooted in or oriented toward objects. Objects themselves are always cogiven. The cogivenness of the objects has its epistemic, linguistic, and logical correlates. Object-bound propositions are themselves interlocked. The languages in which the propositions are expressed and communicated have their common vocabularies. In spite of the indefinitiveness of ordinary language, its use—extensive and repeated use—minimizes the ill-effects associated with its indefiniteness, indeterminacy, or imprecision.

The facts and conditions that make possible our shared life, theoretical as well as practical, leave room for change, whether externally caused, internally initiated, or brought about by both external and internal factors. The situation has been likened by many like Neurath and Quine to our sailing in the same boat. Producers and consumers of ideas and experience, theoreticians and practitioners, persons of common sense, philosophers, and scientists are (in a very important sense) all copassengers in the boat of knowledge. They interact, communicate, quarrel, and cooperate and, what is most important, somehow manage to live and work together.

This boat analogy has some very important lessons to offer us in the context of induction. The first thing to be noted is that theoreticians are obliged at some point or other to learn from the people of common sense. The "intellectual superiority" of the former is to be taken in a *very* qualified way and only in some restricted areas. The underlying truth of the point has to be duly noted; viz., experience and views embodied in common sense has an important say even in deciding the fate of abstract theoretical issues and discourses. It is not at all surprising that we often come across some such expression as "this is counterintuitive," "this seriously offends our commonsense understanding," "this is outrageously inconsistent with our ordinary language," and the like. When I say this and emphasize the importance of common sense and ordinary language, I do not mean that theoretical language and theoreticians are under any special obligation to follow the "dictates" of common sense and ordinary language. The basic point that deserves our serious attention is this. At some stage or other theoreticians and their theories or constructs have to *satisfy* the demand and the will of the people of common sense. Even the so-called uncommon persons, theoreticians, for example, have their own commonsense moments of will and interests that have to be attended to. The same point may be put in another way. Our theories, however abstract, are in some way or other answerable to the tribunal of experience. This tribunal need not be construed in the "narrow" Quinean sense. It may be taken also in the "expansive" Husserlian sense. "The phenomenal" is not necessarily inconsistent with "the phenomenological." The connecting point is to be found in the nature and role of evidence. The evidence of experience, even if suspect in principle, has to be respected in practice, at least provisionally.

The second point of importance in the boat analogy is this. *All* that goes in making the boat possible and mobile is not simultaneously questionable and corrigible irrespective of the quality, positive or negative, and quantity, one or many, of evidence. Because of his "negative methodology," to use Quine's term, Popper attaches more importance to falsifying or infirming evidences.[4] He is understandably interested more in the negative quality of evidence than in the large amount of positive or confirming evidence. Even then, it has been rightly pointed out, one or two pieces of negative evidence hardly prove decisive. Falsifiable theories are rarely falsified and overthrown by one or two examples of negative evidence, however crucial they might be claimed to be.

This indicates, among other things, two very important points. First, the fate of falsifiable theories is intimately related with other well-established theories that, though themselves falsifiable in principle, are in effect not sought to be falsified simultaneously with the former. Second, when negative evidence purported to falsify a theory fails in its mission, that is, proves unsuccessful in overthrowing the theory in question, it lends support to it, corroborates it, and adds to its content. In spite of this "limitation" of his negative methodology Popper sticks to what is called methodological *localism*. According to him, one of the basic *norms* of scientific method requires the working scientists to try to test separately or locally different components, that is, different propositions or parts, of the concerned scientific system. Like a boat, it cannot be overhauled as a whole and at one time. When we talk of the boat as a whole, of the scientific system as a part of and nourished by life, nothing is left external to it. The talk of an external-internal distinction makes no sense when science is taken in its totality, as born out of, sustained, interrogated, and improved by the life-world. It is not Popper's methodological localism and negative methodology that require him to deny the intimate relation between life and science. When it is said that science *qua* boat has to be repaired or rebuilt plank by plank while staying afloat, it is implied that we cannot disembark, there is nothing external to it, a port and a dock, for example, where it can be taken for repair and rebuilding. Smart seems to be perfectly justified in observing that Popper and Quine are in "substantial agreement" in accepting Neurath's boat analogy.

There is still another aspect of the similarity between Popper and Quine in approaching science from a biological point of view. Quine repeatedly draws our attention to the biological basis of epistemology. What we know is due to the sensory input, often very "meager" to start with. The fact that the output, knowledge and its inductive extension, turns out to be large is mainly because of two things. What we learn from experience is retained in our body-mind complex, provided it proves practically helpful, behaviorally satisfactory, and enables us to *adjust* ourselves with our environment. Second, even those elements of our experience found to be unsuitable or unhelpful cannot be *suddenly* disowned or dropped. Like boat repairing the process of building science is also gradual. Notwithstanding our subjective difference in identifying the connection between words and experience, we remain socially united by our communication and beliefs. Our success in social communication keep and make us move together, despite the peculiarity and privacy of our experience as evident at times in our eccentric or unconventional use of language. "Society, acting solely on overt manifestations, has been able to train the individual to say the socially proper thing in response even to socially undetectable stimulations."[5]

Building science is like living life, which requires us to learn from experience, adjust ourselves with the environment, develop "a taste for simplicity in some sense and a taste for old things." Quine is careful enough to draw our attention to the fact that inductive learning presupposes (1) the learner's point of view, (2) the resemblance between the given stimuli and the anticipated stimuli, and (3) simultaneous likeness between distinct points of view of different learners (and teachers).[6]

Quine's epistemology is a part of psychology sustained immediately by biology and mediately by ecology, both natural and social. This probiological view of epistemology as the foundation of science is bound to remind one of Hume's naturalism and the attending predicament. It is not at all surprising to hear Quine saying, "the Humean predicament is the human predicament."[7] Like Hume, Quine recognizes that, in spite of our taste for "old" things, we have to go for "new" things in scientific theory construction and anticipate the unobserved ones. Also like Hume, he is persuaded that even if we have sufficient sensory evidence *for* science at our disposal, "scientific doctrines must fall short of certainty." We are destined to live with probable knowledge.

This biological orientation of epistemology is clearly evident in such later works of Popper as *Objective Knowledge: An Evolutionary View* and *The Self and Its Brain*. "By the biological approach to knowledge" Popper means "an approach that regards knowledge, whether animal or human, as the evolutionary result of adaptation to the environment—to an external world."[8] Popper draws distinction between inherited adaptation and learned adaptation. The two coexist but not always. When the environment throws up new things, a new situation develops around us, particularly of the problematic sort, and our inherited adaptation is found inadequate, it threatens our "old things," old ways of dealing with the world, and we are obliged to explore and learn new ways. Even to learn new things we need the old ones. Another point that Popper reminds us is unconscious knowledge. This is said to be a peculiarly human, not subhuman, phenomenon. The biology of human beings is of such a nature that it keeps informing the human organism of its environment although the organism itself may not be conscious of this process. For us information gathering is not necessarily a conscious enterprise. In the light of this information, continuous and complex information, we adjust and readjust ourselves with the environment. This process is also not always conscious or reflective. But Popper's biological orientation of epistemology is very cautious and qualified. In this respect he differs from empiricists like Hume and Quine. He points out that, although our consciousness and intelligence are nourished by information received from outside, the relation between the two cannot easily be measured. It is argued, with reference to the case of Helen Keller, that on the basis of limited sensory input of information one can develop considerable or even vast knowledge. This argument of Popper might easily remind one of Quine's observation that "meager" sensory input may yield "torrential" cognitive output to humanity. Both Popper and Quine agree that, whatever might be the quantum of sensory input or the extent of the evidential "base" of knowledge, knowledge is bound to remain probable or conjectural forever.

It is well known that Popper wants to maintain a safe and respectable distance from that view of knowledge which is known as probabilistic and closely associated with Hume. His preference for the word *conjectural* in the context of epistemology is also equally well known. In his or her language, the scientist aims at bold conjectures whose probability value is low and content value high. Time and again we are reminded that the scientist is for *im*probable, *not* highly probable, conjectures or hypotheses. In other words, according to this personal interpretation, the scientist is for theories with *high content value* and *low probability value*. In pro-Popperian vocabulary, probability and content of scientific theories are inversely related. He argues that "since a low probability means a high probability being falsified, it follows that the high degree of falsifiability, or refutability, or testability, is one of the aims of science . . . the same aim has a high informative content."[9] High testability and improbability are interchangeable terms to Popper. To increase the empirical content of a theory its explanatory power in relation to known (as well as yet unknown) evidence must go up.[10] Because the structure and substructure of the unknown (or yet unavailable but possible) evidence are not mappable a priori, the truth claim of scientific theories is bound to remain provisional or temporary.

This account of scientific knowledge and its growth does in many respects stand close to Quine's view on the subject. To say this is not to overlook the important points on which they differ. The basic point of their similarity is to be found in the common scientific approach to philosophy. By *scientific* what I have primarily in mind is scientific method. At one stage Quine was of the view that philosophy *is* science, nothing more or nothing less. In fact, what he had in view and later on developed is this: philosophy embodies the method of self-reflection of science. That science as a corpus of empirical propositions cannot stand still long and remain idle has been highlighted by both. In a way this is a recognition or restatement of what science is like and how it works as a going concern. In spite of his methodological holism, Quine never belittles the importance of the scientist's professional obligation to learn from empirical evidence, especially that which is purported to be falsifying, not justifying. It has already been pointed out that, notwithstanding his cautious and sophisticated restatement of falsificationism, Popper has not been able to convince most of us that methodological localism sets for science a more rigorous or demanding standard or criterion that is not satisfiable by the Duhem-Quine thesis and that makes any actual and substantive difference to the career of science. Like Popper, Quine rejects the classical approach to epistemology formulated basically in terms of sense data and the laws of associating them. His observations on the point deserve careful attention.

> Our dissociation from the old epistemology has brought both freedom and responsibility. We gain access to the resources of natural science and *we accept the methodological restraints of natural science.* In our account of how science might be acquired *we do not try to justify science by some prior and firmer philosophy,* but neither are we to maintain less than scientific standards. Evidence must regularly be sought in external objects, *out where observers can jointly observe it.* Speculation is allowable if recognized for what it is and conducted *with a view to the possible access of evidence* at some future stage.[11] (My emphasis.)

Note that in the preceding paragraph Quine mentions, in his characteristic concise fashion, several important points in quick succession. First, when he speaks of freedom, what he has in view is primarily freedom from "abstract ideas" of sense data theorists like Berkeley and Hume. Obviously he is aware of the responsibility of telling us convincingly how to mobilize the resources necessary for the construction of scientific epistemology. Second, what adds to his responsibility is the rigor imposed by the method of natural science. Third, he makes it abundantly clear that he is not a justificationist. He is neither an inductivist in the traditional sense nor engaged in search of a firmer philosophy to justify induction. His notion of induction is explorative and prospective. It is compared to the process of "natural selection." It may be remembered here that Popper also has likened the growth of scientific knowledge to the process of natural selection. Fourth, when Quine speaks of the strength of evidence of external objects and insists on their public character or joint observability, one is reminded of Popper's account of basic statement, "Basic statements are . . . statements asserting that an observable event is occurring in a certain region of space and time."[12] It is to be noticed that in later years Popper recognized that even basic statement "remains essentially conjectural."[13] Once it is recognized that even basic statements are conjectural and testable, rather *easily* testable, their provisional character, by implication, also, is conceded. Finally, on the role of speculation in the scientific epistemology of Popper and Quine one expects that "naturalist" Quine and "rationalist" Popper can never agree. Without denying their difference one must try to realize the significance of why Quine allows speculation to play an important role in science, in scientific epistemology. When speculation remains answerable to evidence, its "pernicious" transcendental trait is purged away. This interpretation of speculation stands close to Popper's notion of "influential metaphysics," metaphysics that influences and is influenced by science. Not every part of the world is available to our knowledge, certainly not to scientific knowledge. We are obliged to speculate about the composition and structure of both the very small and the very large chunks of reality. Though Quine gives us a "psychological" account of how we start knowing the world from the very bottom and gradually go up, he hastens to add that at the commonsense level our starting point is medium-sized objects like table, chair, and building. Both Quine and Popper advise us to keep our "speculation" or "bold conjectures" earth bound, that is, answerable to empirical evidence, so that they do not go wild and aimless.

Committed as he is to the spirit and method of science, Quine finds it impossible to accept science as a definitive system of knowledge that is not open to further questioning and correction. In this respect his difference with the Cartesian and the Kantian enterprises is clear. He does not propose to provide us any rock-bottom *philosophical foundation* or conclusive *philosophical justification* of science. On the contrary, his own "new epistemology," a part of science itself, aims to show that by its very nature it is skeptical and, what is more instructive, this skepticism shows its own merit or strength, not weakness. Science has "demonstrated the limitedness of the evidence for science." It is now for the new epistemologist to show how best to make use of the "limited sources" provided by science. As an empirical psychologist he is now "liberated" from the requirement of talking in terms of sense data or physical objects and can start his scientific investigation into "man's acquisition of science," taking his basic cues from sensory stimulation. The liberated and new epistemology is claimed to be "enlightened in recognising that the sceptical challenge springs from science itself, and that in coping with it we are free to use scientific knowledge."[14] The ancient skeptic, like the modern counterpart, also challenged science from within. Using the scientific principles and with reference to the external scientific reality each showed "the fallibility of senses." Yet, the skeptic somehow failed to realize the strength of this scientific fallibilism. That on the basis of science itself one can fruitfully question the conclusions of science is a new realization of the liberated epistemology. This self-interrogating or reflective enquiry may well be called *philosophy*. But it is not a "first philosophy, firmer than science, on which science can be based." If we get to its "bottom" we would see to what extent it is our own "free creation." The Quinean image of science, as I see it, is similar to the Popperian one in several significant respects. Let us look at it from the latter end.

3. Popper's Antifoundationalism and Quine's Rejection of First Philosophy: Their Agreement and Difference

Popper's antifoundationalism is well known. Neither the basic statements, reports of direct experience, nor any first philosophy, according to him, can provide any firm foundation to science. But science needs both. Metaphysics as "first philosophy" influences or inspires science. Science has no definite and fixed point of *origin*. Popper discounts the genetic approach, if taken in the strict sense. The issue of genesis or origin and that of validation are separate. One may trace the psychological or sociological origin of a scientific theory, yet one cannot be logically sure of its correctness. To say this is not to deny the relevance of the relation between science and its nonscientific background, on the one hand, and that between science and the testing statements, on the other. Such nonscientific elements as myth and metaphysics are very much there in the background of science. Like science, myth and metaphysics are also attempted explanations or understandings of some problem or puzzle of our life of experience. But unlike science, they are not falsifiable in the Popperian sense. Where myth and metaphysics end and science begins nobody can say very definitely. Yet it is in terms of falsifiability that Popper seeks to draw the line of *demarcation* between science and nonscience. It has already been pointed out that despite its undoubted merits, the falsifiability-based negative methodology, leaves many questions unanswered when it tries to claim that falsifying evidence can decisively refute a theory. The gap between falsifiability and falsification is considerable. When the implication of the unsuccessful attempts to falsify a theory by basic statement is fully brought out, the insight of induction reappears in a different guise and captures our attention.

Supported (from behind) by metaphysics and corrigible (prospectively) by basic statements, science is linked with both. No scientific theory, nor even its individual propositional constituents, should be construed separately, still less atomically. Notwithstanding his methodological localism, Popper's view regarding the relation between science and metaphysics shows his proximity to Quine and distance from the logical positivist. In the latter's scheme of scientific things myth and metaphysics find no place at all. In contrast, according to Popper and Quine, they are intimately related. Whereas Quine is interested primarily in their ontological affinity, Popper explores their epistemological, rather methodological, continuity.

If something, whatever that thing might be, is accorded an ontological status, we are somehow obliged to account for it epistemologically. Even if the *proclaimed episteme,* valid knowledge, of fictitious things or posits turn out on scrutiny to be mere *doxa,* popular opinion, the need of the latter, by implication, is acknowledged and affirmed.

On the provisional or probable character of knowledge, Popper and Quine have hardly any differences. Their difference becomes prominent when their respective accounts of the relation between science and its background are studied in depth. On the question, What would be the repercussion of refutation, or even significant modification, of a scientific theory on the related nonscientific theories, logical, metaphysical, and so on? Popper's response would be, "Only local," and Quine's "Total."

This point of difference, instead of being exaggerated, needs to be carefully looked into so that the true human character of scientific knowledge is not lost sight of. According to Popper, when a theory is falsified the whole of its background is *not* affected and that only a part, the part that has direct or immediate bearing upon the falsified theory in question, is damaged and needs repair. This clarification does not seem to be very satisfactory and invites some queries. For example, it is not easy to draw a line of *demarcation* between the so-called immediate background of a theory and what is not so immediate. For Popper never disputes the social or objective character of knowledge. Nor does he deny, in truth he specifically maintains, that the basic statements, the falsifiers of theory, are public in character; that is, sharable by people of normal senses. At this point an undeniable element of conventionalism enters into Popper's otherwise consistently realistic image of scientific knowledge.

This seems to add further complexity to his well-intended methodological localism. Once we concede the point that basic statements are not basic on their own account but because of their sharability by people of normal senses, we open the door of science, though slightly to start with, to the skeptic and our first retreat from realism starts. For this reason, and not for any fun, the early Wittgenstein kept all human factors off his account of elementary proposition.

Another Popperian point to be borne in mind in this connection is the theory ladenness of every statement, every report of direct experience. Theory ladenness of basic statements, openness even of our elementary sense experience to past habits, present dispositions, future expectations, and so forth not only reduce their realistic content claims but also conceptually link them up with other parts of our background knowledge, the background related to our process of acquiring and repairing knowledge.

To take the ambiguity of the Popperian point a little further one can always plausibly argue that "the normal senses" of all human beings over the world are not identical. The abilities of seeing, hearing, and so on of all people living in different environments, hot deserts, cold and high mountains, noisy and crowded urban areas, sparsely populated and calm polar regions are not biologically identical. This factor has its bearing on psychological receptivity of information brought to the human senses by such natural elements as light and sound.

The tenuousness of the relation, if not untenability, between methodological localism and methodological holism may be shown in another way. When Quine speaks of *interanimation* (in principle) between all products of human experience, from metaphysics to mathematics, Popper speaks of causal *interaction,* upward and downward, between Three Worlds, 1 (physical-natural), 2 (psychological), and 3 (theories, problems, arguments, musical forms, and so on). This suggests, among other things, that the world breaks into our languages and the latter present the former to us in different ways. Different are the ways of making the world. Even those who like Husserl believe in "essences" concede different possible ways of representing them; viz., descriptively, eidetically, and transcendentally.

It is interesting to note the reason why Quine wants to remove the curtain that generally is believed to have separated natural sciences, on the one hand, from logic and mathematics, on the other.[15] It is a widely shared empiricist dogma, traceable at least to Hume, that logical and mathematical statements are neither based on nor refutable by the evidence of the senses. In contrast, statements of natural sciences, together with their verifiers and falsifiers, are believed to be questionable. Another way of describing this "one-way screen," that is, the relation between the analytic statements of logic and mathematics and the synthetic ones of natural sciences, is construed in terms of informative content. Whereas the former are said to be devoid of empirical content, the latter are credited with the same. However, the alleged empirical contentlessness of logical and mathematical statements does not degrade their reliability or throw them open to skeptical questioning. On the contrary, their empirical contentlessness is taken to be a proof of their irrefutability. It is to be recalled here that, according to Popper, the statements that are not refutable or falsifiable are not scientific. Viewed from that point, logic and mathematics are not scientific. In other words, logical and mathematical statements are empty. Obviously, here the word *science* is being used in a very special sense. It is in this and analogous contexts that Quine draws our attention to the mistakes arising out of our excessive zeal to follow "the terminological boundaries between sciences."

For example, there is hardly any good reason to maintain that there can be no empirical evidence against mathematical statements, unless we are dogmatically committed to some particular meaning of the term *empirical evidence,* which is applicable only to the domain of "informative sciences" and *not* to that of logico-mathematical statements. Here again we are thrown back to the old problem of ambiguity concerning the word *information.* Must we maintain that logic and mathematics do not tell us anything about the world, that they are uninformative in their nature? This question is bound to invite another question, Why then are logic and mathematics used so extensively in natural sciences like physics and chemistry and even in social sciences like economics and sociology? The correct answer to this question, according to Quine, is to be found in "viewing *empirical evidence* as evidence for the whole interlocked . . . system" of science of which all disciplines—logic, mathematics, physics, chemistry, economics, sociology and psychology, and so forth—are integral parts. By implication what is meant is this. Evidence for or against one part of the system has its impact, more or less, local or global, on other parts of the system. Given this view, evidential experience never goes waste. It is of *consequence* to each and every statement belonging to a system. However, this consequence may not be visible or perceptible. This sort of assertions is bound to disturb Popper and appear to him as metaphysical in the bad sense.

However, Popper is not unmindful of the importance of the issue that not only mathematics but also metaphysics has to be shown related to the basic principles of empiricism. On the question of the relation between science and metaphysics he has written a very well-known paper, "The Demarcation between Science and Metaphysics," with special reference to the view of Carnap on the subject and to which I have already referred earlier.[16] To the other equally important question, that of the relation between reality and the calculus of logic and arithmetic, he also addresses himself very ingeniously. Anyone who, like Popper, is committed to both realism and empiricism is required to answer this question very carefully. In this context, Popper wants to reject the Millian-Rylean view of calculus as a formal science.

Both Mill and Ryle are of the view that we infer particular statements from particular statements and that the universal or the general statements often used to "justify" the inference is nothing more than an "inference ticket." This "inference ticket" is an expedient device for moving from one statement to another. It has nothing to do with the real world. It is not even a purported description of the latter. This view is bound to remind one of Hume's *logical* problem of induction. Logically speaking, Hume found no passage from the general or universal truth to a particular one. On the basis of what happened in a domain of experience in the past we cannot predict definitely what would happen in it in the future. But this *logical* difficulty does not prevent Hume from *psychologically* believing, believing rather firmly, that the laws of nature or the course of events are more or less sure to remain constant. It is mainly with reference to this line of Hume's argument that his critics have spoken of the Humean fork, of the division between what is logically warranted (probabilism or skepticism), on the one hand, and what is psychologically warranted (belief in the external world or naturalism), on the other. Thomas Reid is one of those early thinkers who critically pointed out the simultaneous presence of these two "incompatible" trends in Hume's thought. In recent years Strawson also somewhat approvingly draws our attention to the same.[17]

Both Popper and Quine have tried, in their own ways, to do away with the Humean distinction between the logical and the psychological. Whereas Quine tries to achieve this end by drawing our attention to the untenable distinction between "the analytic" and "the synthetic" drawn on the basis of some muddle-headed notions of meaning, synonymy, and so forth, Popper moves toward this end by pointing out that both "scientific" theories and "nonscientific" myths and metaphysics (falling on the two sides of the changing and unclear line of demarcation) are more or less provisional and conjectural. It is to be highlighted here that even after solving the problem of induction Popper does not claim to have achieved the objective of antiskepticism; that is, certain knowledge.

Popper first tries to draw an important distinction between the Rylean interpretation of the "logician's hypotheticals" or rules of inference as rules of procedure and his own interpretation of the rules of inference as somehow applicable to the real world The former are purely procedural; that is, ways of inferring, having nothing to do with the states of affairs in the world. According to the Millian-Rylean interpretation, rules of inference remain always undisturbed or unaffected by the outcome of the inference. Rules of inference as procedures, like the rules of chess, are only rules of performance and never intended to be applicable to the world. The world has no say on the question whether the rules of inference do or do not apply to it. To Ryle it appears to be an ambiguous question and, therefore, merits no definite answer.

Popper, without disputing the possibility of this interpretation, offers himself another, which, he claims, helps us to understand why certain rules of logic prove good or useful and others do not. One of the basic purposes of all logical activities, making statements, drawing inferences, and so on, is to get to the truth, avoid falsity, detect errors, proofs, and criteria of truth and falsity. For example, in logical inference we try to see how, given true premises, true conclusions follow. Given true premises, if the conclusion that "follows" turns out to be false, we try to find out the reason why; viz., whether the premise or one of the premises is false or in the process of drawing inference some rules have been violated.

One may be justified to a certain extent in affirming that rules of inference are world neutral. But this justification is somewhat limited. Though we can form different types of logic and language purported to speak about the world or a part of it, all of them may not be equally "powerful." A logic or language may be powerful enough "for describing all the facts we wish to describe" but it "may not permit the formulation of the rules of inference needed to cover all the cases in which we can safely pass from true premises to true conclusions."[18]

In other words, the rules of inference found in a language about the world may not have *in* that language the rules in question. For achieving that purpose we may have to form an appropriate metalanguage. Alternatively, we may dispense with the distinction between object language and metalanguage and remain content with the resources and infirmities of natural language. However well-formed a language might be, the rules of inference of *all* valid intuitive inferences cannot be formulated within it.

The question of application of the logical calculi to reality may be viewed in two different ways—"internal-syntactical" and "external-semantical." As semantical systems the calculi are intended to be descriptive (of the world). Viewed thus, they are open to refutation. Viewed otherwise, that is, as uninterpreted, the calculi as systems of well-formed formulas do not face the tribunal of experience and, therefore, the possibility of refutation. But this distinction loses its sharpness when we bear in mind the fading distinction between the analytic and the synthetic. Once we leave behind the approach of the naive realist, the question of the relations between the logico-linguistic means of description and the describable facts of the world brings its deeper implications to our notice. Words and objects cannot be clearly separated. In a sense words are also objects. To view the matter from the other end; objects enter into the life of the words and their use. It has been aptly observed by Popper that "facts are something like a common product of language and reality."[19] When we fully grasp this insight into the interdependent or interpenetrated relation between world and language, we start seeing why our knowledge or description of the world is bound to prove more or less inadequate, inviting new investigation, new questioning, and needing correction.

4. Inductive Ascent: Description of State, Structure, and Constituent in Wittgenstein, Carnap, and Hintikka

The question of justifying induction or the sort of knowledge we have through inductive method is in a way indicative of the elusive character of truth unalloyed by anything of *human* experience. It is not at all surprising that probability has often been defined in terms of range *between* truth and falsity, the range between unity, 1, and nullity, 0. Those who believe in the sharp analytic-synthetic distinction ascribe the unity truth value, 1, to analytic statements or tautologies and the nullity truth value, 0, to self-contradictory statements. Naturally those who do not subscribe to the said distinction speak of probability (or improbability) of all sorts of informative statements. To them every description of reality is bound to be more or less incomplete. This is due to the lack of saturation of the concepts used in the predicative position for descriptive purpose.

Those who, like the early Wittgenstein, believe that we can speak about the "truth grounds" of a statement and that those truth grounds are themselves either true or false are not prepared to recognize that there is nothing in the world corresponding to a probability statement.[20] A statement, in itself, cannot be said to be either probable or improbable. What makes it true either happens or does not. "There is no middle way." About the middle way or range between 1 and 0 we do not have more detailed knowledge—more detailed than that of the circumstances attending the occurrence of particular events or its lack. Strictly speaking, a probability statement does not pertain to an event. It is "a general description of a propositional form." We are obliged to make a probability statement only because our knowledge of the concerned fact is incomplete and we cannot be certain about it. However, Wittgenstein points out, even while we are to remain content with probability statements we do know *something* about their form. In relation to a certain situation a picture may prove incomplete but, he points out, it is always "a complete picture of something," "a sort of excerpt" from other statements.

Whenever we speak of description in the context of probability what we have in mind are *complex situations* and not nameable objects.[21] Only in terms of the truth value of the constituent statements of a complex situation can we come to know of its probability, if any. We are informed of reality only in two exclusively different ways, affirmatively or negatively. Descriptions of nameable objects are available in terms of their *external* properties and that of reality in terms of its *internal* properties. "A proposition constructs a world with the help of a logical scaffolding so that one can actually see from the proposition how everything stands logically *if* it is true."[22] According to Wittgenstein, there is nothing compulsive in our description of reality. Whatever is describable could be described otherwise. "There is no apriori order of things."[23] The propositions of logic are indeed the descriptive scaffolding, nothing more than scaffolding of the world. That is, they represent only the form of the world but not its "subject matter." Because logic claims only to give the form of true description of the world, it cannot tell us about the form of probability description itself. In other words, probability logic is logic only by courtesy. The best possible logical description of the world is provided by some *model* of mechanics (Newtonian or Einsteinean, for example). Laws of mechanics are of help, only of use value to us in describing the world; they themselves are not descriptive of the world. Only the possible *plans* of description of the world are provided by different models of mechanics. Somewhat unlike the laws of mechanics, those of physics attempt to describe the world though indirectly. But, according to Wittgenstein, the physical description of the world is confined to what can happen in the world. The limit of description is set by the possibility of happening. Whatever is *prohibited* by the laws of physics cannot be described, remarks Wittgenstein.[24]

Wittgenstein's *describable* world, describable in physical science, is amazingly, almost incredibly, simple. Or, one might say, the complexity of the world is logically or "artificially" simplified by Wittgenstein so that he can put them into the "manageable" scaffolding of the *Tractatus*. The elementary propositions used by Wittgenstein to describe the world are equiprobable in their evidential values. Everyone of them, probabilistically speaking, is like the rest. Consequently, every *state description* is of equal confirmation value. A true state description confers equal probability or confirms equally a hypothesis as any other true state description. In brief, their confirmation values are equal. Let this Wittgensteinian confirmation function be called *C*(*W*). This isomorphism of state description or the statistical equiprobability of different individual state descriptions has been defended by philosophers like C. S. Peirce and J. M. Keynes.[25]

It was rightly perceived by Carnap and his followers like Bar-Hillel and Jeffrey that, given the complex and disorderly (at least at the local or the micro level) character of the universe, *C*(*W*) cannot tell us how to learn from our experience. If the experienceable universe, at some level or other, turns out to be unpredictable or surprising (and it seems to be the case as evident both from common sense and science), it puts in question the rationality and usefulness of *C*(*W*). Collaterally, this character of the experience shows that experience is educative and not necessarily repetitive. We seek additional information about the world not only to confirm but also, admittedly more so, to test our accepted hypotheses and beliefs. To take care of this important aspect of instructive experience and eliminative induction, Carnap developed a new and more comprehensive confirmation function, which may be called *C*(*C*), that is intimately related to his notion of *structure description*.[26] Structure descriptions are deemed superior to state descriptions in a very important respect. In addition to the common feature of the different states falling under a structure, structure description keeps specific information about the concerned states. Hence it gets out of the restrictive character of *C*(*W*) associated with state description. In this way structure description shows us the way of higher-level inductive generalization or, what Popper would say, bold and comprehensive hypothesis. At the same time, it also enables us to reconcile the positive feature of enumerative induction with the negatively instructive features of eliminative induction.

Taking his cues from Carnap's notion of structure description, Hintikka proposes a more general theory of inductive generalization.[27] Hintikka, though primarily influenced by the Carnapian ideas of inductive logic, is very mindful of the Popperian objections and requirements on the subject. "In spite of the enumerative character of [his] basic assumptions" Hintikka tries to develop a notion of *constituent description* that tries to show that the degree of confirmation of a generalization is "not increased by a mere addition of new instances if they are similar to the old ones." On the contrary, according to Hintikka's formulation of his confirmation function, let us call it *C*(*H*), repeated confirming instances eventually decrease the confirmation value of the concerned generalization. In this respect *C*(*H*) tries not only to obviate the restrictions of *C*(*W*) but also of its improved variants proposed by Keynes in terms of his Principle of Limited Variety of Nature, which I have discussed earlier. Hintikka affirms that "the requirement of the variety of instances . . . in our inductive logic [is] so strong that it overrules a form . . . of a principle which has sometimes been put forward as being obviously correct and which says that the degree of confirmation of a generalization is never decreased by a positive instance of the generalization."[28] Keynes is not specifically referred to here, but he, among others, does attach importance to positive instances. However, in fairness to him, it should be mentioned that he has explicitly said why enumerative induction, addition of positive instances, strengthens a generalization. Mainly because of the *difference* in the attending circumstances the concerned positive instances do not turn out to be merely repetitive. They have something of additional significance in them.[29] It seems that Hintikka is not completely convinced by this argument. According to him, "the first and primary inductive task is elimination: mere enumeration begins to play a significant role only after elimination has been carried as far as it can possibly go."[30] When we closely look into Hintikka's *C*(*H*), it becomes gradually clear that he owes a lot to the main confirmation theory of C. D. Broad and J. M. Keynes. Evidently it is stronger in its confirmation. And in this respect the influence on him of G. H. von Wright's "Principal Theorem of Confirmation" is very clear. According to Wright, "if the initial principle of generalisation is not 0, then its degree of confirmation is added to by each new observation which is implied by the generalisation but which is not maximally probable relative to the previous observations." Hintikka's formulation statement of this theorem of Wright is claimed to have been satisfactorily taken care of by *C*(*H*).[31]

From what has just been stated it is clear that state descriptions are more "concrete" or "earth bound" than structure descriptions, and that constituent descriptions are more "abstract" or "heaven bound" than structure descriptions. Constituent descriptions do not inform us how many individuals belong to the different predicates used to describe them. They only inform us of all the concerned predicates whether they are instantiated in the world or not; that is, of empty predicates. Like Gödel, Carnap feels that we can inductively ascend beyond constituents; from constituent description we can go to *constituent structure description.* "Instead of saying of each predicates whether it is exemplified or not," remarks Carnap, "we simply say how many of them are instantiated in the world and how many are *empty.*"[32]

Perhaps there is an implied criticism of *C*(*H*) in Carnap's view. If the constituent structure descriptions proposed by Hintikka have an added advantage of simplicity and comprehensiveness, it is because of their *in*difference to the concrete particulars of experience. When a particular confirmation theory wants to attain the Goedelian height, it has to pay a price for it. It cannot be fair to the details of experience. It seems that Hintikka's *C*(*H*) is addressed to the region between the "logician's heaven" and the "empiricist's earth," aiming to develop a theory of induction that is prepared to learn from experience without forgetting the need to go beyond it. He is reconciled to move within the "inductivist's limbo in which we all are doomed to live."

5. The Popper-Carnap Controversy on Induction and Probability and Conflation Issues

If for the time being we keep aside the technical aspect of *C*(*C*) and *C*(*H*), we are thrown back to the old philosophical question; On the basis of experience how far can we go beyond experience? The plausibility of the answer to this question largely depends upon the qualitative richness or the cognitive capacity of human experience *and* the extent of disorder that the universe has in it or can tolerate without ceasing to be what it is. It seems undeniable that in some way or other each one of us, irrespective of his or her area of work, business, or politics, domestic life or public life, is objectively required to go beyond the limited area of personal experience and make intelligent guesses and decisions when evidence for or against them is not conclusive. The problem of the working scientist or the theorist of knowledge is not altogether different. On the contrary, the issues they are concerned with are basically similar.

Any informed researcher in the field of scientific method and epistemology is entitled to feel that much of the controversy on what is ordinarily known as the problem of induction is somewhat avoidably confused. The classical example is provided by the Popper-Carnap controversy.[33] Time and again it has been affirmed by the Popperians that there is nothing like induction and the question of using it as the (or even as a) method of scientific enquiry does not arise at all. Repeatedly it has been pointed out by Carnap and his friends that the terms *induction* and *probability* have more than one meaning and, unless we have initial agreement as to their meaning, our arguments and counterarguments are destined to prove misdirected and, therefore, useless. Popper is perhaps right in affirming that the history of inductive logic is not very encouraging. The very fact that the problem of induction formulated by Hume is yet to receive a universally acceptable solution may well be interpreted, as it has indeed been, that there is no induction at all. What goes by the name induction is (implicit) inverse deduction. The general question of confirmation referred to earlier is not of any interest to the anti-inductivist. To him, the successive attempts to formulate different confirmation functions, *C*(*W*), *C*(*C*), *C*(*H*), and so on, are not of much consequence to scientific method. The scientist is basically concerned with problem solution. Scientific knowledge at every stage of its growth shows some problematic features.

Problems are due mainly to incompatibility between accepted theories or beliefs and relevant experiences. Problems also are due to imprecision, inadequacy, or lack of comprehensiveness in the theories. By induction, as it is ordinarily understood, none of the problematic features just mentioned can be effectively removed or tackled. When facing a problem, the scientist frames some hypothesis or conjecture purported to sort out the problem—a conjecture open to some testing or falsifying evidences. But Popper admits that good hypotheses, though in principle falsifiable, in fact are not so easily or frequently falsified. This partly explains the stability and continuity of scientific systems or world-views. It has already been indicated how the falsifying evidence, when it fails to overthrow the concerned hypothesis, adds to its empirical content. By implication Popper is obliged to recognize the additive-enumerative significance of evidence. However, this does not in any way minimize the importance he attaches to the intended eliminative role of evidence. We have already noticed how inductivists like Wright and Hintikka are also emphatic in their recognition of the importance of eliminative induction. Given this problem situation, one feels like asking why Popper is so opposed to all sorts of confirmation functions—*C*(*W*), *C*(*C*), and *C*(*H*). This question is all the more pertinent in view of Popper's repeated affirmation that the scientist's interest is not only in constructing theories to solve problems but also in corroborating and correcting them in the light of testing evidences.

Popper's response may be summarized in the following way.

> [A] sharp distinction [has to be] made between the idea of the probability of a hypothesis, and its *degree* of corroboration. It is asserted that if we say of a hypothesis that it is well corrobo-

> rated, we do not say more than that it has been severaly tested . . . and that it has stood up well to the severest test we were able to design so far. And it is further asserted that *degree of corroboration cannot be a probability,* because it cannot satisfy the laws of the probability calculus. For the laws of the probability calculus demand that, of two hypotheses, the one that is logically stronger, or more informative, or better testable and thus the one which can be *better corroborated,* is always *less probable*—on any given evidence—than the other, . . . [This] shows not only that we must distinguish sharply between probability . . . and degree of corroboration or confirmation, but also that the *probabilistic theory of induction, or the idea of an inductive probability is untenable.* The impossibility of an inductive probability is illustrated . . . by a distinction of certain ideas of Reichenbach's, Keynes's and Kaila's. One result of this discussion is that *in an infinite universe . . . the probability of any (non-tautological) universal law will be zero.*[34]

Popper's disregard for different confirmation functions, indicated earlier, and interest in what he calls *corroboration* are directly related to his notion of empirical content of scientific theory. From Popper's works it is clear that he does not recognize the distinction between the critical and the constructive tasks of inductive theory. The constructive significance of the efforts to develop the different systems of inductive logic or inductive probability has been berated because of their failure to solve Hume's problem. The critics of inductive philosophy, from Whewell to Popper, in their criticism of induction and probability have paid attention only to the criticizable inadequacies of the constructed systems and neglected the positive aspects of the systems that apparently failed. Popper, who otherwise commends the methodological rule that we must try to learn from our failures, practical as well as theoretical, has not followed his own precept in his critique of induction and probability.[35]

6. Self-Corrective Induction of Peirce and Reichenbach and Popper's Concept of Truth Approximation through the Method of Trial and Error

Fallibilists like Peirce noticed the advisability of the self-corrective character of inductive policy. Placed in a bewildering situation marked by different tendencies, we are ordinarily inclined to go by some "fair sample" or inductive policy. Every generalization we make in an uncertain situation is gradually corrected and qualified by the contending possibilities inherent in the situation, compeling us to include as many of those possibilities as we can. It is a form of convergence to the limit of permissible possibilities—permitted by the concerned generalization. The classical example of this view is provided by the case of tossing coin marked by two possibilities, heads and tails. The proclaimed self-correcting character of induction, mentioned by Peirce, has found favor, perhaps unknowingly, with Reichenbach. Let us assume that a potential infinite sequence called *S* has a certain property *P*. First we observe an initial segment of *n* numbers of *S*. The observed proportion of *P* in *n* is *p*. Then we inductively generalize that *p* is the limiting frequency of *P* in *S*. This generalization is called the *posit of the first order.* This posit is appraised and possibly corrected in and by a posit of the higher order. There may be a hierarchy of posits of generalizations. We are inductively advised, to posit *that value* as the limiting frequency of *P* in *S*, which is *most frequently* the limiting frequency of *P* in a sequence, of which a set with the observed relative frequency of *P* is a selection. It is clear that the higher-order posit is intended to correct the immediate lower-order posit. But the correcting posit is itself in need of correction. Thus we go up and up in an infinite hierarchy of posits and appraisals. The "going up" needs to be "accelerated" or "speeded up" so that this inductive up-going takes us to truth. But it has been rightly pointed out by von Wright and others that no degree of "acceleration" is a good enough premium to ensure our getting to truth.[36]

Like von Wright, but for different reasons, Popper also is not impressed by the self-corrective claim of Reichenbach's frequency theory of probability and induction. Popper's *propensity* interpretation of probability, in spite of its numerous difference with the frequency interpretation defended by von Mises and Reichenbach, has rightly been characterized as a variant of it. It is true that Popper has in him a strong realistic tendency and, in the later years, stands firmly committed to Tarski's correspondence (semantic) theory of truth. But, nevertheless, his commitment to truth or, more precisely speaking, truthlikeness ("verisimilitude"), is more an ideal (to be followed) than an achievement. If Reichenbach's posits-appraisals take us toward but does not to the truth, one might say, the same point may be raised against Popper's conjecture-refutation journey toward the truth. Gradual enrichment, enrichment in terms of empirical content, of a theory is no guarantee that it is conclusively, not provisionally, true. In either case, Reichenbach's or Popper's ideal truth seems to be elusive. However, this is not to deny the importance of the refinement to be found in the propensity interpretation of the frequency theory of probability.

7. Popper and Carnap on Empirical Content

To come back to another important aspect of Popper's claim regarding his approach to the problem of induction and probability: this is about his account of empirical content. He has been repeatedly complaining that Carnap and others of his persuasion have failed to take due note of this view. The latter have repeatedly denied the charge and, what is more, pointed out that their approach to induction and probability is different from Popper's not because they do not recognize the importance of the latter's view but because they are interested in some other substantive issues like the construction of inductive logic and laying the foundation of a satisfactory probability theory. According to them, their concerns are not antagonistic to the Popperian epistemology nor the aim of science. Additionally, they claim that Carnap, like Popper, is also deeply interested in the question of empirical content.

To start with, Carnap emphasizes his agreement "to a large extent with Popper's views on *general* questions of the theory of knowledge and the methodology of science."[37] Also he recalls his own, Neurath's, and the majority of the other members' (of the Vienna Circle) agreement with Popper that all knowledge is basically guessing, that certainty is beyond the reach of the scientist, and that on re-examination scientific guesses could be modified and improved. But Carnap rejects the Popperian thesis that probability and content measure of hypotheses chosen by a good scientist or a reasonable business person are bound to be inversely related. Carnap offers arguments and examples to show where lies the basic mistake of the Popperian thesis. The term *probability* may be construed in either way, absolutely, Popper's $P(x)$ and Carnap's *m-function,* or relatively, relative to given evidence, Popper's $P(x)$ may be read in either way, "initial probability of x" or "a priori probability of x." Given e, the total evidence currently available with the observer, we may read $P(x, e)$ as the present probability of x. Popper defines content measure as the reciprocal of the initial probability:

$$1.\ Ct(x) = df. 1/P(x)$$

According to Popper, as noted earlier, if initial probability of x is larger than z, then its content would be less than that of z:

$$2.\ \text{If } P(x) > P(z), \text{ then } Ct(x) < Ct(z)$$

But the scientist or the business person is concerned with the conditional or relative, not absolute, probability of *x*. Bearing that in mind, the point may be schematized in this way:

3. If $P(x, e) > P(z, e)$, then $Ct(x) < Ct(z)$

Carnap maintains that 3 is not generally valid. It is valid only in the special case when the logical implication obtained between *x* and *z* is unidirectional:

4. If *z* L-implies *x*, but *x* does not L-imply *z* then
   a. $[Ct(x) < Ct(z)]$
   b. $P(x) > P(z)$
   c. If any evidence *y* (not L-false), $P(x, y) > P(z, y)$

Now we face two alternatives:

5a. Choose the hypothesis with the higher content measure.

5b. Choose the hypothesis with the higher probability.

Popper is clearly in favour of 5a, whereas Carnap is said to be committed to 5b. But Carnap rejects this imputation and refers to his earlier supporting reasons.[38] Carnap is not prepared to accept 5a except in some special cases where "the content-measures of the two hypotheses differ but the other relevant factors are equal and among them, in particular, the probabilities." Therefore only a modified form of Popper's 5a is acceptable to Carnap:

6a. If two hypotheses have different content measures while their probabilities (and other circumstances) are equal, then choose the hypothesis with the higher content measure.

Carnap's own counterpart to this rule is claimed to be equally valid:

6b. If two hypotheses have different probabilities but their content measures and other circumstances are equal, then choose the hypothesis with the higher probability.

Carnap points out that if the Popperian insists on the acceptance of 2 in an unmodified form, then both 6a and 6b turn out to be impossible.

It is to be noted here that Carnap focuses his attention on the relation between content measures and "other circumstances." Further it has to be pointed out that his construction of the relation is "synchronic" and not "diachronic." That is, his concept of the content measure is to be understood as constant and not in the changing context of growing evidence and his concept of probability is to be taken in the sense of present probability. Thus the rules he arrives at are shown to be complementary and not incompatible.

7a. If $Ct(x) > Ct(z)$ and $P(x, e) = P(z, e)$, then choose x

7b. If $P(x, e) > P(z, e)$ and $Ct(x) = Ct(z)$, then choose x

Although Carnap is persuaded that both high content measures and high probability are compatible as features of scientific hypothesis, he cautiously adds that it would be unreasonable to increase as much as possible one of the said two magnitudes. He is for a compromise in which each requirement is *partially* satisfied. If the content measure of a hypothesis is pushed to the infinity, its probability would be zero; or, in other words, every prediction derived from that hypothesis would be false. *Per contra,* if the probability of a hypothesis is taken to be very high level, there is left little room for learning from experience in terms of that hypothesis.

It is well-known that the scientist is not concerned only with the choice of a particular hypothesis having in view its content measure or probability or both, but often he must choose one among many particular methods of experiments that have direct bearing on the choice of the hypothesis. Good inductive logic is required to help us not only in *theoretical* sphere but also in the *practical* sphere with or without relation to the former. It is not practically easy to choose the particular point at which the preceding two magnitudes are to be found in balance. Through choice and related action we want to gain (or avoid loss of) some such things as psychological peace, physical health, and money. In this sort of situation whatever course of action we propose to follow leads to a different outcome. We can never be sure which of the possible outcomes will be actualized. In a relatively uncertain situation we naturally want that action whose outcome has the maximum utility. Here the maximum utility of the outcome, because of its possible and uncertain character, is a matter of expectation. "The expectation value is the weighted mean of the utilities of the different outcomes, each utility value of an outcome being multiplied with a factor which is to be taken highter the more reason we have to expect just this outcome to occur."[39]

It is to be noted here that Carnap's shift of emphasis from probability and content measure to utility is a recognition that methodology of scientific theory cannot satisfactorily take care of the needs of our practical sphere of life. True, there is a relation between the two. For example, science through technology enters into our practical lives in very many ways. Even then it has to be realized that the logic of *decision making* in the practical sphere has some peculiarities of its own. It is in this context that one has to understand the necessity of the introduction of the rule of *utility maximization*.

Some interesting attempts have been made to narrow the gap between the sphere of scientific methodology and that of logic of decision. For example, Hempel thinks that the adoption of scientific hypotheses and theories may well be understood as the process of maximizing some such "epistemic utilities" as *simplicity* and *explanatory power*.[40] Hempel's program, successfully worked out, could show the convergence of the methods of decision theory and those of the philosophical problems of induction. But it has been critically pointed out that, although in decision theory our main concern is utility maximization, in inductive construction of scientific theory we are primarily interested in truth. Hintikka and Pietarinen observe: "This purely theoretical search of truth or of approximations to truth seems to be something a decision-theoretic approach cannot accommodate."[41] In support of their view they argue that if Hempel's rule of expected utility maximization is interpreted in their way as maximization of expected information, then some headway could be made toward the desired direction; viz., bringing a theoretical search of truth closer to (b) practical search of success or utility maximization. It is to be seen that this is one more attempt to reconcile information measure (defined in terms of inductive probability) and content measure and corroboration as defended by Popper.

Earlier we have noticed von Wright's belief in supplementing the critical task (of induction) by the constructive one. It is not enough to find out where the traditional approaches to induction have gone wrong. What, in addition, is necessary to show is how to construct inductive logic that stands closer to both theoretical enterprises like physics and chemistry and practical enterprises like business and betting. In recent years interesting work has been done in this area, among others, by Isaac Levi[42] and Amartya Sen.[43]

Historically speaking, it is interesting to note that Popper chose, among others, Kaila, von Wright's teacher, as a target of his criticism for espousing inductivism. Admittedly, Wright's own works provide cues to remove some of the shortcomings of induction and bring it closer to the view of scientific method defended by Popper himself. Hintikka, a student of Wright, not only criticizes Carnap and Popper, especially the latter, but also develops a "compromise" inductive logic drawing elements from both of them. This limited historical context of the problem has a larger historical and conceptual background that deserves our careful attention.

8. The Popper-Carnap Controversy as a New Case Study of the Ongoing Theory-Practice Dialectic

The Popper-Carnap controversy is not a new book, neither methodologically nor substantively. It is a new chapter of an old book authored by great thinkers down the centuries. This can be illustrated both from the history of Western philosophy and that of Indian philosophy. To a simple but basic truth this intercontinental and interepochal parallelism may be traced: the problems that we face in *practical* life and those we encounter in *theoretical* sphere are not quite separate. Their common human roots on scrutiny may be clearly perceived. Theory provides guidelines for taking up and pursuing the practical courses of action. The latter, in the process of their execution, encounter different problems not envisaged by the theoretical plans behind them. In the light of the problems theories must be reviewed, reconstructed, and reapplied. This *theory-practice dialectic,* not dualism, should not be viewed as a vicious circle. If the circle metaphor is to be retained, it is to be interpreted as virtuous or increasingly instructive. This is not to suggest that the history of human theories and practices is unidirectional and has no occasional sign of degeneration or setback. It has been rightly pointed out by Lakatos, for example, that the history of science has two sorts of shifts, progressive and regressive.

A clear analogue of the controversy between the Carnapian and the Popperian in this century may be easily found in the nineteenth century. For example, the controversy between Mill and Jevons shows that both their substantive and methodological interests are comparable to what we are discussing today under different terminologies. The difference between Peirce and Whewell on this issue may also be represented in a similar way. Conceptually speaking, the entire gamut of problems relating to induction and probability may be taken back to the earlier century, to Hume and Kant and their respective followers.

Although we emphasize the dialectic unity (and continuity) of theories and practices of the philosophy of induction and probability, it is instructive to note that Hume's skepticism, though primarily theoretical, has not gone in vain in convincing a die-hard rationalist like Kant that there is something very practical to learn from experience. From the other end, it might be pertinently pointed that Hume's followers like Mill did take serious note of Kant's epistemological reconstruction of the Newtonian paradigm of science after it was assailed by Hume. If Hume's work was primarily *critical,* Kant's was *constructive.* If the former's work was basically diachronic or dynamic, the latter's was synchronic or static. Rightly understood, Kant's critiques are intended transcendental rebuttal of empirical skepticism.

Equally instructive is to recall that interpretations of inductive philosophy offered by Mill and Peirce, in spite of their common empiricist inspiration, are considerably different in their formulations. Peirce's fallibilism may be tolerated by the empiricist Mill, but his Kantian inspiration and inclination are not likely to be accepted. From the other point of view, it seems very insightful that Whewell, notwithstanding his Kantian inclination, is diachronic or historical in his approach to the understanding of science. As early as in 1860 Whewell not only wrote *On the Philosophy of Discovery,* but also pointed out that science owes its origin to some or other "happy guess" and its growth to testing evidence.

The influence of Kant was not confined only to Whewell. It also reached his junior contemporary Jevons. Following the ideas of De Morgan and Boole, he visualized the possibility of constructing a logical machine of calculation, a forerunner of the modern computer. The inductive machine envisaged by him could display the combinations consistent with the information fed into it. To him, as to Whewell, induction was the inverse operation of deduction indicating the general truths or laws underlying a given set of phenomena available to us. The anti-Baconian and the anti-Millian tenure of the Whewell-Jevons approach to scientific inference is unmistakable. Like De Morgan, he subscribed to the subjective view of probability. Also he offered a comparative measure of rational expectations on the basis of information and its lack in the light of the available evidence. Jevons's methodological theory of inverse probability designed to choose from among the possible hypotheses based on intuitive methods, though unsatisfactory in detail, anticipated some features of Popper's method of conjectures and refutations.

The same story may be illustrated as well from the varieties of proinductive and anti-inductive theories of science that have been developed with care and ingenuity and having in view the legacies of Hume and Kant, severally or jointly.

The most radical form of anti-inductivism would straightaway rule out the possibility of justifying induction on the simple ground that there is nothing like induction to be justified. The less radical formulation of the view is this: induction knows no inductive or probabilistic justification and the only possible justification that could be construed for it is bound to be deductive; but in this case induction as a separate method of scientific inference loses its separate identity and gets assimilated under deduction. There is another modest and realistic approach purported to solve the problem of induction. In effect, it concedes that a scientific hypothesis, though not acceptable inductively, adds to its empirical content in this way. Popper's own view that failed falsifiers of a hypothesis contribute to the increase of its content has itself been plausibly construed along this line.

Popper himself made the precept current that there is no *logic* of discovery. Still more difficult is he on the *psychology* of discovery. In both cases his objection seems to be the same: there is no logic of rigorous test. Putnam's remark is worth quoting on the point: "All the formal algorithms proposed for testing by Carnap, by Popper, by Chomsky, etc., are to speak impolitely, *ridiculous:* if you don't believe this, program a computer to employ one of these algorithms and see how well it does at testing theories."[44]

Earlier I spoke of the virtuous and constructive relation between theory and practice of science. Theoretically or even methodologically speaking the genesis or discovery of a scientific hypothesis can never be pinned down. It owes to a sort of intuitive-practical ingenuity. However, negatively speaking, the creative spirit of a scientific discoverer, of an Archimedes or a Newton, is not to be ascribed to an isolated and fleeting experience. Why it seems to be so to the outside observer, on scrutiny, is found to be rooted in the background, theoretical and historical, of the concerned scientist. But this rootedness is too complex to be logically demonstrated. But when we say, *a la* Popper, that a theory (together with its initial conditions) can be tested only by a very carefully designed experiment or a surprising phenomenon to be found in a specified reason of space-time, we are implicitly highlighting the primacy of the scientist's *practice.* Putnam has rightly pointed out that Popper's failure to recognize the importance of practice has led him to attach exaggerated significance to his criterion of demarcation separating "falsifiable" science from "criticizable" myths, metaphysics and ethics. Without *practical* wisdom the philosopher or the scientist is unable to choose the initial conditions and the potential falsifiers without which their hypotheses cannot be tested. But, unfortunately, no logic, either inductive or deductive, could unfailingly enable the concerned inquirer to perform the task in this difficult and complex area. This is not to say, even remotely, that either deduction is useless or induction is unjustified.

9. Induction, Probability, and Deduction: Relation without Confrontation and None beyond Question

The question of the relation between induction and deduction must not be viewed through the eyes of the confrontationist. To speak of "deduction *versus* induction" is to misconstrue the scope of these two areas of logic. If induction is in need of justification, so is deduction. As logic they are at a par. It is with considerable justification to his credit that Carnap maintains that probability logic as logic is not to be distinguished from other forms of logic. Logical formalization or systematization of cognitive as well as formal (noncognitive) discourse is in need of some sort of principles. However, this is not to deny the possibility of the alternative ways of formalization or systematization. After all it affirms that all logic and epistemological theories are human-made constructs, drawing the elements from the real world. The difference between deduction, induction, and probability becomes worth talking about and even important only when we are engaged in epistemological discourse or its underlying methodology.

The empiricists from Bacon, via Mill, to Reichenbach and Quine have not only remained engaged in knowing the world experientially but also in systematizing their inductive logic. True, most of them did not try to develop a speculative theory of "universal learning machine." But that they were groping their way to it in some way or other is now evident. There is no reason why a Popper, who favors bold speculative venture, will not welcome it, despite Gödel's caution.[45] On this point Putnam's "compromise" observation is bound to remind one more of Popper than of Carnap.

My brief survey shows that scientific research has two different but very intimately related aspects—explorative and systematic. The explorative aspect is guided primarily by a sort of *in-formed* realism and the historical view of knowledge. Realistic orientation proves somewhat corrective of the excessive zeal for formalization. The forms we frame and use to formalize our knowledge, knowledge of the world, are required to take note of the forms of the world itself. Unfortunately we have no a priori valid ways of grasping the latter. In this respect the Kantian-Copernican Revolution overshot its target. In the process of refuting the Humean skepticism it claimed to have found a final (and yet scientific) view of the world. Kant's claim, his excessive systematization, leaves little room for exploration and arrogates to itself the task of unilaterally giving *forms* to the world of science.

In our own time when Popper has remained preoccupied with the normative methodology for promoting scientific revolution, revolutionary exploration of the world of sense-experiences, Carnap confined himself primarily (but not exclusively) to the task of formalizing the practice of scientific enquiry. Carnap might have failed in some respects, but that should not make us blind to "the crucial role conceptual models play in our understanding and design of a good inductive logic."[46] In fairness to him it has to be admitted that Carnap has shown the intellectual courage not only in recognizing his "failure" but also in trying painstakingly to undo the same. His progressive shift from confirmation-function theory, *C*(*C*), to the decision-theoretic approach (to epistemic utility estimate) illustrates the point.

Somewhat similarly, I would say the criticism aired by persons like Carnap, Ayer, and Hintikka have also persuaded Popper, at least partially, to recognize the inadequacy of the initial statement of his falsification marked by a form of strong anti-inductivism. The point has been elaborated both by Watkins[47] and Lakatos,[48] especially the latter. They are among Popper's leading disciples. Unfortunately, contrary to his own profession, he refused to learn from the criticism of his own critics. One might say that Popper the theoretician of scientific method did not show convincing willingness to learn from Popper the practitioner of scientific method. This shows, among other things, that the convergence of theory and practice, though desirable, is not very easily achieved. Many of us tend to develop a sort of vested interest in *our* scientific theories and get easily provoked and involved in bitter quarrels with our critics. Then how would we be able to learn from others' criticism and experience?

10. Different Senses of Knowledge and Different Forms of Its Presentation

There is a widespread misconception that the deductive forms of knowledge are certain and the inductive ones uncertain or skeptical. This misconception persists despite the well-known view, mentioned earlier, that inductive logic as logic may be presented in a deductive or even axiomatized form. Those who favor the idea of these two forms of logic and refuse to admit that it is a misconception highlight the point that it is not the *forms* of logic that are most important but *knowledge* that is sought to be presented in those forms. Knowledge may be questioned, whatever may be its form of presentation, mathematical or physical. There are radical empiricists like Mill and Quine who try to show how even mathematics is constructed on the basis of experience. Carnap uses logical techniques to show how we move from the autopsychological to the heteropsychological elements of the world and then out of those elements construct the world of the physicist. But none of these empiricists claim that any form of knowledge is beyond question and correction.

More interesting in this connection is to recall that those thinkers, especially metamathematicians, who are opposed to empiricism, have constructed rigorous foundations of mathematics and yet in the end concede themselves the weakness or inadequacy, if not the outright paradoxical character, of their systems. This shows, among other things, that mathematics is not merely a form or mold in which knowledge can be cast in any way one likes. Mathematics itself is a branch of knowledge, no matter whether it is self-contained or derived from logic or psychology. To think that it is merely a set of techniques and that its value is only instrumental is to misconstrue its scope and nature.

Another important point to be considered in this context is to remember the different construals of the concept of knowledge itself. Ordinarily we believe that knowledge is of this person or that person; that is, it is a part of one's biography. It may be taken as the *psychological* sense of the concept.

But on reflection we easily understand and agree that wherever may be the origin or locus of knowledge it is more or less available to other persons. That explains why one's knowledge can be communicated to and questioned or confirmed by others. This sense of knowledge is studied in and as *epistemology.*

There is a third sense of knowledge which is *ontological*. Knowledge need not necessarily be taken as a part of someone's biography. Its genesis or locus has nothing to do with the determination of its truth or otherwise. When a Hegel affirms that knowledge and reality are identical at bottom or essentially same, he defends the impersonal and objective character of knowledge. Similarly, when a Popper likens his concept of objective knowledge, a citizen of his world3, to the Hegelian concept of objective mind, one gets a clearer idea of what the ontological sense of knowledge is.[49]

In the third or ontological sense of knowledge one is bound to hear the echo of Platonism. However, in fairness both to Plato and his modern "enemy" Popper one should draw a distinction between the *strong* ontological sense of knowledge and the *weak*. If the former, basically ahistorical in nature, is closely associated with the name of Plato; the latter is associated with Popper's evolutionary and Whitehead's organismic view of knowledge.[50] Scientific knowledge is inductive in the sense that its theories are end seeking, truth approximative, and adaptive. Whitehead observes: "the modern evolutionary view of the physical universe should conceive of the laws of nature as evolving concurrently with the things constituting the environment . . . the conception of the Universe as evolving subject to fixed, eternal laws regulating all behaviour should be abandoned."[51]

The distance of the weak sense of knowledge from the strong one is clear. One is *static* and speaks only of the fixed structural relationship obtained between the objects of knowledge; the other is dynamic and highlights the *dynamic* equilibrium between the knowing subject, on the one hand, and the known and know*able* objects, on the other.

I find a comparable line of thought in Husserl. He speaks of the self-fulfilling nature of knowledge. What "fulfills" knowledge, that is, evidence, is anticipated by it. But evidence is not necessarily apodictic or even adequate. This explains why the Husserlian *horizons* of knowledge are ever-expansive, prove elusive to the certainty seeker, and yet the evidence that makes the expansion possible is not "eternal" to what is expanded, knowledge. The horizons are "regulative" but not fixed, elusive and not a priori.

All these views on the nature of "objective knowledge," despite their difference from one another in many respects, have one thing very important in common. The proponents of these views try to show that *objectivity* and *changeability* are not only not antagonistic but in a significant way mutually supportive and necessary to one another. The evidence that shows the objectivity of knowledge is also responsible and accounts for its change. Additionally, what brings them somewhat close to each other is the common anxiety over the possible threat posed by the evidentially warranted change in the course or career of scientific knowledge. If the change exhibits no pattern, is full of surprises and imponderables, and renders inductive anticipation unreliable or impossible, the threat assumes the skeptical proportion. But does it actually happen in real life? Or, is it the result of the theoretician's rootless anxiety and heated imagination?

The answer of the realist Nyāya-Vaiśeṣika philosopher, like that of the pragmatist, is, "consult your own and other people's practical experience, *lokācāra*." Interestingly enough, even the skeptical Buddhist recognizes the importance of what is called *causal efficacy, arthakriyākāritva*. If water (as a cause) can quench one's thirst, has one any good reason to doubt its identity and efficacy, particularly when the empirical reports of other drinkers of water consistently corroborate it?

a practical Buddhist recognizes the importance of what is called causal efficacy or arthakriyakaritva. If water (as assumed) can quench one's thirst, has one any good reason to doubt its identity and efficacy, particularly when the empirical reports of other attributes of water consistently corroborate it?

5

What Is Wrong with Skepticism?

1. Different Senses of Skepticism

The main difficulty that one encounters in dealing with the themes I am addressing is rooted in the notorious vagueness of the different associations of the term *skepticism*. For example, when the Buddhist (of the Mādhyamika School) affirms that nothing in the world has an intrinsic nature of its own, this is not meant to deny the existence of nonintrinsic worldly things as possible objects of human perception. The Buddhist is even prepared to recognize the conventional truth as a way to reach the absolute truth.[1] It has been argued, at least partly in support of Nāgārjuna and referring to Ayer's characterization of "philosophical skepticism," that the skeptic's aim is rather modest; that is, he or she is content with showing that our standards of proof are logically questionable.[2] Hume, who is often quoted to buttress the basic point of skepticism, is himself on record to have said that philosophical skepticism is a kind of "academic joke." "'Tis evident, that so extravagant a Doubt as that which Scepticism may seem to recommend by destroying *every Thing,* really affects *nothing,* and was never intended to be understood *seriously,* but was meant as a *mere* Philosophical Argument."[3]

It is interesting and also instructive to note that in a roundabout way Russell, who is generally known as a strong defender of realism, also supports Hume's "jocular" characterization of skepticism. According to him, "scepticism, while logically impeccable, is psychologically impossible, and there is an element of frivolous insincerity in any philosophy which pretends to accept it."[4] This view of Russell is bound to remind one of Hume's observation: "When [a Pyrrhonian] awakes from his dream he will be the first to join in the laugh against himself, and to confess that all his objections are mere amusements."[5]

If the skeptic's "arguments" are nothing more than "jokes" or a mere matter of "philosophical amusement," that is, "never intended to be understood seriously," what is the point in trying to refute or rebut them? If these arguments are "logically impeccable," then why do we, like Russell, talk of the "frivolous insincerity" of the philosopher who accepts skepticism, Besides, the Russellian is obliged to explain why skepticism, though "logically impeccable," is condemned to be "psychologically impossible." By implication, what Russell affirms is that the sceptic's case, in spite of its *logical* look, is psychologically impossible. In other words, what he means is that the logic and psychology of skepticism do not sail in the same boat. Obviously this view presupposes particular notions of logic and psychology that were not there when, for instance, Sextus Empiricus, Nāgārjuna, and Hume wrote on the subject. For the logicians like Russell *psychologism* is a pejorative term and if we insist strongly on following it we are often landed in an indecisive position. In his later life, it is true, Russell took a more charitable view toward psychology. In the context of human knowledge he draws a distinction between (1) "What do we know?" and (2) "How do we know it?" Sciences like astronomy and physics provide answer to the first question and the nature of the question turns out to be "as impersonal and as de-humanised as possible." But, in relation to the question (2) psychology is recognized by Russell as the "most important of the sciences." The data of our scientific knowledge are psychological. But the "apparent public" character of our scientific world "is in part delusive and in part inferential."[6]

It is clear that Russell finds the scope of psychology much more comprehensive than that of logic. Only a part of psychology can be neatly captured and formulated in logic. In logic, one is interested in what makes a sentence true or false, whereas in psychology one is interested additionally in the state of mind of the person using the sentence with belief. "In logic, '*p*' implies '*p*' or '*q*', but in psychology the state of mind of a person asserting '*p*' is different from that of a person asserting '*p*' or '*q*'."[7] What Russell submits is that the meaning of *or* is subjective; it is a state of (someone's) mind; in the world of science there is no place for *or*. "In fact, we only employ the word 'or' when we are uncertain, and if we are omniscient we should express our knowledge without the use of this word—except, indeed, our knowledge as to the state of mind of those aware of a greater or lesser degree of ignorance." Russell's main problem is not so much with the believed "facts" disjoined by *or* as with the "form" of [their] partial cognition. The term '*p*' as such does not pose a problem to him, nor does '*q*' as such. But what is the "logical form" of the epistemic state of mind underlying "'*p*' or '*q*'"? This seems to be indefinable. Here, Russell feels, logic is baffled by psychology and cannot capture the latter rightly.

True to this basic empiricist position, Russell affirms that human knowledge, in its best possible form, is confined to particular facts that are available to us either through perception or memory; that is, through experience. He is inclined to maintain that this sort of knowledge "calls for no limitation." But neither deductive logic nor experience allows us to be cognitively sure about deductive logic. Nor does experience allow us to be cognitively sure about "inferred particular facts." This is not to deny the practical possibility of generalizations from and on the basis of experience. This principle of inference may be called knowledge only in a qualified sense. We have in us a "propensity of inferences." Our "inferential habits" are promoted by and stand near the laws of nature. With his characteristic candidness Russell concedes that the term *knowledge* does not lend itself to precise explication. "All knowledge is in some degree doubtful, and we cannot say what degree of doubtfulness makes it cease to be a knowledge, any more than we can say how much loss of hair makes a man bald."[8]

From Russell's account of skepticism it appears that in his early years, he had in him a strong streak of logicism leading him to try to assimilate "psychology" of knowing under the principles of logic. At that early phase of his epistemology, to be found in *The Principles of Mathematics* (1903) and to a lesser extent in *Problems of Philosophy* (1912), Russell used to believe that corresponding to knowledge is a definite objective configuration of referents. If the logical framework of a sentence expressing knowledge had in it such words as *the, is,* and *or,* it must be assumed that the cognitive content of the sentence was, at least partly, due to the corelatable extra-linguistic "referents" of those logical words. Subsequent reflection persuaded Russell to refine and substantially abandon this strong realistic theory of knowledge in favor of a sort of constructionism. In his philosophy of logic he encountered a number of problems to arrive at maximally intelligible and indubitable entities. Right from the time of Descartes persistent doubt has been expressed about the view that we can directly perceive the physical objects. The emergence of physical theories framed in terms of such unperceivable entities as electromagnetic fields, energy-quanta, protons, and forces exerted at a point provided the encouragement to leave commonsense realism behind. Unless the principles of constructionism are followed, it is, Russell felt, not possible to obtain and explain publicly available scientific objects on the basis of sense data.

When we start speaking of the *construction* of scientific theories, one might say, we have left behind commonsense realism or faith-based realism. The problems of correctly translating theoretical language into observational language are indicative of distance of scientific knowledge from its objects. Kindred to these problems are what is called the *theoretician's dilemma*[9], "underdetermination of theories," the gap between theoretical skepticism and practical certainty, the difference between scientific skepticism and philosophical skepticism, and the like. It is instructive to note that all skeptics, in some way or other, are found to be realists, and that all realists, in some sense or other, entertain skepticism. In the Indian context Nyāya philosophers like Vātsyāyana and Uddyotakāra, who try to refute the skepticism of the Buddhists like Nāgārjuna also recognise the possibility of initial doubt (*saṁśaya*). However, they think that by and through the means of knowledge (*pramāna*) we can reach certitude (*nirṇaya*). The *initial* doubt of Nyāya, in some respects at least, is different from *provisional* doubt of Descartes. The Cartesian is methodologically determined to doubt whatever appears as dubitable. Unlike the Naiyāyika, the Cartesian is not initially committed to vindicate the claims of some ways or means of knowledge (*pramāṇas*).

The "deceitful demon" of Descartes is allowed to obstruct all ways, perception, inference, and so on, leading to (possible) knowledge. When he doubts all means of knowledge, strictly speaking, Descartes as a scientist does not lose faith in the validity of science. Without ceasing to be a "scientific realist" he feels methodologically or philosophically entitled to question or doubt all sorts of knowledge, including the very means, perceptual, inferential, and so forth, that make knowledge possible. Here it is evident, in the case of Descartes, that he recognizes the distinction between scientific skepticism and philosophical skepticism. The former is not only provisional but also dependent upon the outcome of the latter. Philosophical skepticism cannot be logically rebutted, scientific skepticism cannot be fully vindicated. The "working" scientist is admittedly free not to doubt his knowledge. The *philosopher* of science, though refraining from questioning the skeptical attitude of the "working" scientist, tries to show that *in principle* the latter's attitude as well as beliefs can well be questioned. Philosophers feels interrupted or obstructed in their skeptical journey and come to a stop only when they "find" (both cognitively and morally) it impossible to proceed further.

A universal skeptic, Sextus Empiricus or Nāgārjuna, for instance, may raise a pertinent question at this stage: What exactly does the Cartesian finds whitch makes it *really* impossible to proceed further in the skeptical journey? The Cartesian may well refuse to see in the *cogito* principle (*cogito, ergo, sum*) of Descartes anything *really* preventing questioning it. In principle, one might say, and in fact it has been said, that the *cogito* principle and the *dubito* principle (*dubito, ergo, sum*) are at a par. And in that case, logically speaking, self-doubt is as good or as bad as self-knowledge. If by doubting one's own self one can come to the end of one's skeptical journey; what is the point in trying to seek self-knowledge to refute skepticism? If by doubt itself we can destroy doubt, why this question at all? Its significance lies neither in knowledge nor in doubt but in something else that makes both modes of consciousness (cognitive and sceptical) possible and meaningful. But to admit this is not to vindicate the primacy of knowledge over that of doubt.

When someone takes knowing and doubting as only two modes of consciousness and sustained by it, one really does not disprove the limit of doubt or knowledge, but merely expresses one's *unwillingness* to go to the limit of either of the two. To fall back on self-consciousness and to claim that in it is to be found the refutation of skepticism is not to prove either the limit of doubt or the knowledge of self in the form of self-consciousness. It is for us to note here the well-known Cartesian view that the root of our errors, false (claims of) knowledge, is *will*. If consciousness could be purged of will, it would be free from doubt and error and not inclined to false (claims of) knowledge. In other words, freedom from inclination is stated to be the clinching answer to skepticism.

But is it really so? Can we have consciousness without will or inclination whatsoever? Is it an available form of human consciousness? Or, is it a mere presupposition or a postulated form of it? Suppose one doubts it, what happens? To say that in and through thinking the existence of thinker cannot be doubted or, to change the example, in and through doubting the existence of doubter cannot be doubted is not to prove, at least not logically, the existence of the thinker or the doubter, because to prove that we need certain presupposition(s) and rule(s) of inference. The only way to circumvent this problem is to affirm, as some Cartesians do, that our certitude about self-existence, self-existence in the form of consciousness, is intuitive, not inferential. But one can still raise the question, Why must self-existence be in the form of consciousness and not otherwise? Why this presupposition of existence-as-consciousness? What in consciousness makes it free from all presuppositions?

To say that it is a deliverance of intuition is not enough to put a stop to the critical journey of the skeptic. Again the question may be raised, Is there anything special in intuition that makes its deliverance so unassailable; that is, unquestionable? Do not some of our intuitions turn out to be erroneous? If to remove the possibility of error from intuition it is defined in a particular way, viz., "intuition by its very nature is error-free," are we not buying certitude, an antiskeptical "conclusion," at a very high price? Is it not a mere strategy to have certitude *by definition,* taking "definition" as a sort of dogma? For other notions of *intuition* do not claim that it unfailingly gives certitude or veracity. Why do we reject those notions? Why do we accept this particular notion of intuition that is antiskeptical only (by definition)? Is it not a matter of will or *decision?* If *definition* can be credited with the capacity to deliver certitude, why can *decision* not be similarly credited? In fact there are views that affirm the supremacy or finality of decision, of free decision, against the verdict of which there is no appeal to any "higher tribunal." By implication, the existence of higher tribunal beyond decision is denied.

The limitations of definitional and decisional strategies to put a stop to the skeptic's journey are merely illustrative in nature. There are other possible strategies to stop the skeptic. To stop the skeptic is not to refute skepticism. When we are told by philosophers like Nāgārjuna that universal skepticism is irrefutable, what they have in mind is *philosophical* skepticism and not *scientific* skepticism. The Cartesian skepticism, strictly speaking, is scientific; that is, methodological, both in inspiration and explication. In other words, the Cartesian decides rather "arbitrarily" where to stop and still has options of interrogating these conclusions. The scientist does not interrogate all that he or she possibly can. Because of his practical satisfaction or decision the scientist stops somewhere and does not go on raising questions and expressing doubts, but the philosopher goes on.

The Nāgārjunite skepticism, engulfing even self-existence, is much more radical than the Cartesian one. Nāgārjuna, the Buddhist skeptic, finds no reason why consciousness or self-consciousness should be accorded a privileged epistemic (or ontic) position. Even its intuitive "proof" framed in terms of *clarity* and *distinctness* does not appear to him probative at all. The self-evidentialist's (possible) contention that consciousness is self-revealing, that is, self-proving, is also not beyond doubt. If consciousness is claimed to be self-evident or self-proving, how could we possibly make sense of the demand for, or necessity of, proving consciousness or seeking evidence *for* it?

The radical skeptic takes note of a thesis of the antiskeptic realist, in this case of the Naiyāyika, to the effect that everything of the world, including proofs or *pramānas,* are *prameyas;* that is, *probanda* (to be proved), and then uses it against the latter. Self-evidence as proof, it is argued, is itself in need of evidence. If what is other-evident and what is self-evident are both equally in need of further evidence, all the basic concepts of logic and epistemology—proof, probandum, probans, arguments, and evidence—are thrown into the vortex of doubt. Given this confusing situation, if the radical skeptic is asked to state his own thesis and spell out his argument(s), if any, he "disarmingly" confesses that he has no thesis of his own to defend and, therefore, is in no need of proof or argument. Pressed further to state what induces him to disapprove all proofs, arguments, and so on, the radical skeptic humbly submits, rather reiterates, that he "proposes" only to show the emptiness (*śūnyatā*) of the proclaimed "essence" or "existence" (*bhāva/sva-bhāva*) of all the so-called ways (*pramāṇas*) of knowing and proving. But then he hastens to add that what he "proposes" to show is not itself a probandum or provable either and is empty (*sva-bhāva-śūnyatā*). There is *nothing* to be proved or known or said. What the radical skeptic thus says, in his own admission, is *nothing* (empty or *śūnyatā*).

At this stage the critic of the radical skeptic might intervene and ask, "By saying nothing what does he achieve?" and "What sort of saying is 'saying nothing'?" To this the skeptic might reply, "There is *nothing* to be achieved by saying *nothing,* saying *nothing* is itself also *nothing.*" This game of question and answer may be played endlessly, at least by the skeptic. But, in fact, the skeptic does not propose to play this game indefinitely and, what is more important to note, to him, this game, even to the limited extent it is played, is serious, not frivolous.

Though the Nāgārjunite "style of philosophizing" has been characterized as "at best a distortion and at worst an illusion," yet it has been recalled that through its "constant practice" one gets "an insight into the nature of what is ultimately real (*prajñā*)."[10] What, perhaps, motivates the radical skeptic to theorize about his position is to demonstrate the futility or, at best, only limited utility of theories and arguments in gaining knowledge in the truest sense.

The radical skepticism of Nāgārjuna is clearly philosophical in its nature. It takes the skeptic position to such an extent that it is bound to appear as sophistry to the person of common sense as well as to the working scientist. It is hard to believe that such an elementary truth was not clear to his keen intellect. Even then the main reason for which he went on stating and restating his position with literally devastating logical acumen seems to be primarily *practical.* His view, noted earlier, that constant practice of negative dialectic leads to the realization that "everything" is *nothing* (*śūnya*), empty or void. *Śūnya* is the ultimate reality, says Nāgārjuna.

2. Skepticism: Radical, Philosophical, Practical, and Universal

The radical skepticism of the Buddhist is bound to remind the student of the Western philosophy of Pyrrhonism as stated by Sextus Empiricus. Pyrrhonists, like the Buddhists, are clearly conscious of the nonprobative nature of their arguments. Although they point out, for example, the circularity of reasoning (*diallelus*) or *regressus ad infinitum* in the argument of the antiskeptics, they are not oblivious of the possible charges of fallacy that might be raised against their own view, however indefinite or elusive it might be.[11] The *diallelus* (the wheel) argument, when spelled out, shows its highly abstract or philosophical character. What it says in brief, is this: one cannot establish a position or thesis unless one has already at one's disposal a standard or criterion (in terms of which the *claim* of the concerned establishment can be judged). This argument has been variously stated by various philosophers like Bacon and Montaigne, for example. To distinguish between true and false arguments one needs a *method* that can adjudicate between the rival (truth or falsity) claims and settle them in favor of one or other truth value. The method in question itself is in need of *justification;* that is, justifying argument. The circle becomes full and vicious when it is pointed out that the necessary justification is in need of that very method by which the justifiability of the method in question is to be settled. In brief, it is again a case of (1) making a *claim* backed up by some justificatory or falsificatory *argument* in support of it; (2) searching for a *method,* criterion, or standard for showing the sustainability of the claim; and, then, again (3) to produce some argument in support of the *method* in question. But the *method* [of (2)] cannot adjudicate between the claims of arguments [of (1)] and at the same time, remain dependent on arguments [of (3)]. The same point may be made out differently. For example, juridically speaking, the tribunal whose jurisdiction or constitution or both is under question can hardly be expected to deliver the *right* judgment.[12]

In spite of its clarity, Sextus did not attach much *theoretical* importance to his criteriological circularity argument. He, like Nāgārjuna, could clearly see that this argument not only destroys the opponent's opposition but also destroys his own position. Yet he was impressed by its *therapeutic* value. Skeptical arguments have been likened to purgative medicines, which are expelled together with what they expel from the body. The dictum "nothing is true" is applicable to all dicta, including itself.[13]

My reason for saying that a skeptic's *theoretical position* should not be confused with a *practical position* is very important. Practically speaking, the skeptic is as hardheaded as we, people of common sense, are. For example, Sextus himself, referring to the position of the skeptic, reminds us that they themselves do not heed the strictures of skepticism "universally about all matters." It is confined to only to the objects that are "nonevident" and those of "dogmatic inquiry." Reminded of their cautious attitude, the skeptics state that their assertion is limited to what "appears to them" and that they pretend to make "no positive declarations about the real nature of external objects."[14]

It may be pointed out here that the word *practical* is rather ambiguous. Eating, thinking, and scientific theory construction are all "practical" but, carefully speaking, not in the same sense. Peeling potatoes and thinking theorems are not ordinarily taken to be acts of the same sort. So when Sextus assures us that he is not so "impractical" as to universalize the scope of doubt and that he points out only the *possibility* of doubting what is not sufficiently evident and the *necessity* of questioning what is dogmatically accepted, no one should be surprised. But when the same Sextus questions the conclusive or clinching claim of every bit of evidence, including that of practice as evidence, or fails to spell out the sufficient conditions of admissible evidence and, what is worse, claims the universal applicability of the dictum, "nothing is true," one seems to have good reason to feel intrigued and puzzled. The skeptic's refusal to make "positive declarations about the real nature of external objects" is not sufficiently assuring. More disturbing is his remark "We assert that which appears to us." Given the universal applicability of the dictum, "nothing is true," even his reference to "the real nature of external objects" is bound to raise the critic's eyebrows. Limitation of his assertion to "which appears to us" is not very comforting either. For so many presuppositions are implicit both in his refusal to make "positive declarations" and in the "assertions" of appearances; viz., (1) the existence of "the real nature," (2) that of "appearances," (3) the difference between the two, (4) that between what is external and what is internal, and (5) possibility of assertions and positive declarations. Very many other implicit presuppositions are also there. To the radical skeptic both sorts of presuppositions in (5) are also there. To the radical skeptic both presuppositions and propositions should be equally suspect because of the unavailability of sufficient evidence, whatever that might be, and the universality of the dictum, "nothing is true." What is stated after *because*, that is, (1) and (2), lacks in probative force.

If this is so, what is the point, if any, in saying anything. Is it to propagate or defend a particular position or to oppose all positions, including that of the skeptic? It makes hardly any sense to speak of position *and* opposition if the difference between the two cannot be indicated even minimally. But to do that one is required to presuppose so many things.

This is a requirement that, strictly speaking, can never be fully satisfied. It is just not practically possible for one to be aware of all presuppositions involved in taking a decision, for example, regarding the rationality of a proposed course of action or the truth value of a particular proposition. To decide or to act "rationally" the conditions we are called upon to fulfill are in effect unfulfillable. Action or praxis hardly goes, at least not logically, with *skepsis*. Yet we all act, rather are obliged to act, under uncertain circumstances. If we are advised to act only when we are certain and if we decide (but how?) to follow that advice, we are perhaps destined to "perish," as persuasively pointed out by both Locke and Hume. Locke points out: "He that will not eat till he has a demonstration that it will nourish him; he that will not stir till he infallibly knows the business he goes about will succeed, we have little else to do but to sit still and perish."[15] This view has been substantially endorsed by Hume when he states: the skeptic "must acknowledge . . . that all human life must perish where his principles universally and steadily prevails. All discourse, all action would immediately cease; and man remain in a total lethargy, till the necessities of nature unsatisfied, put an end to their miserable existence."[16]

It is clear that according to both Locke and Hume, praxis and *skepsis*, theoretically speaking, do not go together. But the persistent fact that we act, live, and ordinarily do not perish shows that the skeptic's position, though conceivable, is not practicable. Even the possibility of its *coherent* conceivability has been doubted.

Pragmatists, classical and modern, do not feel much disturbed by this logic or theoretical or philosophical skepticism. By *logic*, they mean (consciously or unconsciously) inductive logic; by *theory*, some vision or view that is itself practicable or has at least some practicable consequence; and by *philosophy*, general principles abstracted from sense-experience and not speculative, that is, not empirically rootless or inapplicable. In a way a refutation or answer to skepticism, if any, has to be sought within our own animal nature and practical life. Both animals and human beings take their basic lessons from experience. By experience they enlarge or abridge the scope of their experience. When Pascal, for example, remarks that "nature comes to the help of impotent reason," his target of attack is deductive or ratiocinative reason.[17] Reason of our animal nature, that is, the ability to learn from experience, is not here under attack. This is supported, though from a different point of view, by Santayana when he writes that it is by "the animal faith [that we] live by from day to day."[18] That we go beyond experience by experience is a deliverence of experience itself. In other words, inductive extrapolation of experience, though inductively unjustifiable, cannot be helped. More positively speaking, it may be probabilistically supported. Here *probability* is to be taken in a weak inductive sense. The criticism of induction cannot actually proscribe our inductive moves. Inductive conclusions or decisions, though questionable in principle, are humanly unavoidable. Hume is worth quoting on the point:

> Shou'd it here be ask'd me, . . . whether I be really one of those sceptics, who hold that all is uncertain . . . I shou'd reply, that this question is entirely superfluous, and that neither I, nor any other person was ever sincerely and constantly of that opinion. Nature by an absolute and uncontrollable necessity has determin'd us to judge as well as to breathe and feel.[19]

The universal skepticism of Sextus and Nāgārjuna, on ultimate analysis, takes the form of what Rescher calls the "no criterion" argument.[20] Time and again attempts have been made to meet it in very many related ways; viz., in terms of animal faith, practice, some or other form of pragmatism, inductive probabilism, or self-justifying induction. When I say these are the kindred ways of *meeting* the skeptic's position, I recognize, by implication, the strength underlying it. I also recognize the important distinction between the "theoretical" and the "practical." When I subject these concepts to rigorous analysis it becomes increasingly clear to me that these two concepts are not as antithetical as is often thought. In an important sense, theory construction is a practical action. Or, one might say, theory can hardly be constructed without making use, directly or indirectly, of practical experience. Practice may be animal or refined. But underlying every practice we may detect some theory, articulate or inarticulate; that is, the result of distilled or undistilled experience. The interdependence of theory and practice need not be construed as circular, still less fallacious. Their relation seems to be dialectical. One may enrich, enlarge, or constructively abridge the other. One may also bring out the inadequacy, imprecision, and criticizable features of the other. In some of our moments of life we exhibit a pronounced bias to what is practical or practicable. There are also moments of life when we go by the primacy of theory. Once we remember the interfusive and interpenetrative, that is, dialectical, characters of the relation, we need not be unduly disturbed by the alleged circularity of their relation.

3. Different Forms of Skepticism and Different Ways of Meeting Their Challenge

The skeptic's challenge cannot be met in an omnibus manner. The challenge itself takes different forms. Before one thinks of meeting the different forms of skepticism, one has to take note of the peculiarities of each form. The "no criterion" argument of the skeptics assumes an intractable form only when it is viewed as an undifferentiated whole. When the specific features of the different forms of the argument are discerned, one feels that these may be met or answered. From the long history and vast literature of skepticism it is clear that all skeptics do not follow the same form. For this reason we often hear of (1) initial skepticism, (2) methodological skepticism, (3) *epoche* type of skepticism, (4) philosophical skepticism, (5) scientific skepticism, and (6) universal skepticism. Carefully analyzed, every form is found to have subforms within it. But, strictly speaking, it seems to me that it is difficult to find out a concrete and credible case of universal skepticism. In spite of the radical tenets of Sextus and Nāgārjuna, it can be shown that they have some definite positions of their own and some definite forms of logic and argument to support them. If they appear very humble in putting across their positions and sound highly self-critical, it is only because of their decision to fight dogmatism and not to take any dogmatic stand in the fight.

The two main arguments of the skeptic, of "circularity" and "infinite regress," can be met by the same practical-dialectical strategy. In fact, these arguments are being regularly, though silently, followed or practiced by us in our daily lives. Our theoretical assertions and practical decisions are validated not *prospectively,* in advance, or a priori. The validation or invalidation of what we assert and what we decide becomes evident to us only *post eventum* or a posteriori or in and through practical experiences.

To speak of a form of skepticism that it is *initial* or *methodological* is to give only a sort of *post eventum* description of it. Unless one knows the *final* outcome of an inquiry to refer only to its *initial* stage as skeptical makes very little sense. Somewhat similarly, to characterize a given form of skepticism as merely methodological or provisional and thereby to imply that its substantive nature or final aim is something beyond is, in intention, first, to discover the ground of doubt and then, second, to show why the method leading to the substantive or final discovery should be left behind. It may be likened to Wittgenstein's ladder. Once it takes one to the top, the realization of truth, to the concerned one it loses its use value, appears as a nonrenewable resource, and invites some such cheerless and post facto description as "merely methodological."

Cartesian skepticism is both initial and methodological. In either case it may be designated as provisional. Given Husserl's interpretation, it is also of the *epoche* type. For the final judgment regarding its true character is held back in suspension unless the regressive exploration is taken to its end point. Its inspiration is unmistakably philosophical and final aim, in his own words, to "establish [a] firm and permanent structure [of] the sciences."[21]

Born out of distrust toward "outer" or superficial sense-experience his skeptical inquiry takes him step by step to the "inner" consciousness of self-existence. Both "external objects" of sense-experience and "internal" ones of consciousness are available within the "inner space" of consciousness marked by different layers or levels. Some levels are open to more deception and some to less deception. At least one level is said to be "here," which detects all possible deceptions without itself being vulnerable to any deception whatsoever, externally or internally induced. On the contrary, the undeceivable core of consciousness in the form of self-existence is claimed to be assured of its invulnerable position by some "external" authority. Clarity and distinctness of innermost consciousness, self-existence, by themselves could not assure Descartes of the *absolute* indubitability of his own self-existence. Consequently he had to invoke or fall back upon the idea of *veracitas dei,* the undeceiving God. And on this ground, one recalls, Descartes's epistemology invited Arnauld's criticism of "circular reasoning." It has been repeatedly pointed out that "in the order of reasons" the proof of self-existence precedes that of God's existence. But the probative force of the former is parasitical upon or derived from the latter. That is, unless the existence of an undeceiving God is proved first (in the order of knowledge), the proof of self-existence can hardly be taken as absolute. For, absolute truth, in the order of existence, is first to be found in God and then it is God's undeceiving nature that lends it to self-existence through clear and distinct ideas.

As is well known, the charge of circularity has been met in various ways. First, as indicated earlier, by drawing the line of distinction between the "order of existence" and the "order of reasons." Second, it has been argued that certainty of self-existence is not derivative or God-lent, but intuitive and self-evident, and therefore the question of proof being derivative (and, in the process, circular) does not arise at all. Third, it has been affirmed that *without* God's guarantee the claim of absolute truth *is* sustainable. Given this interpretation, the Cartesian notion of *veracitas dei* seems to be dispensable. Certainly this is not the impression one gets from the texts of Descartes. Fourth, some are of the view that the rule or test of truth, closely related to the third argument, and (framed in terms of clarity and distinctness) by itself can guarantee the absolute truth of self-existence and that no guarantee of God, or that sort of strong transcendental argument, is at all needed.[22] Finally, whereas philosophers like Alexander and Hintikka propose an absolute truth claim of self-existence, others like Frankfurt are content with the *relative* truth-proof claim of self-existence. If we propose to follow, as Beck[23] and Kenny[24] do, the texts of Descartes, we are inclined to conclude that Descartes's own formulation of the rule of truth framed in terms of clarity and distinctness is of such nature that the very possibility of absolute falsity is ruled out. Needless to add, the modest interpretation of this sort offered by Frankfurt and others is *not* rendered invalid thereby. On the contrary, it strengthens the proskeptic argument, the one advanced by Unger, for example, that *ignorance* can never be totally removed.[25] This possible threat of ignorance, it seems, persuaded Descartes to tighten his (transcendental argument) belt rather strongly, if not wrongly as well, making it difficult for himself to explain the *growing* character of scientific knowledge.

To me the main question in this context is not so much to identify and establish the absolute truth of self-existence but to identify those actions, decisions and their expressions, linguistic or behavioral, that could be rationally or evidentially defended as true or correct. When I say this I do not claim to have closely followed the Cartesian texts. My limited intention here is to indicate how the Cartesian argument could be presented as a route of transition from philosophical skepticism to scientific skepticism. I get the impression that Descartes got stuck up in philosophical skepticism and therefore it proved difficult, if not impossible, for him to reach the more promising area of scientific skepticism.

As I have indicated earlier, quoting Descartes himself, that his original motivation for examining the nature of doubt was methodological and to vindicate the claim of scientific knowledge against radical skepticism. But later on, it seems to me, he tried to be *substantively,* not merely methodologically, antiskeptic. And, in the process, he was led to a view of science that is not only antiskeptic but also antirelativistic and antihistorical, refusing to take note of the growing character of science. If he had suspended commonsense view of things to review them critically later on in the light of scientific findings and logical analysis, one could have agreed with the *epoche* type of interpretation given of his view. Here I am alluding to the weaker form of phenomenological interpretation of Descartes and not exactly to Husserl's own understanding of it marked by a strong transcendental streak.[26] To recall Husserl's quest for certitude in the realm of transcendental subjectivity and without God may be construed as an attempt to show the "forgotten" relation between the life world and science. If the transcendental argument purported to beat back radical skepticism is a bit "weakened," realistically formulated, and a serious attempt undertaken to relate science to life, it enables one to give a satisfactory account of the growing nature of science and also to save us from the tortuous journey to the end of philosophical skepticism. I describe the journey as "tortuous" not because it is not logically refined, which, as we have noted earlier, it is, but largely because of its practical superfluity or near futility. To claim, as the Buddhist does, that constant practice of skeptic dialectics leads one to a sort of spiritual illumination seems to me unconvincing, if not out of place.

If Nāgārjuna, for example, had ever realized the ultimate reality (*śūnya*), it is mainly by following the positive path of meditation and contemplation as recommended by Buddha and not *via negativa,* not by mentally undergoing skeptical dialectical experiences. A comparable criticism may be raised against the Cartesian skepticism. For, he gives me an impression that he is not content only with beating back scientific skepticism but also determined to remove the ground of philosophical skepticism as if refutation of scientific skepticism is not enough to remove the threat of philosophical skepticism. Scientific skepticism stands above commonsense skepticism. And above the scientific stands philosophical skepticism. It has also been noted that one could be scientifically skeptic without being so at the commonsense or the down-to-earth level. Proponents of the strong transcendental argument do not fail to see the intimate relation between these three levels of skepticism. So they want to ensure steadiness and stability of "knowledge" at all levels. Though they are foundationalists, their foundation being essentially deductive in structure, its strength or steadiness is injected at the top, by a veracious God (of Descartes), a transcendental unity of apperception (of Kant), or a transcendental subjectivity (of Husserl). It is not that foundationalism of the other type, built up brick by brick, on the basis of basic statements or protocol statements of the logical positivists like Schlick and Carnap, is not known in the history of scientific epistemology. But concerned, as I am here, with the Cartesian sort of skepticism, I want to show that though the Cartesian feigns to doubt what lies at the bottom of the structure, the foundation of this doubt lies, in fact elsewhere, at the top. It is *essentially* transcendental in nature. The "deepest" foundation of doubt, doubt of all objects, both of common sense and of science, essentially speaking, is an undeceiving God. He is the limit of all forms of ignorance, including self-ignorance. Self-ignorance, speaking paradoxically, has been rendered impossible by God-given knowledge of self-existence. Once *this* God is removed from the picture, the seeming steadiness of nonskeptical structure gets badly shaken. More radically speaking, without God the *Cartesian* structure is bound to collapse. But does it really? I wonder.

To take Descartes's own example. Although he himself, sitting by the fire with a piece of paper in his hand, doubts whether this is the case, he has as much right to doubt as not to. It depends on him, the case being the same. If he has decided to trust his senses, he has no right (or rather has given up the right) to doubt the case. If however he decides to wait, neither believing nor doubting the verdict of his own senses, and waits for God's certificatory or nugatory voice on the matter to reach his consciousness clearly and distinctly, nobody can make him decide otherwise (except perhaps by unbearable force). The process consists of the double—progressive and regressive—movement of his "indubitable" self-consciousness, (1) reference of the concerned sense verdict by self-consciousness to God and (2) God-certified reference back of the same via the same self-consciousness. Even then if Descartes decides not to accept as true what has been certified by God, nothing *cognitive* on earth or even beyond it can coerce him to decide otherwise. An all-powerful malicious demon may *perhaps* make him decide or dream otherwise by, say, suppressing the nature of the case. But how can the *veracious* God possibly act that way? Unless Descartes, motivated by his love for truth or knowledge, is himself prepared to accept the verdict of

God, nothing possibly can force him to do so, taking away his freedom to say "no" or decide otherwise.

Descartes's example appears to be neither persuasive nor his argument probative. At the moment I am not deliberately raising some otherwise pertinent questions like, "How can he *rationally* form the idea of an all-powerful malicious demon?" One feels that Descartes's doubting game is a make-believe game, and that it is not being played seriously. He is feigning to play the doubting game both at the level of perception and at that of dream, violating right from the beginning the rules of the concerned games and remaining secretly committed to certain rules that do not recognize at all these games and are not applicable to them.

If from the area of sense-perceptual game we move to that of dreaming one, the Cartesian strategy of not playing it seriously, following its own rules, does not basically change. If by sitting by the fire with a piece of paper in his hand, that is, having been perceptually informed of the case in question, he feels entitled to say, can say, or just decides to say "I am doubtful whether this is the case" or "I do not know that this is the case," what can possibly prevent him from doing so? It would be wrong to imagine that by referring to the seemingly proskeptic case of malicious demon (as a thought experiment) Descartes epistemologically improves his antiskeptic case. If perceptual verdict can be doubted, the verdict of dream (whatever might be its scope, particular or universal) can also be doubted, though differently. However, the points of difference (between the games of sense perception and dream) are of minor importance in this context. Descartes, remaining, as he is, committed to the rules of the game of transcendental argument, "finds" that, to start with, these (rules) do not apply to either of the two games. He fails to see that the rules of transcendental argument are blind in relation to different forms of empirical games. Transcendentalism is not clear that this appears to be so only because of its dogmatic insistence that the rules of perceptual synthesis and those of dream synthesis would not be recognized at all unless and until they could be shown to be derivable from the "higher" rules of transcendental synthesis (embodied *absolutely* in the *veracitas dei*).

Strangely enough, the acute mind of Descartes is led to believe by a malicious demon that all information of the world given by senses to him can be devored by, fit well with, dream. But must one debunk all dreams *a la* Descartes? Have we not been told by Moore of a Duke of Devonshire who not only dreamed once that he was speaking in the House of Lords but also on waking up found that he was indeed speaking in the House of Lords?[27] Rigorous parasychologists' observational findings on dream cases are not to be lightly dismissed in a scientistic vein. If all perceived objects are treated epistemologically at a par with the dreamt-of objects or claimed to be assimilable under the latter, one suspects that the commitment which impels the transcendentalist to "advocate" this sort of radical theses is indubitably dogmatic. Whatever is corrigible, not self-evident, is, to him, dubitable—dubitable from the highest point of view, which itself, of course, stands above all doubt.

Clearly the implication of Descartes's refusal to accept the verdict of sense-experience and to liken it to dream experience is very disturbing. It extends the scope of skepticism to experience of all other persons of all times and places. From this universal skepticism, Descartes rightly concluded, no one, except God, could save him or, for that reason, any other person. The implication of this sort of skepticism for a nonbeliever is bound to appear absolutely hopeless. I am inclined to endorse Stroud's critical observations on the matter: "The consequences of accepting Descartes's conclusions as it is meant to be understood are truly disastrous. . . . [T]he whole idea is simply absurd. . . . [U]ltimately it is not even intelligible."[28]

Stroud's words of criticism against Cartesian skepticism are undoubtedly strident. However, it would be wrong to say, as Stroud does, that Descartes's position is not even intelligible. It is not clear to me why he goes to the extent of describing it as "simply absurd." To my mind the Cartesian position is intelligible and not absurd. The whole strategy of Descartes is intended to vindicate transcendental philosophy by pointing out the provisional and methodological inadequacy of the empirical sciences. When we recall that Descartes as a professional scientist was interested in unifying terrestrial mechanics and celestial mechanics, it is hard to maintain as Stroud would that his philosophical skepticism made it difficult for him to recognize the reality of everyday life or the commonsense view of the world. I think it is primarily because of his philosophical commitment that his scepticism vis-à-vis the external world, comprising both everyday life and scientific view of nature, assumes an acute problematic look, inviting undue harsh criticism. However, this is not to deny the perennial attraction and importance of the Cartesian skepticism.

What Descartes questions from one standpoint is defended by him from another. From the empirical standpoint he finds nothing wrong in questioning the existence and objectivity of the external things studied by scientists. But once he feels that he has succeeded in showing us the irrefutability of self-existence and the indubitability of God's existence, he returns to the world of everyday life and science and without any practical doubt about it in his mind. In brief, he starts with doubt and ends up with indubitable self-existence affiliated to and guaranteed by God's existence. His skeptical journey from the dubitable to the indubitable and his demonstrative return journey from the indubitable to the dubitable are two complementary halves, regressive and progressive, of one transcendental argument. In this way I find the Cartesian philosophy of science clearly intelligible, although we can well differ about its correctness.

As an antiskeptical strategy it is not new. For example, Plato's (1) attempt to show the shadowy or copy character of the mundane objects of senses and (2) to prove that there are original(s), Forms, of these shadows are familiar elements of a transcendental argument (though of a different sort). The common point of interest between Plato and Descartes is that both are opposed to explaining the empirical by the empirical. For they are afraid that this empiricist approach condemns us to some or other form of skepticism and that as philosophers their task is not only to go beyond the empirical but also to explain it in terms of the transcendental so that the former is lifted out of the rut of doubt. Obviously this is primarily a philosophical move. One might say, this is a move of the *philosopher* of science who, on the one hand, recognizes the importance of science and, on the other hand, is interested in providing it a philosophical justification so that the skepticism associated with it (due to its changing or fallible fortune) is removed, at least theoretically, for all time to come. It seems to me that this approach to science and philosophy, rather to a philosophy of science, is evident in all leading transcendentalists like Plato, Śaṁkara, Descartes, and Kant. None of them denies the reality of the empirical world recognized in science, but each of them speaks of "higher" reality or realities in terms of which, according to them, the best possible understanding of science is possible.

4. Two Ways of Meeting the Challenge: Cartesian and Kantian

Kant, like his distinguished predecessors, was deeply concerned with the problem of how to be sure of the existence of things outside us. He was not prepared to accept it "merely on faith."[29] To concede that the existence of external things studied by science has to be accepted merely on faith and without any "satisfactory proof" appeared to him a "scandal to philosophy" and an embarrassing victory for idealism and skepticism. He, therefore, not only was opposed to "the problematic idealism of Descartes" and "the dogmatic idealism of Berkeley"[30] but also to those "geographers of human reason" who, like David Hume, thought that the sweep of our scientific causal knowledge, the horizon of reason, is characterized only by a "subjective necessity."[31] Kant was not prepared to hold, *a la* Descartes, that all empirical assertions, except "I am," are uncertain. Equally unwilling was he to accept the Berkeleyan view that the external things in space are "merely imaginary entities." Whereas the Cartesian confesses an inability to prove the existence of external things only on the ground that they are given to our sense experience, the Berkeleyan is led to believe that external objects, situated in outer space, cannot be taken as given to sense experience. Kant felt obliged to refute both these forms of idealism. Additionally, he took upon himself the task of proving that our belief in the existence of external objects, contrary to Hume's contention, is not merely subjectively necessary or "customary."

In fact Kant's investigation was intended to broaden the horizon of human knowledge, brushing aside the external and unwarranted "censorship" of reason, and vindicating "the transcendental employment of the principles" of reason. The initial censorship of reason zeroes down to dogmatism. Its second step promotes doubt or skepticism. In between the two there is room for play of the Cartesian *cogito* principle, "physiology of inner sense," which can inform us only of what seems to be immediate and indubitable but is actually incapable of apprising us of the "properties [of] . . . possible experience" and "yielding any apodeictic knowledge regarding the nature of thinking."[32] Kant aims at the *criticism* of reason, not the censorship of it, so that its bounds and limits, not its *self-ignorance,* could be fruitfully explored. Skepticism is accorded a limited recognition by Kant as "a resting place for human reason." It is somewhat corrective of its own "dogmatic wanderings." The skeptical reason "make(s) survey of the region in which it finds itself, so that for the future it may be able to choose its path with more certainty." Obviously, Kant is not content with any temporary dwelling place. He wants "a permanent settlement" that is obtainable "only through perfect certainty in our knowledge, alike of the objects themselves and of the limits within which all our knowledge of objects is enclosed."[33]

It is clear from Kant's stated scope of investigation that he wants the sort of knowledge that is permanent and certain in character. The question is whether this is the sort of knowledge we really aim at in science. Comfortingly enough, Kant is opposed to "the permanent peace" of the skeptic, a virtual acknowledgment of the impossibility on the part of human reason to get out of the allegedly vicious circle of doubt.

Kant's defense of human reason, meant for achieving scientific knowledge, leads him to search for a permanent settlement. It is not surprising that Kant's image of scientific reason has been likened to his "sheltered life" in his native town, Könisberg, beyond the limits of which he never traveled.[34] Given Kant's "critical" confidence in human reason, especially in the transcendental use of its a priori principles, he thought that as a geographer of human reason he, in spite of being physically confined to a particular spot on the earth, was quite capable of surveying the broad (but not unlimited) horizons of possible experience. His basic strategy to get out of the zeroed circle of skepticism (or *madhyamā* of the Buddhist) is to show that there is no division between external bodies and our representations or perceptions of them. The spatially orderable bodies are said to be immediately perceived and not the result of inference. Though, in principle, we are liable to be mistaken in our perception of external things, we are in a position to have them as representations or contents of our consciousness. We are inferentially aware of them. The relations between the representations are so connected that, when discovered according to appropriate empirical laws, we are not liable to be mistaken.[35] But this formulation of Kant, unless further explicated, does not make clear how his position could be considered as empirical realism. For, he is obliged to maintain that, in some sense or other, external objects are there independent of their being perceived by us. Otherwise, he can hardly distantiate himself from the Berkeleyan's idealism.

The metaphysical realist believes that the external world is there irrespective of whether we know it or not. The metaphysical realist, strictly speaking, may be said to have "animal [or dogmatic] faith" in the independent existence of external things. Kant is not prepared to be convinced by this "scandalous" reason (of faith). To avoid the scandal, dogmatic or skeptic, one is required to look into the *depth* of metaphysic of experience, because metaphysics as such will not do. Differently speaking, Kant's proof of realism is epistemic or experientially metaphysical. Whether the external world (outside our consciousness or knowledge) does exist or not cannot be meaningfully formulated, discussed, and decided unless we look into the issue where it figures or arises. It arises *in* our knowledge or consciousness. This use of the terms *knowledge* and *consciousness* is not to be taken in a *general* way. That is one of the reasons why Kant uses the term *representation*. *Representation* conveys the sense of *objective* content. Human consciousness is to be taken as consciousness of objects. Similarly, human knowledge is to be taken as knowledge *of* objects. Whereas the metaphysical realist has "animal faith" in the existence of the external world populated by objects of different kinds, the empirical realist, in this case Kant, is not prepared to infer that world "of faith." He wants to show that in human consciousness the object is *already* present in a sense and that it is the human task to explicate that immediate sense. The explication would be a proof of external (objects of the) world. To put the matter differently, humankind does possess knowledge of external objects but this possession is to be legitimized; that is, shown to be rational and not accepted merely on the basis of faith.

Where Descartes goes wrong is that he doubts external objects in a precritical manner, without showing us what leads him to doubt them. Unless in some way or other the very possibility of the objects doubted is indicated, the doubt turns out to be as bad or dogmatic as faith; that is, based on faith. Even appeal to the cases of illusory perception does not improve the case of the Cartesian skepticism. For there we are thrown back to our own experience, sense representations for detection as well as correction, if possible, of the concerned illusory perceptions. One who denies this line of argument is obliged to fall back on a sort of dogmatic idealism, according to which sense representations of things are prior to and independent of external objects. Illusory or erroneous perception is undoubtedly corrected by other perceptions. However, the point to be remembered is that the latter, corrective perceptions, are not mere appearances of objects but objects represented, object-ward contents, of our consciousness. The problematic idealists, like Descartes, it seems, have gone wrong at this point. Apparently they fail to realize adequately the intentional and object-ward character of consciousness. The fancied divide between consciousness and the world of objects is the source of problematic idealism. In other words, the gulf between "inner experience" and "outer experience" is imaginary. Therefore, the question of bridging it, strictly speaking, does not arise at all. According to Kant, one has only to *see* the bridge, and the seen would then constitute a proof of empirical realism. To quote him, "we have *experience,* and not merely imagination of outer things . . . our inner experience, which for Descartes is indubitable, is possible only on the assumption of outer experience."[36] Even imagination is objectively rooted. Nothing or nobody can make one imagine of one's own existence *in abstracto*. The consciousness of one's existence is at the same time an immediate (not inferential) consciousness of the existence of other things outside the concerned one.[37]

Thus we find that Kant's proof of realism consists in showing that the very nature of our experience has in it the necessary imprint of and reference to independent objects around us. Objects are not mirrored in our experience. *Per contra,* objects conform to our cognitive experience. This conformity, according to Kant, is not explainable unless an element of *transcendental* ideality is recognized at this point. This ideality is said to be transcendental because the *capacity* of our experience that makes it possible for us to say that objects conform to our experience is not experience itself. It is something "higher" and not "borrowed" from the empirically real world of objects. Denial of transcendental ideality in the constitution of possible objects of experience leads us to skeptical idealism. The ideality Kant speaks of has a priori and necessary character in it. It steadies our knowledge of the objective world, the world of senses. It is necessary also for refutation of both dogmatic idealism and problematic idealism.

The critic may point out that if the "problematic" idealism of Descartes collapses without the transcendental assumptions of self-existence and the existence of the undeceiving God, Kant's own version of empirical realism also breaks down without the transcendental capacity that he ascribes to the knowing (human) self. The "dogmatic" idealism of Berkeley was subjected precisely to this sort of criticism by Hume, who was not *empirically* persuaded of the existence of either self or God. The resulting skepticism of Hume, as we know, is not at all to the transcendental taste of Kant. To have the best of both the worlds Kant combines, or, at any rate, tries to combine, the "appeal" of experience with antiskeptical "steadiness" of "transcendental metaphysics." Second, the critic does not find it at all easy to reconcile "empirical realism" and "transcendental idealism." The uneasiness is due mainly to the fact that one and the same world of objects is claimed to be both, empirically *real* and transcendentally *ideal.* We have already noted how Kant tries to prove the existence of "external" objects in and through "internal" representations by delving deep into the latter and claiming to have found in them their objective reference. In brief, he claims to have succeeded in going *beyond* experience *through* experience itself. But once we look carefully into the structure of Kant's metaphysic of experience the use of such words as *in, through,* and *beyond* makes very little sense. All the moves, analytic and strategic, that are made to draw the distinction between "the empirical" and "the transcendental," between "the internal" and "the external," take place *within* our "experience." Besides, the critic may point out that the catch lies in the word *experience* as used by Kant. He for his own antiskeptical purpose metaphysicalizes it or gives an unduly loaded metaphysical interpretation to it.

One suspects that through the route of the metaphysical Kant surreptitiously brings in the transcendental and all its proclaimed antiskeptical benefits. Kant's transcendental investigation into the nature of experience or knowledge engages him "not so much with objects as with the mode of our [experience or] knowledge of objects in so far as this mode of knowledge is to be possible *a priori*."[38] His investigation reveals that in all our empirical knowledge there are a priori or nonempirical elements. The latter are said to be responsible for the availability (from experience) of the lesson that a thing must be so and so and not otherwise. Kant feels impelled to undertake this sort of investigation because he finds that "the nature of things . . . is inexhaustible" and bewildering, whereas the judgments passed on them are of limited extent and a priori in origin. Kant sounds pretty confident, if not ambitious, when he claims that our a priori possessions of the understanding, on survey, are found to be comprehensive and complete.[39]

Kant's critique of pure reason aims at establishing a *complete* transcendental philosophy that proposes to be fair to the empirically real. And this fairness is sought to be ensured from within the nature of knowledge, freeing itself from all sorts of idealism: problematic, dogmatic, and skeptical. The anatomy of Kant's argument reveals that the empirically real is available to us "because" we can supply ourselves the necessary conditions "hidden" in us and discoverable by a priori reflection. First we are told that the true nature of our knowledge cannot be grasped unless and until the objects of the concerned knowledge are grasped as well. And, then, we are further told that the necessary conditions of our knowledge (of the objects) are to be discovered by a priori reflection alone. I feel certainly puzzled at this point of his argument. If empirical realism is not sustainable without the aid of transcendental idealism, how could the latter itself be understood without the latter. Of course I am not pleading for a *vicious* circle, one half of which is constituted by "the empirically real" and the other half by "the transcendentally ideal." My basic point is that Kant's proof of the existence of the external objects does not mean much if it is not accorded due importance in (or at the level of) transcendental philosophy. This is a point that, later on, drew serious attention of Husserl, among others. In a more radical manner it has been dealt with subsequently by Merleau-Ponty, who was vehemently opposed both to "distortion from above," of the empirical by the transcendental, and "distortion from below," of the transcendental by the empirical. If the skeptical idealism of Hume is accused of the latter, the transcendental idealism of Kant may be accused of the former. However, in fairness to Kant, it has to be admitted that his *First Critique* is an extremely ingenious and earnest enterprise of reconciling the empirical with the transcendental. The critic might even complain that Kant's realism is not realistic either in the ordinary sense or in the scientific sense.

It is not that Kant was not aware of this possible criticism. But, given his main strategy, he could not perhaps oblige the critic on this count. For, it seems, he was afraid that if transcendental strategy is sought to be extended to the realm of external objects, to that of realistic realism, we might unwittingly walk into the trap of empirical idealism, if not its worse ally, skeptical idealism. Once external things are viewed as totally detached from their internal representations within us, the relation between the two, the "inner" and the "outer," stands snapped and we are caught up in the "inner"; that is, confined to our own representations, without any credible access to the "outer."[40] The dilemma of Kant's transcendental philosophy becomes very clear at this point. He is unwilling, on the one hand, to commit himself to empirical idealism and the resulting skepticism and, on the other hand, stubbornly refuses to accord the status of things-in-themselves to external objects. The empirically real, which, in a way, owes its existence to the transcendentally ideal, cannot be given a transempirical or misplaced realistic recognition.

Kant's apprehension is apparently based on a Cartesian legacy, according to which external things are external only by courtesy and that all things *deemed* to be external are external to, or external in relation to, some or other consciousness; human, divine, or both. This Kantian strategy of upgrading the transcendental and relatively downgrading the empirical is likely to offend one's not only common sense but also scientific sense. At the same time, because of the extreme ingenuity of its formulation one finds it very difficult to refute.

It is somewhat like the Vedāntin position of Śaṁkara, which also goes all the way to recognize the importance of the reality of scientific world but, simultaneously, affirms that from the ultimate standpoint of the unity of pure consciousness the scientific world of diversity of objects is simply *non*existent. Painfully conscious of the puzzling nature of his view (or of the very nature of reality itself?) Śaṅkara was sober enough to confess that it is unspeakable, *anirvacanīya*. But the interesting point to be noted is that even then language is used and theory constructed to "speak" about "the unspeakable." When they are very cautious the defenders of "the unspeakable" submit that their words "about" it are nothing more than sparks of fire, glimpses of truth, and are not intended to be ingredients of any theory. To speak of "sparks," "glimpses," makes no sense unless it is claimed, maybe on the basis of "postulation" or "realization," that there is something "supremely transcendental." To vindicate what is empirically real, that is, the scientific as well as the prescientific world, are we obliged to fall back on what is transcendentally ideal? The expression "what we are obliged to fall back on" seems to be obscure, if not misleading, and, therefore, begs elucidation. The vindication of the world of science and prescience may be achieved in different ways; for example, by dogmatically assuming that it is there, that is, taking it as a matter of animal faith, or by skeptically-theoretically questioning it but then practically accepting it. There of course are other ways of beating back skepticism and vindicating realism.

5. On the Alleged Uncritical Character of the Antiskeptical Methods

One might say that each one of these antiskeptical methods, on critical scrutiny, shows its own uncritical or antiskeptical presupposition. If skepticism is sought to be criticized on some dogmatic ground, presupposition or assumption, the skeptic may rightly claim an initial victory, pointing out that the task, that is, the uncritical method of the dogmatic, has to be recognized as a serious cognitive undertaking. In other words, in the fight against skepticism, the company of dogmatism can hardly be accepted as welcome. For this reason philosophers like Descartes favored *a* (if possible, *the*) presuppositionless method. Whereas Descartes claims that his methodological skepticism purported to show the untenability of skepticism and vindicate the possibility of genuine scientific knowledge is indeed presuppositionless, Husserl tries to point out that it is not so, that it is not sufficiently radical.[41]

The critique of skepticism, broadly speaking, takes two forms: (1) justificatory, directly or obliquely, wittingly or unwittingly; and (2) genuinely critical, if not self-critical. It becomes halfhearted and unconvincing, if it is obliged itself to depend on some unquestioned presupposition. Husserl's complaint against Kant is that it happened with the latter. According to him, Kant accepted many things without question; viz., naturalistic (pro-Humean) psychology of his time. He accepted the mathematical and geometrical theories of his time and, above all, the transcendental principles of his philosophy, for which he did not offer any convincing proof. If the grounds of proof are not themselves provable or transparent or are granted immunity against interrogation and criticism, their use value in the fight against skepticism becomes questionable. Another point to be noted here is this. The presuppositions or assumptions underlying the so-called antiskeptical proofs may be of two types: pertaining to the existence of the world itself and its experientially available messages or pertaining to the structures and capacities of the "inner" and the transcendental worlds (psychological or psychical). If Husserl is right, a third alternative cannot be ruled out. For, according to him, the world around us, is already within us in the form of the pregiven lived world; and, conversely speaking, the transcendental world of our subjectivity, unlike the Kantian things-in-themselves, is both rooted in and answerable to the lived world. This dialectical-critical character of the third alternative, Husserl claims, enables him to see "the empirically real" and "the transcendentally ideal" of Kant in a unified way and, what is more important, without resorting to any presupposition, subjective or objective, inner or outer. This is the proclaimed beginning of the antiskeptical and phenomenological method of a really radical philosophy of which science is a part.

6

Husserl and Popper on Skepticism: Some Problems

1. Antipresuppositionalist and Antireductionist Program of Phenomenology

Antipresuppositionalism as an antidote to skepticism has been engaging Husserl's careful attention right from his relatively early days of *Logical Investigations*.[1] The fundamental attack that the skeptic might launch against a theory of logic is to question the very self-evident conditions of its possibility. The conditions of possibility may be approached from two ends, subjective and justificatory. To be antiskeptical a theory is required to be "inwardly evident," free from blind prejudices, and "luminously certain." Besides these noetic conditions, an antiskeptical logical theory must be based on an objective unity of truths or propositions, bound together by the relation of ground and consequent and founded upon the notions of truth, proposition, object, property, relation, and so on that "enter *essentially into the concept of theoretical unity.*" According to Husserl, a theory, logico-objectively speaking, would be self-destructive if its contents violate the laws without which it cannot acquire any "coherent sense." A theory may be logically vulnerable, both noetically and objectively, if its presuppositions offend against the logical laws that alone can ensure its coherent structure.

Noetical skepticism owes its origin to the violation of the basic laws of logic and the resulting false conclusion that "there is no truth, no knowledge, no justification of knowledge." According to Husserl, empiricism, irrespective of its form, moderate or extreme, is skeptical, essentially "non-sensical"; that is, lacking in the said coherent sense.

Husserl's antiskeptical sweep is very broad and not confined to logical theories only. It also wants to show the sound possibility of such "precious sciences" as metaphysics, natural science, and ethics. If skepticism is to be decisively defeated, it should be fought on both the fronts, epistemic and metaphysical. It is not enough to vindicate the noetic conditions of knowledge. In addition, the pro-Kantian *metaphysical* thesis that denies the knowability of "things in themselves" also has to be refuted by pointing out that arguments and proofs, paralogisms and antinomies, which give rise to apparent puzzles and perplexities, are not really sustainable. Husserl's main interest in what he calls *metaphysical skepticism,* which "wrongly favours epistemological scepticism," lies in drawing our attention to the misleading ambiguity of the subject-object terminology. To show logically not only the possibility of knowledge but also the soundness of its foundation the dichotomy between the subjective and the objective, the so-called gap between things within our mind and those beyond it, has to be done away with. This phenomenological undertaking evident in logical investigations has been carried forward relentlessly by Husserl throughout his life. One might say that in *The Crisis* this research program, unfinishable in principle, reaches a very crucial point.

The antiskeptical phenomenological method of Husserl, as we have noted, rests on two main planks; viz., phenomenological transcendental idealism and the noematic sense of the objective world, comprising other things and beings. The transcendental idealism, which explains the intersubjective sharability of the existential sense of the real world, is not metaphysical in any ordinary sense, despite its intermonadic look of the Leibnizian origin. Husserl's transcendental idealism is both antisolipsistic and antiskeptical. It is sustained by transcendental reduction and on the basis of what is called "the most originary evidence" lending legitimacy to all sorts of knowledge. The process of reduction is phenomenologically explicative, not metaphysically constructive, and intended to peel away the dubitable given, inner or outer, and unbracket the bracketed. This presuppositionless explication seeks to bring out the *sense* of the world for us. It precedes every sort of theorizing or philosophizing. On the contrary, theories and philosophies are engaged in uncovering, not altering, this noematic sense.[2]

Husserl's antiskeptical program derives its strength not only from the primitive noematic sense of the objective world but also from the *criticizable* nature of transcendental experience and knowledge. Transcendental knowledge in its bid to rise above its initial *naivete* of apodicticity interrogates itself or turns out to be reflexively self-critical. In principle, this criticism cannot successfully emanate from any Other. It is bound to be self-originary and self-addressed. Yet it is not psychological because of its alleged grounding in "an absolute foundation." Husserl is thus led to believe that it is he who is giving a concrete shape to the Cartesian idea of philosophy as an all-embracing science on an absolutely sure foundation. It is claimed to be free from both the Kantian constructivism and the Humean *naivęte* of the daily life. It is critical without being skeptical. To Husserl, the explication of the Cartesian idea of a universal science is an "endless program." Its self-justification in the form of *self-criticism* is equally endless. Its "radical self-investigation" is engaged in uncovering the intentional horizons of all phenomena, the existence-sense of objects. But the recognition of the necessity of endless self-criticism shows that the progress achieved in the process of the said uncovering is never beyond the ken of criticism and that explains the presence of continuous self-criticism. Transcendental phenomenology, though claimed to be grounded in an absolute foundation, evidently has in its superstructure certain cognitive elements that invite interrogation and question; that is, criticism. Because Husserl's transcendental philosophy is not affiliated to any (Cartesian) veracious God or (Kantian) transcendental unity of apperception (of an enclosed sort), it remains always open ended. Though it has been characterized as "the path of universal self-knowledge," one gets the impression that it is a neverending path and that it can never be fully traversed by ordinary mortals like ourselves.

Husserl's program, to start with, appears very confident and optimistic. Its claim of universality as a science is reminiscent of the Kantian-Einsteinean ideal of a single unified science. Although for Kant and Einstein it is a substantive science, for Husserl it is a programatic one. Husserl is not a fallibilist in any ordinary sense, neither in the sense of Popper nor in that of Ricoeur. Yet his recognition of the programatic character of the unity of science and the necessity of its endless self-criticism reminds me of Popper's view of philosophy. True, Husserl is likely to reject Popper's "naturalistic attitude" to science and other-induced criticism of knowledge. A deeper understanding of the Popperian philosophy, which is also primarily methodological, and *of* sciences brings out certain elements; viz., a primitive *sense* of the existence of worldly objects, the assimilation of objects under the eidos (eidetic reduction-variation), and the neverending criticizable character of transcendental knowledge. These elements are bound to remind one of the affinity of Popper's thought, at least in some respects, to Husserl's.

Interestingly enough, Husserl's self-critical attitude took a rather dramatic turn toward the end (1930s) of his long creative life. He was deeply distressed to find many negative symptoms of the crisis of the European sciences, for example, skepticism about the possibility of metaphysics as science, the collapse of the belief in the universal philosophy as the guide for the new person, the collapse of the belief in reason and a steady emergence of a sense of disillusionment in man regarding his total "Existenz." In the descending gloom humankind was looking around, forgetting itself, and overtaken by skepticism.[3]

It is a long way for Husserl to travel from the *Cartesian Meditations* to *The Crisis of the European Sciences:* in the former he recalled the Delphic motto, Know thyself! and now he finds that the European mind has become engulfed by skepticism and pessimism. He feels disappointed to see that even the practitioners of rigorous sciences like mathematics are unwilling to look into the philosophical foundation of their disciplines and dismiss them as metaphysics. What is worse, many of them are returning to the "paradoxical skepticism" of Berkeley and Hume, giving rise to a widespread belief that naturalistic psychology is competent enough to take care not only of epistemology but also of philosophy of mathematics. Epistemology is being eulogized as theory of knowledge, and transcendentalism is given an indecent farewell. Reductionism and positivism are pushing the antipresuppositionalism of Descartes out of the center of philosophy. Dogmatic objectivism, praise for *external* testability and criticism, is displacing the spirit of self-criticism. Humanity is becoming increasingly other-oriented, alienated from its own true self.

At times, Husserl gives the impression that even Kant is partly to be blamed for this crisis. For, he resorted to the method of constructive inferring (*schliessende*) and gave up the more promising method of intuitive disclosing (*erschliessende*). With the passage of time Kantian epistemological constructivism yielded to psychological or (superficial) logical constructivism. The logical constructivism of the first half of this century appeared to Husserl "transcendentally unfounded." Unfortunately, Kant was accepted from the wrong end. His transcendentalism was forgotten and constructivism practiced without looking into its transcendental roots. Besides, the science of the time was detached not only from the inner self of humankind but also from its outer life world. Rightly understood, this life world is compellingly present in the world of transcendental mind, transcendental subjectivity, creating the world of science. When science, which is essentially an expression of humankind's true self-knowledge, ceases to be human and gets doubly alienated, both from its own roots and from its surroundings, the crisis becomes inevitable and deepens. The crisis of the European science is the crisis of humankind's self-forgetfulness. The truth of science, rightly understood, is humanistic, universal in its scope and historically rich in its legacy.

The main feature of Husserl's antiskeptical phenomenology that appears to me most important is its open endedness, both empirically and transcendentally. The recognition of the necessity of continuous self-criticism of what is cognitively achieved by the phenomenological method is a consequence of this open endedness. Its programatic character, especially its endlessness, is another very important aspect of the growing character of our knowledge. Little reflection is called for to understand that Husserl, in spite of his early Platonism, antipresuppositionalism, and foundationalist inclinations, is a product of his time (1859–1938), both historically and conceptually. Historically speaking, one finds in his philosophy the profound influences of Descartes, Hume, and Kant. But, when we look into the nature of the influence that these three thinkers had on him, their conceptual aspect becomes very clear. Conceptualization of history, pushing the details of history to a back bench and highlighting their structural feature, is typical of Husserl's antinaturalistic and transcendental methodology. Simultaneously, he takes note of the sense of the pregiven of the life world and its assimilability under the transcendental subjective structure. Neither the sense of the pregiven can inform us conclusively of the properties and relations of the concerned object(s) nor the transcendental subjectivity that makes use of this sense can provide us the final truth about the nature of the object(s) in question. Objects become available to transcendental subjectivity through sense. Transcendental subjectivity, which can operate only in association with the senses, makes use of them to articulate their objective reference. But this articulation is not done in one stroke; it is a continuous neverending, though progressive, process. The sense as phenomenon has a constructive duality in it. Because of its intentional object-wardness it is noematic and because of its grounding in transcendental consciousness it is also noetic. The reflection in the sense is made possible because of the noetic ability of our consciousness, the ability to turn all the way back upon itself.

The first point to be noted here is that Husserl's phenomenological method is thoroughly antireductionist. His noematic ontology is an unmistakable proof of it. Without it, he feels afraid, phenomenology tends to degenerate into phenomenalism either of the psychologistic or of the positivistic variety. Without noematic ontology phenomenology is pushed to phenomenalism. Husserl rightly apprehends that his position may then be close to skepticism. But he is not content only with fending off skepticism at one end. To be doubly sure he introduces the notion of noetic ontology. According to Husserl's mature version, noesis is the *real* phenomenological content of our mental acts. The noetic phase of an act contains in it the different phases of experience that have in them the specific character of intentionality. Husserl takes pain to distinguish the notion of noesis from that of "the psychological." The real content of the act is the instantiation of its intentional essence, or the intentional content of the act. By drawing the distinction between "the psychological" and the real content of the noetic, Husserl tries to defend the second antiskeptical flank of his philosophy. Neither the noetic nor the noematic aspects of our mental acts are vulnerable to skepticism. Between the two, the noesis and the noema, the real content and the ideal content, a unity as well as continuity is ensured by the intentionality of consciousness. All these concepts and their relations, rightly understood, may be construed as preventive of skepticism and expansive of the meaning horizon of knowledge.

2. Kant's Influence on Husserl and Popper: Their Points of Agreement and Difference

The unity and continuity of subject and object, of being and meaning, are explicated by Husserl mainly in terms of his notion of transcendental subjectivity. Whereas Kant's notion of transcendental unity of apperception is primarily, if not exclusively, an overarching principle of unification and systematization, Husserl's notion of transcendental subjectivity combines in it both unifying and creative-explorative activities. This point of comparison between Kant and Husserl and the related issues have been carefully looked into by, among others, Kockelmans.[4] The unity and continuity of being and meaning may be viewed from both ends, empirical as well as transcendental. This unity may be construed naturalistically, *as if* it is there in the world. Of course, it may be looked at otherwise; that is, from the end of transcendental subjectivity or synthetic unity of apperception before which the world itself is represented or present as ideal content.

The main distinction between Kant and Husserl becomes clear when we concentrate our attention on two points: (1) the unity of subject and object that, to Kant, is *enclosed,* giving the impression of completeness; and (2) Kant's (notion of) reason as finitistic and restrictive, unduly apprehensive of paralogisms and antinomies and equally scared of the possibility of skepticism. Obsessive fear of skepticism prevents Kant from overstepping the bounds of reason. Unless he is certain that the empirical is neatly tied up with the transcendental, he does not find it possible to explore something more than what is given to or represented in him. He does not find any meaning in the talk of the things that are not objects of *possible* experience. Added to it is his restrictive stipulation that nothing can be empirically possible, objectively possible, unless it is transcendentally sanctioned or affiliated. Thus one finds that given the Kantian framework, a noetically dualistic framework, the Husserlian program of endless quest for scientific unity and continuity turns out to be a futile undertaking.

That this undertaking is *not* futile is evident particularly from the history of the philosophy of science from the days of Kant. Whereas Kant and Laplace were committed to complete determinism, modern philosopher-scientists like Einstein keep on reminding us that the unity of the world of knowledge is always a point of departure for the quest of a larger unity. No unity is final and complete. Every unity is a phase of continuity.

In the context of studies in skepticism, for and against, two thinkers of our time, Husserl and Popper, readily come to my mind. Both have drawn heavily on Kant but, curiously enough, reached apparently different conclusions. It is clear from Popper's work that he has studied Husserl carefully. But his references to the latter are not particularly complimentary or appreciative. Popper is critical mainly of Husserl's essentialism[5] and antipresuppositionalism.[6] It is not clear to me why Popper fails to note Husserl's rejection of Kantian dualism and attempt to vindicate the growing unity of scientific knowledge. Further, Husserl's antipresuppositionalism and foundationalism must be considered with circumspection. For example, when he speaks of "radical beginning," he does not gainsay the "rootedness" even of "the beginning point" in the "life world." After all, Husserl with all his antinaturalist tendencies cannot arrogate to himself the role of a canvas cleaner. Popper is required to bear in mind that "radical beginning" is not to be confused with canvas cleaning. Besides, the objective critic can hardly miss the interesting parallelism between Husserl's notion of changing and growing horizons of knowledge and Popper's notion of the growth of scientific knowledge. In this respect, both are anti-Kantian and against the "finitistic and complete image of science. However, it has to be admitted that Popper's view on and sympathy for scientific fallibilism is somewhat incongruous with the Husserl's idea of *rigorous* unified science. The difference between the two on this crucial point partly accounts for their different attitudes to skepticism. Fallibilism is opposed to skepticism; but it can be, and in fact has been, successfully exploited by the skeptic. Somewhat similarly, notwithstanding his crusade against skepticism, the double-ended openness of Husserl's phenomenological method, tends to take away much of its claim of rigor and lends itself to "softer" interpretations. Husserl's own quest for deeper and deeper transcendental unity of science and Heidegger's endlessly forward-looking transcendentalism are to be remembered in this connection.

Another point in Popper's criticism of Husserl, that the latter's essentialist method is not acceptable either to science or to philosophy, does not appear to be entirely fair or charitable. For Husserl's insistence on noematic ontology and the uneven structures of phenomena does not go well with essentialism. Further, it is to be borne in mind that Husserl, unlike Brentano, draws a distinction between "fulfilled" intentionality and "frustrated" intentionality, between "real" object and "fictional" object. True, Husserl's position on this point is not free from ambiguity. However, if the basic point is duly recognized, the pro-Aristotelian essentialism that Popper attributes to Husserl appears strained and suspect.

If we closely look into Popper's own philosophy, we will find that in spite of his protestation to the contrary, it, like Husserl's, is primarily methodological and that it is also antiskeptical from both the ends. Popper's emphasis on anti-inductivism and the theory-laden character of the given shows unmistakably his antinaturalistic orientation. Second, another Popperian thesis that buttresses this point is that all observations are meaningful or informative only on satisfaction of the condition that they are relatable in terms of some or other hypothesis or conjecture. Negatively speaking, he maintains that observations as such are not cognitive and meaningful unless they are theory related. In this thesis one can hear the echo of Kant's celebrated Copernican Revolution; that is, the reversal of the relation between subject and object. The role assigned by Popper to theories that are essentially hypothetical or conjectural in character is also reminiscent of the view of Kant's categorical framework. In spite of his repeated claims that he is *not* a methodological holist, one often comes across his reference to the "well-entrenched" character of a good scientific theory. In other words, a scientific theory can hardly be good entirely on its own account. It is required to be comprehensive, comprehensively related to other well-entrenched theories.

However, a word of caution needs to be uttered at this stage. Otherwise another important thesis of Popper's, particularly important from the antiskeptical point of view, is likely to be misunderstood. Although his accent on the theory-laden character of observation is likely to be taken as an antinaturalist and protranscendentalist aspect of his epistemology, his equally important reference to the role of crucial experiments, the potential falsifiers, highlights his strong realistic orientation. That the objects of the real world have a very crucial say in molding or modifying our theories is an important point that was apparently missed by Kant. Looking at the two aspects, the subjective and the objective, of his theory of scientific knowledge it is clear that Popper's fallibilism, though different from Kant's critical theory, on the one hand, and from Husserl's phenomenological theory, on the other, has some basic similarities with both.

Popper's theory of scientific knowledge demands our close and critical consideration mainly for two reasons. First, his transcendentalism, unlike Descartes's and Kant's, is open ended. There is no terminal transcendental level. When he speaks of the "depth" of scientific theory, all he means is its comprehensiveness or its contentual richness. The other terms used by him in this connection are *simplicity, coherence,* and *beauty.* The simplicity, coherence, and beauty of a theory are dependent on its depth—in relation to its "depth" one can talk of these qualities.[7] For example, one finds a new beauty in Newton's dynamics because of its depth, a depth that can encompass both Galileo's terrestrial physics and Keplers' celestial physics. The findings of Galileo and Kepler assume added significance in the Newtonian framework. The resulting theory also becomes more *testable.* The quest for depth does not necessarily entail *correction* of the facts already found in the theories unified. To take another example, Maxwell's electromagnetic wave theory incorporates within it Fresnel's wave theory of light without correcting its factual findings. The depth dimension of the theory, contrary to the popular understanding of it, is achieved in terms of its higher, increasingly transcendental levels. To put it paradoxically, depth is explored from the (transcendental) height of increasingly comprehensive theories. The height of theory knows no bound. In other words, unification, to recall Husserl's phrase, is an unfinishable program.

As an uncompromising antifoundationalist Popper recognizes no rock bottom base of scientific theories. In this respect, he rejects the empiricist's thesis that certain basic observational reports, unquestionable and incorrigible, lay the foundation of science. Equally opposed is he to the rationalist contention that there are in us some innate ideas that (from above) provide the self-evident foundation of science. One might say, borrowing and varying an expression of Merleau-Ponty, that Popper is opposed to both dogmatism from below and dogmatism from above. Antifoundationalism is a close ally of antipresuppositionalism. Neither any proposition nor any presupposition can be taken to be unquestionable forever. Because the horizon within which the scientist is obliged to work is set by some or other specific problem situation, no proposition or presupposition can be unconditionally declared as basic or indispensable. The presupposition, transcendental or otherwise, and the proposition, empirical or metaphysical, that the scientist needs to construct a theory are influenced by and oriented toward the problem in question. Antifoundationalism and antipresuppositionalism are to be understood in the context of the concerned problematic issue(s).

Popper's anxiety to show the openness, both methodological and substantive, of his theory is understandably motivated to support his claim of cognitive fallibilism not amounting to skepticism. In this respect, as I mentioned earlier, he stands closer to Husserl than to Kant. Unfortunately, he himself does not appear to have realized it clearly. What is interesting to note at this point is this. His departure from Kant has been a subject of criticism by the orthodox Kantian. His departure from the logical positivism, his preference of falsificationism to verificationism, has attracted severe criticism from the logical positivist quarters. The orthodox Kantian and the Cartesian highlight the diluted (Godless) form of his transcendental argument and the resulting weakness of his knowledge structure. If there is no *ultimate* certification of the *unity* of knowledge, it is argued, the structure remains vulnerable, if not destined to collapse. If the empirical findings are always allowed to shake the edifice of knowledge, it is bound to crack and break down. In view of this alleged threat from below, the empirical, the transcendentalist tries to find a close-ended structure of this cognitive system.

Popper's problem vis-à-vis the empiricist seems to be equally, if not more, serious. Although he states that he does not believe in the *ultimacy* of observational reports (in the forms of protocol statements or basic statements), he is obliged to recognize the ultimacy of falsifiers. Falsifiers are obviously stronger than verifiers in their efficacy to test a hypothesis or to overthrow a theory. The problem arises mainly out of Popper's claim that he is a realist, anticonventionalist, and that the basic statements, which work as falsifiers, are *public* in character. It is clear that Popper's twin claims of realism and the public character of falsifiers are essentially antiskeptical in their motivation. But the question remains how he can assure himself that *his* observational reports are identical with *my* observational reports in relation to a particular given object. To say that it is publicly available is to beg the question itself. For, experience or observation by its very nature is not public. What is public is the referent of our experience. But to establish this we are obliged to fall back upon our experience itself. Given this problem-situation, one is reminded of the importance of the recognition of some such things as "elementary picturesque propositions" (Wittgenstein) and "protocol statements" (Neurath). To a professed antifoundationalist this sort of "indubitable" statements are not available. Popper's own "basic statement," though close analogues of "direct observational reports," are to be distinguished from both Wittgenstein's elementary picturesque propositions and Neurath's "protocol statements" views on the subject. It is not easy for Popper to defend realism at the level of "basic statements" and, at the same time, demand that they be publicly available. His problem is compounded by another favorite thesis of his that there is nothing like "direct observation" or "simple apprehension" and that every observation or apprehension is theory laden and problem oriented.

3. Popper and Watkins on Basic Statements: A Short Critique

To say this, however, is not to deny that Popper has made a heroic effort to show that even his basic statements are open to question and correction. On this issue I find Watkins's analysis very incisive and instructive.[8]

Popper's antipsychologism is evident right from his first published work.[9] His antifoundationalism obliges him to deny that basic statements are foundational rocks and chips. At the same time, as an empiricist he cannot deny the crucial and very important role of experience in scientific investigation. Finally, as a rationalist he has to tell us why only certain basic statements (and not others) are to be accepted as basic. To put it differently, the epistemologist is *logically* required to provide reasons for his acceptance of basic statements. Popper proposes to show that his account of the empirical basis (constituted by basic statements) is free from the triple blemishes of dogmatism, psychologism, and infinite regress. It is irrational to accept basic statements dogmatically; we must be able to justify or support them by some rational arguments. But the problem is that rational arguments or logical proofs lead us to a sort of infinite regress. If we accept basic statements on the grounds of their alleged immediacy, that is, on the grounds of their being "based on" immediate experience, we invite the charge of psychologism. If we are anxious for rational arguments we land ourselves in infinite regress. To cut the matter short, if we accept certain statements as basic simply because we think to decide that they are so, we are dogmatic. So to avoid the trilemma, dogmatism vs. infinite regress vs. psychologism, we must try to produce some reason in support of our acceptance or choice of basic statements.

In both Neurath's "protocol statements" and Carnap's "basic forms of scientific language," Popper finds "the fundamental ideas of the psychologistic approach."[10] It is a disputable question whether Neurath or Carnap is psychologist in any ordinary sense. However, from his earlier work, *Aufbau* (1928), where Carnap used such terms as *auto-psychological* or *hetero-psychological,* one gets the impression that even his "formal mode of speech" is not totally unrelated to our psychological experience.[11] In his earlier writings "protocol statements" appear to be ultimate and "touchstone" statements with reference to which every scientific assertion had to be judged. Neurath's view on "protocol statements" seems to be more acceptable to Popper. For, according to the latter, "protocol statements" that record experiences are not irrevocable but at times may be rejected. Neurath's view at least does not smack of dogmatism. But Popper wants to give a more rigorous, objective, and rational account of the empirical basis.

First, he wants to remove the possible misconception about the alleged relativity of basic statements. According to him, "any basic statement can . . . in its turn be subjected to tests, using as a touchstone any of the basic statements which can be deduced from it with the help of some theory, either the one under test, or another."[12] In principle, basic statements are questionable and corrigible but obviously not in isolation. But, in practice, the working scientist does not go on questioning the basic statements simply for the fun of doing it, for carrying on this game. He stops somewhere. The point where he stops is usually accepted by others also as the point to stop. It is indicative of the public character of where we stop and of successful communication. In a technical sense one might say that the point where we decide to stop is arbitrary and we resort to a sort of *dogmatism* to avoid *infinite regress*. But this sort of dogmatism, says Popper, is "innocuous." The kind of infinite regress to arrest which the scientist becomes dogmatic is also said to be innocuous. Finally, the charge of psychologism is rejected by Popper by highlighting only the *causal* connection of our experiences with basic statements showing how the former "motivates [our] decision" (regarding the *satisfactoriness*) of the latter. But, curiously enough, Popper refuses to admit that his use of the terms, *experience, motivation, decision, satisfaction,* has anything to do with psychologism. The basic plank of his antipsychologism is *testability* (in principle) of the basic statements.

Second, these statements play the role of falsifiers in scientific theories. This role they can play because of their ability to satisfy two conditions: (1) their nonderivability, without initial conditions, from a universal statement, and (2) their ability to contradict and be contradicted by a universal statement. Condition (2) is satisfiable because it is possible to derive the negation of a basic statement from the theory that it contradicts. Now it follows that a basic statement must have a logical form such that its negation itself cannot be a basic statement in its turn.

In brief, Popper affirms that the basic statements have the form of singular existential statements. Apart from "the formal requirements" he refers to "a material requirement" that has to be satisfied. This is the *observability* requirement, observability of the space-time regions that the concerned singular existential statements describe (or purport to describe). "Basic statements are therefore—in the material mode of speech—statements asserting that an observable event is occurring in a certain individual region of space and time."[13] But Popper here stubbornly resists the psychologistic interpretation of his "observability." In defense of his view, he is prepared to take "an observable event" as "an event involving position and movement of macroscopic physical bodies." Because Popper is not a physicalist in any ordinary sense and because he repeatedly speaks of "empirical basis," it is difficult to see how he can totally exorcise the ghost of psychology or human experience from his what he calls (later on) *empiricism.* It also seems to be incongruous with his antireductionism as evident from his theory of three worlds, world1 (physical), world2 (psychological), and world3 (theoretical-transcendental).

Finally, it is to be noted that even his account of "public observability" is not totally free from the traces of *psychologism.* So, here the critic may understandably maintain that Popper's basic statements are not entirely free from the blemishes he himself finds in Carnap's observation reports and Neurath's protocol statements. The more radical critic committed to cognitive psychology may raise the question, What is wrong with psychologism? Can we eliminate psychologism without restoring to dogmatism or uncritically accepting an unexecutable program of reductionism or both?

It seems that neither Popper nor his followers is prepared to accept psychologism, for they are all more or less equally afraid of skepticism, which, they think, is an inevitable corollary of psychologism. Popper has offered his own strategy of combating psychologism and the resulting skepticism. The basic concept used by him in this context is *testability,* testability of basic statements. Once he succeeds, if he succeeds at all, in showing that even the basic statements are testable in conjunction with some other theory or assumption, he is in a position to claim that both psychologism and dogmatism have been successfully contained. But Watkins has argued painstakingly that Popper's strategy does not work. In spite of all that Popper claims, Watkins feels that following the Popperian strategy of "testing" the basic statements, "all [that] we are getting . . . is a lengthening chain of derivations: no *tests* are being made."[14]

Having rejected the Popperian interpretation of the basic statements, Watkins offers his own. He proposes to show that "perceptual experiences are both causes of and reasons for the acceptance" of basic statements. If the basic statements are regarded as causal (motivationally or decisionally), the charge of psychologism sticks to them. Therefore, it is understandable that Watkins should try to find a rational basis for the acceptance of basic statements.

The testing process of the basic statements is brought to a stop not because of any causal decision or motivation, but because, it is claimed, we are able to produce reasons why we have so stopped the said process. Watkins rejects both Ayer's interpretation of the Popperian basic statements and Popper's own view on the subject. According to Ayer, the basic statements fail to work as *conclusive* verifiers but they have in themselves the "adequate ground for accepting them." Basic statements are, in a sense, self-justifying.[15] Popper himself is prepared to characterize our perceptual experiences underlying the basic statements as "inconclusive reasons." "They are reasons because of the generally reliable character of our observations; they are *inconclusive* because of our fallibility."[16]

Here it seems to me that Popper reiterates his old position that now appears as quasi-inductive to Watkins and not significantly different from Ayer's view on the matter. The expression *generally reliable* sounds inductive to Watkins because of its informative claim, providing information about the external world. Although we appreciate Watkins's anxiety over the unwelcome threat of inductivism, it is not clear how the denial of the informative character of the basic statements would be of help to us in distancing ourselves from inductivism. On the contrary, it appears that if the basic statements are regarded as totally incapable of informing us of anything about the external world, we are obliged to retreat from realism. In the process of trying to avoid psychologism and inductivism, are we not shying away, may be unintentionally, from realism?

"No," I presume, would be Watkins's answer. He formulates the problem very ingeniously indeed. First, he tells us that he is not engaged "in any kind of quasi-inductive inference from inner experience to outer reality."[17] What he thinks he is doing is to frame an explanation of 0-level sensory reports in terms of statements of other (higher) levels. This shows that we may need explanation even of our sensory reports because of their puzzling or problematic nature. The statements of higher levels, 1 (observational reports about things), 2 (empirical generalizations about things), and 3 (exact and universal laws of mathematical physics), because of their more or less theory-laden and problematic character, may be "wastefully" explanatory. But ordinarily we find that the theoretical load of lower-level statements is less than that of the higher-level ones. The point that Watkins insists on is that problematic 0-level statements (or sensory reports) though "autopsychological" in character, are accepted because of the rationality of their explanatory hypothesis. Negatively speaking, their acceptance is not dogmatic or psychological or dictated by what he calls *quasi-inductive reasons*. The explanans of 0-level explanandum consists of a suitable set of level-1, level-2, and level-3 statements. This suggested explanation is a conjunction of level-1 (basic) statement concerning the object(s) of the explanandum and other statements of the attending circumstances and regularities. Clearly, the form of the proposed explanation is hypothetico-deductive. The structure of the explanation, as formulated by Watkins, is related to the problematic nature of the concerned 0-level statements. Watkins is of the view that if *this* explanation of ground-level, that is, 0-level, sensory reports is available, then these reports may be accepted until a better alternative explanation of the concerned sensory report is available. By this formulation of the temporarily acceptable "empirical basis" of scientific theory, Watkins thinks, he can solve the problems of dogmatism, psychologism, and quasi-inductivism at one stroke.[18]

A pertinent point that may be raised at this point: "the empirical base" need not necessarily be a problematic base. Watkins gives no arresting reason to show that only his 0-level statements constitute the empirical basis of science. One is free to regard the level-1 observation reports about things as *the* empirical basis of science. It depends on the scientific discourse, rather its level, one has in mind. The difference between 0-level statements and 1-level statements is not *intrinsically* fundamental here. The most important question is whether they *are* problematic. Or, are we ourselves *choosing* the problematic level? Once we agree, as apparently Watkins does with Popper that "science does not rise upon rock-bottom . . . rises . . . above a swamp . . . [and has no] natural or 'given' base," the question of choosing the empirical base cannot be dictated from without. The problematic level is a matter to be decided by the concerned persons or the working scientists. Popper tells us that our choice is free despite not only the *problematic level* but also the *explanatory level*. "We . . . stop when we are satisfied that they are firm enough to carry the [scientific] structure, at least for the time being."[19] It is to be noted here that what Watkins has to say on this point is strikingly similar to Popper's own words on the subject. "Having arrived at . . . an explanatory hypothesis, it will be rational to retain it unless [we] afterwards find a better alternative to it or later evidence tells against it."[20]

4. Popper and Quine on Realism and Instrumentalism: Where Their Agreements Are Substantive and Differences Secondary

It is very difficult to detect any fundamental difference between Popper's approach to the trilemma of dogmatism vs. psychologism vs. infinite regress and that of Watkins's. The only difference that seems to me noteworthy is Watkins's extra anti-inductive zeal. I do not know how Popper with all his life-long record of crusade against inductivism can possibly differ from Watkins on that point. It may, however, be possible if Popper maintains that Watkins's strategy of avoiding the infinite regress is uncalled for and that the latter's extra anti-inductive zeal has been well taken care of by the former himself.

For the time being let us forget the difference between Popper and Watkins on their interpretations of the basic statements. The point that keeps on disturbing us is whether anything in us can make us certain about our knowledge of the external world. A further question may be raised regarding the character of that something which claims to have the capacity to give us certain knowledge about the external world. Is it psychological, or is it epistemological? Another question to be taken into account here is, Which of these two sciences, psychology and epistemology, is more respectable or comprehensive? According to some thinkers like Quine, psychology is more comprehensive and dependable than epistemology as ordinarily understood. Others like Popper take epistemology more seriously and are afraid of psychology because, among other things, of its psychologistic undertone. It is clear that although the former are proreductionist the latter protranscendentalist. The common point between the two is the recognition that science is a going and growing concern with no finalistic aim. In different ways both are realistic. Quine's realism, as we see it, is very liberal, but Popper's realism is relatively austere. Neither of the two proposes to offer a proof of the external world for the sake of vindication of his commitment to realism. One might say that both of them are internal realists. I am not using the term here in Putnam's sense.[21] As distinguished from these moderate views against skepticism and for realism, I recall Moore's famous "Proof of an External World."[22] "*An* external world?" or "*the* external world?" Can *external* take care of the distinction between *an* and *the*? Do we not, rather can we not carve out "different external worlds" from "one and the same external world?" To my mind, both Popper and Quine answer the question in the affirmative. Let us see briefly how they formulate their answers.

Quine finds no *fundamental* difference between particular things and universal beings, physical objects and abstract objects, existence and essence, and the like. Intersubjectively observable physical things prove basic in our *language learning* and for the purpose of *communication*. Because of their continuity and homogeneity medium-sized physical objects like tables and chairs appear more real than the objects of highly inferential or constructed sorts. Out of sensory objects Quine finds how physical objects could be obtained without postulating sense data. The superiority claim of physical objects over that of sense data lies mainly in direct association of sentences, not of individual terms, with sensory stimulation. Therefore, compared to abstract terms, common terms for physical objects are found to be more useful for cognitive systematization. However, Quine is not altogether opposed to the admission of abstract objects into theory because of their efficacious role. Positively speaking, in his canonical notation of quantification, neither physical objects nor abstract ones are accorded a special ontic status or privileged position. There is no universal "object-positing pattern" of different languages. Ontic commitments of words and sentences are conceded to be of "limited polemical power."[23] In regimented language the economy achieved is purely theoretical and not claimed to be objectual.

Quine tries to show that all sorts of objects, physical, metaphysical, logical, and mathematical from *this* and *now,* tables and chairs to attribute, proposition, intention, class, number, and unactualized possibles are ontologically alike in ultimate analysis. The Occam's razor used by Quine cuts deep into the realm of ontological entities reducing their number to the minimum; viz., "value of a variable."

Possible objects and unactualized possibles are due to defective nouns and perplexity over identity. He explains the point in terms of stimulation and stimulus meaning. Even when a particular position is ostended, as in "the possible new Taj Hotel near the Gateway of India in Bombay," the identity of position does not make the possible object identical. It remains as elusive as Quine's "Gavagai" and "Rabbit." By using a modal operator the sort of identification of a possible object that can be achieved does not make sense to him. Rather, he would prefer some neighborhood relation to the existing Taj Hotel, Gateway of India, and Bombay. Thus the possibles are sought to be brought closer to the actuals.

Second, propositions are dismissed by Quine on the grounds that their admission rests on an unreasonable wish for finding out some "eternal truth-value vehicle independent of particular languages."

Third, admission of facts and negative facts is also strongly discouraged by Quine. He thinks that our talk in terms of facts is due to our anxiety to answer the question of how the truth values of sentences are to be decided. Another motivation underlying the fact-oriented discourse is to believe that sentences, like names, must posit some facts and that without the latter they would be awkwardly hanging in the air. Quine's point is that absurd or troublesome objects like facts and unactualized possibles may be banished from the domain of values and their variables without impairing the regime of logical language.

In this context, he refers to the concept of infinitesimal and how the problem regarding its definition has been resolved. Newton and Leibniz introduced this concept in their differential calculus to indicate quantities infinitely close to zero and at the same time, strangely enough, distinct from one another. Though the idea itself of infinitesimals seemed absurd, yet in the differential calculus in which they were recognized as values of the variables yielded useful results. The problem of defining the classical concept of infinitesimal was resolved by Weierstrass by introducing his theory of limit and showing thereby "how the sentences of the differential calculus could be systematically reconstructed so as to draw only on proper numbers as values of the variables, without impairing the utility of the calculus."[24]

A comparable strategy is resorted to by Quine in parallel cases like *ideal objects;* namely, mass points, frictionless surfaces, and isolated systems. These objects of mechanics appear to be contrary to physical theory. But for the purpose of theory construction these ideal objects are profitably exploited. If one states that mass points behave in some particular ways, what one means is this: particles of a given mass behave in some more or less predictable ways when their volumes are smaller. Exact behavior of *actual* particles or their systems can never be correctly anticipated. In effect, what is conceded is that the *ideal* gas particle is never identifiable individually but only aggregatively.

To Quine, ideal objects of physics are "symbolic" in character. In this sort of case the rich resources of language is used by the physicist somewhat like the literary critic, the psychoanalyst, the philosopher of language, and the anthropologist. Symbols are used because of their explanatory use value, vividness, beauty, and simplicity. Strictly speaking, one cannot draw a sharp line of demarcation between the primitive's myth and the scientist's hypothesis. In this respect Quine seems to have rejected, at least significantly departed from, the Popperian criterion of demarcation, that is, falsifiability, that seeks to highlight the difference between myth and metaphysics, on the one hand, and rigorous sciences like physics, on the other. From classical mechanics to quantum mechanics, Quine tries to impress upon us, ideal objects have been liberally used. Their utility has been borne out even for the rigorous purpose of computation or calculation. To Quine, this approach of science is in no way inconsistent with ontological realism. Idealization of actuals entails no retreat from the actuals of the real world. On the contrary, actuals *qua* actuals, as empirical givens, cannot be effectively captured and systematized except within the network of theory. I am sure the Popperian would interpret this ontology as very thin and its accompanying methodology as objectionably instrumentalist. But I do not see why those rationalists who reject the myth of the given and recognize the explanatory powers of myth and metaphysics should embrace illiberal realism. The distinction between instrumentalism and realism, in this context, is more idiomatic than substantive.

To brand Quine a realist or physicalist without qualification perhaps would be unfair and prove inconsistent with his careful texts. Of the alleged weakness of his theory of meaning he seems to be quite conscious. To preserve the flexible referential character of referring names his strategies are undoubtedly ingeniuous. His formulation of the concept of ordered pair, $\langle x, y \rangle$, following the cues provided by Peirce, Wiener, and Kuratowski is intended to undo the miseffects of defective nouns. The ordered pair has been characterized by Peirce as a mental diagram consisting of two images or symbols of two objects, one actually attached to one member of the pair, the other with the other. By this device two objects are treated at a time as if objects of some sort were being treated one at a time. It is purported to assimilate relations to classes. It is a key analytical tool for the purpose of philosophical explication resulting in dissolution or elimination of seeming problems and paradoxes. Here Quine claims to be in the line with Wittgenstein. Vague or perplexing objects "denoted" by defective nouns are eliminated and claim to be replaced by the analytical device of the ordered pair.

Given the Quinean device of eliminative analysis, one of the well-recognized but vague objects that gets eliminated is mind. Mentalism does not find favor with Quine. However, his preference for physicalism does not entail total rejection of mental terms. He is not prepared to admit that his theory entails total rejection of the mental states. The reductive interpretation of his view seems rather drastic to him. What he highlights is that there is "no unbridgeable differences in kind between the mental and the physical."[25] What Quine is quite unwilling to accept is that the mental states can exist independently. Positively speaking, he affirms their parasitical character, parasitical upon bodily states. Understandably he is unenthusiastic about the ambitious program of radical reduction, a sort of metaphysical behaviorism.

Unlike Nagel and Quine, Popper refuses to accept the view that the distinction between instrumentalism and realism or between physicalism and mentalism is merely idiomatic. He gives the impression that his own variety of realism is indeed uncompromising.

At the risk of being repetitious let me briefly mention the points why I somewhat discount the strength of his realistic claim. First, the dramatized form of his anti-inductivism and fallibilism is primarily indicative of his rejection of metaphysical realism, be it of Kant's transcendental sort or Moore's commonsense sort. One who starts from problem and hypothesis and is engaged only in testing the latter to solve the former can never reach the terminal point, if any, of epistemic certainty. The basic statements or falsifiers by which the hypothesis is tested are themselves testable in principle and their shared acceptance is a sort of convention and inductive decision.

Second, both Popper and Quine are for *simplicity* of theory and *austerity* of ontology. Excess baggage of ontological presuppositions is left behind by them. Equally notable is the similar look and content of their mixed bags of *truth* and *convention*. In determining truth both word and object, language and reality, have their says. Tradition and context are given due recognition. Possible world semantics and glorification of rigid referential theories of meaning entailing essentialism are scrupulously eschewed by both. Putnam's attribution of "conventionalist essentialism" to Quine seems to be trivial.[26] To think, as Putnam does, that Quine's translation manual (rules and constraints) seek to determine referential *meaning* "exhaustively" and uniquely rests on a misunderstanding and is untenable. It is due to his commitment to the Kripke-type of causal theory of reference.

Popper's espousal of methodological nominalism is also bound to be resented by realists with essentialist leanings.[27] Otherwise a consistent realist, Popper's concession to nominalism in the context of definition is purely methodological or practical. Elsewhere Putnam sounds critical of Popper on the ground of inadequate practical orientation.[28] It is not difficult to understand the critical disposition of a pro-essentialist realist toward ontological austerity and methodological liberalism, one promoting another.

Third, because of their respective methodological positions neither Popper nor Quine claims that our knowledge can be conclusively true or that one could be rationally certain about it. Though a staunch defender of Tarski's theory of truth, Popper realizes the difficulty of accommodating it fully in any scientific theory. Therefore, he concedes that what we achieve in theory is truthlikeness or, to use his term, *versimilitude*. This moderate and practical aim of scientific theory stems from another basic point in Popper, the concept of *growth* of scientific knowledge. The question of growth makes no sense unless it is recognized that in our quest for knowledge we can never possibly reach a point that could be rightly regarded as the end of the journey. Incidentally this explains why Popper characterizes scientific theory by truthlikeness rather than by truth as such. To him, truth is a *regulative ideal* that can be more or less approximated but never actually reached. In a sense it is ever elusive yet very useful. In the process of growth both theory and the problem to which it is addressed undergo change or revision. Through testing the theory is changed and some of its errors are eliminated. But the resulting new theory encounters new problems requiring new tests. And this process, marked by nonlinear variation, progressive and regressive, goes on.

A similar, but not identical, view is found in Quine. According to him, our theoretical universe in the face of new and relevant experience undergoes change. The change is felt maximally at the periphery and minimally at the center. As the result of theoretical universe's test encounter with "external" sense stimulations the truth values of its constituent sentences undergo redistribution. In other words, the universe is repeatedly revised. The so-called analytic center consisting of logic and mathematics is least affected and the peripheral or frontier disciplines like psychology and biology are most affected. The most important thing to be noted here is that in the Quinean universe nothing is completely impervious to the effects of experience. Empirical information never goes totally vain.

Quine's *revisability* and Popper's *fallibility* are kindred concepts and indicative of their empirical bias. The Quinean universe, as formulated by him, is really one and unified. Though Popper speaks of three worlds, physical, psychological, and logico-mathematical, they are causally interrelated and interactive. It is not difficult to find the affinity between Quine's concept of *interanimation* and Popper's concept of causal *interaction*. The difference between the two is one of degree and not of kind. The basic point that emerges out of their views is the revisable, changing, and fallible character of human knowledge.

Finally, from these points one may conclude that Popper and Quine objectively encourage skepticism, no matter what their subjective views about it may be. What Popper's anti-inductive and error-eliminative method delivers remains fallible and uncertain. It is not without any justification that, despite his protestation, he is often accused of encouraging methodological anarchy and epistemological uncertainty, the two main faults attributed to Kuhn and Feyerabend. Most of his basic theses—antifoundationalism, questionability of basic statements, free framability of explanatory hypotheses, verisimilitude, and open endedness of knowledge—are pointers to practical skepticism and promote a critical spirit. To affirm at the theoretical level such ideals as objectivity and realism, though welcome, is not enough to banish skepticism. In practice fallibilism is not far from skepticism. The nonconclusive character of falsificationism makes the point additionally clear. It is instructive to note that even after highlighting the importance of falsifiability, Popper has to admit that good scientific hypotheses are expected to survive falsifying tests. Even more interesting to note is Popper's failure to emphasize another extremely important point that he admits only parenthetically; viz., the tests designed to falsify hypotheses and do not succeed in doing so add to the content of the concerned hypothesis. In effect, having survived its tests a hypothesis gets inductively rich in terms of content. In spite of Popper's professed preference for the "eliminative (inductive) test," he cannot in practice avoid the additive or enumerative aspects of induction. This aspect of his otherwise negative methodology brings out its hidden positive or additive aspect. By implication this shows that every scientific theory, whatever may be its level and scope, remains open to further testing and questioning, addition and elimination.

On the comparable issues Quine's position appears to me more simple and less circuitous. He straightforwardly recognizes the importance of induction for both learning and communicating. Whatever information, true and false, we accept we do so inductively and from limited accepted bases we try to construct or infer theories of wider scope. Although the basic building blocks remain the same, the structure gets inductively larger and taller. But we never reach a stage where we can justifiably claim that our cognitive structure or system of knowledge is so perfectly built that it does not need any repair or renovation. Unlike the foundationalist Descartes, Quine is of the view that there is no basis, *cogito*-like basis, a sort of first philosophy, that can ensure permanence and perfection of scientific knowledge. In vain we ask for *objectivity, reality,* and *truth* that would insulate our cognitive structure from all sorts of wear and tear due, metaphorically speaking, to atmospheric effects like questions, experiences, and experimental tests. All forms of fallibilism, Quinean, Popperian, and Kuhnian, deeply disturb those who style themselves as really realist. Therefore, the latter strongly feel that some or other stronger form of realism or transcendental idealism must be defended and relied upon. Otherwise, they are afraid, the threat of skepticism cannot be adequately met. This is not a historical question peculiar to our age and culture. Time and again this conceptual question has appeared and reappeared in different philosophical cultures and in different forms.

To speak of staving off skepticism by realism is one thing and to achieve it is a different thing. Forms and formulations of realism are endlessly diverse. It is not easy for one to be sure which one would be able, if at all, to contain skepticism.

5. The Idealist and the Realist Ways of Meeting the Challenges of Skepticism: Some Western and Indian Views

At the very beginning of this chapter I tried to show how Husserl, for example, tried to refute skepticism from the standpoint of transcendental idealism, like Kant before him. Whereas Kant's main concern was to rebut the Humean form of skepticism, Husserl was engaged primarily in exposing the weakness of psychological logic and epistemology of his immediate predecessors like Mill, Erdmann, and Lotze. One finds a comparable attempt by the Nyāya-Vaiśeṣika school of logical realism to contain the sceptical Mādhyamika trend of Buddhism.[29] Buddhist skepticism also has been criticized by the Vedantic thinkers like Śaṁkara.[30] Śaṁkara, anticipating Kant, tries to reconcile empirical realism with transcendental idealism.

For the purpose of my exposition I leave aside, for the time being, the views of Kant and the Nyāya-Vaiśeṣika and return to Husserl's criticism of skepticism. Husserl, as noted earlier, speaks of different forms of skepticism: logical, noetic, metaphysical, and epistemic. His antiskepticism is rooted in a form of profoundationalism. He proposes to provide a rational account and a foundation for knowledge. Without defending self-evidentialism, without a grasp of some noetic conditions, the threat of skepticism can hardly be removed. True knowledge, for Husserl, is not merely a matter of subjective unity; it is also objectively based. In fact, he is trying to discover the objective unity of the laws founded upon the concepts of truth, proposition, object, property, relation, and so forth. This theory becomes open to doubt when its presuppositions cannot be conclusively shown to be "luminous certainties." When the foundational notions are phenomenologically discovered, presuppositions cease to be presuppositional in ordinary sense and get logically assimilated into the unity of foundational theory. Similarly, noetic or subjective conditions also have to be shown to be nonarbitrary.

The basic weakness of empiricism, as Husserl views it, is that it leads to denial of truth, knowledge, and a lack of justification of knowledge. Referring to Mill, he argues that the former's formulation of the law of contradiction is a facile generalization from experience. If the basic logical laws like contradiction are trivialized by and allowed to degenerate into empiricism, how can knowledge, Husserl wonders, possibly be shown to be well-founded? "Extreme empiricism is as absurd a theory of knowledge as extreme scepticism. It *destroys the possibility of the rational justification of mediate knowledge, and so destroys its own possibility as a scientifically proven theory*" (emphasis in original).[31]

Empiricism is said to be based on the primacy of a singular judgment of experience. Espousal of this Humean legacy makes it impossible for the empiricist to offer any rational foundation of logic, arithmetic and other higher order cognitive theories. This also explains its inevitable drift toward skepticism. In Husserl's scheme of thought subjectivism, relativism, anthropologism, and skepticism are strongly interrelated and sprout from the same seed of empiricism. Their genesis is traced by him to the Protagorean formula: man is the measure of all things. He tries to show, step by step, the untenability of individualistic relativism, anthropological relativism, and the resulting forms of uncertainty. His conclusion is clear and very authoritative.

> What is true is absolutely, intrinsically true: truth is one and the same whether men or non-men, angels or gods, apprehend and judge it. Logical laws speak of truth in this ideal unity, set over against the real multiplicity of races, individuals and experiences, and it is of this ideal unity that we all speak, when we are not confused by relativism [and skepticism].[32]

To substantiate his bold claim the basic strategy followed by him is, in brief, this. Taking cues from Descartes and, unlike Quine, he tries to show by detailed analysis that philosophy, as distinguished from physics, for instance, is the most basic foundational science. Whereas the latter is primarily a third-person discipline, that is, has to be methodologically proved to be *public* in character, the former, at least to start with, is a first-person investigation, that is, *inward* engagement for discovering what is truly objective and real.

For Husserl, one's starting point in philosophical inquiry is one's own consciousness. This consciousness is not taken as the quality or power of some particular ego or self that, interestingly enough, is said to be elusive. The basic character of consciousness is its intentionality, objectwardness, and disclosive propensity. Negatively speaking, consciousness is *not* self-enclosed or solipsistic in the least. To *intend* is to point to, to be directed toward, to be about, to aim at. Though the presence of sensory or hyletic elements is not denied in phenomenology, these figure only as constituents of complex intentional phenomena, especially at the perceptual level. Intentional phenomena, perceptual, judgmental, or inferential, are to be taken as *acts*. In terms of intentional acts the objects that are grasped or constituted are not physical as ordinarily understood but ideal. By this formulation of intentionality the body-mind dualism and the external-internal dichotomy are sought to be done away with.

Two other concepts, *noema* and *noesis*, are central to the understanding of what intentionality is. *Noesis* is real content of an intentional act, and *noema* is an ideal content of it. The *noesis* of an act imparts specific character to intentionality and every act having a *noema*, or "sense," is the foundation of all intentionality. *Noesis* and *noema* exhibit phases of intentional acts. The noetic phase excludes the sensory contents, or hyletic phases, of perceptual acts.

Noesis and *noema* are complex or structured entities.[33] The noetic aspect of the composite structure is sense (*Sinn*)-giving and what is given is the *Sinn, noema* itself, an ideal unity. The *same* sense may be *differently* given. To understand this differentiation or specificity of what is given *noesis* plays its crucial role. The same rabbit, for example, may be differently given in perception. It is not easy to give different *proper names* to different modes of the givenness of the same rabbit.

This problem of properly naming of the modes of givenness of a particular object is interestingly encountered by Husserl. Criticizing nominalists like Mill in this context he argues to show that the object "given" to perceptual intentional acts is, strictly speaking, an ideal unity, and its specific aspects may be differently grasped in and through different "single-rayed" noetic acts. To name properly what is grasped through "single-rayed" positing acts, Husserl thinks, we need to make a judgment. Only in terms of perceptual judgments can different real senses of (the ideal unity) of the object can be grasped.[34]

In the later phase of his thought Husserl has paid understandable attention to the relation between concrete experience and abstract judgment. How the latter is formed out of the former has been persuasively argued at length in *Experience and Judgement*.[35] The subtitle, *Investigations in a Genealogy of Logic,* of this posthumous publication merits careful attention. Opposed to the genetic-empirical methods of Mill and his successors like Quine, Husserl speaks of "genealogy of logic." Delving deep into the sense of prepredicative and prelinguistic experience, he tries to show, step by step, how higher forms of knowledge, judgmental and inferential, for example, are constituted. Though he speaks of prepredicative and prelinguistic stuff of experience, yet its intelligible forms are brought out and made explicit by intentional acts and from within the framework of predicative and linguistic judgments and inferences.

In the phenomenological structure of thought, perception and judgment, receptivity and activity, are deeply intertwined. The sense of doubt and uncertainty inherent in the lower modes of givenness of the objects goads *interested* human consciousness to look for higher levels and broader *horizons*. Disappointing perceptual apprehensions explain the percipient's search for and closer inspection of what is given incompletely or illusorily and why, for instance, perceptual anticipation proves frustrating.

Doubt is the joint outcome of perceptual disappointment and continued interest in the search for valid knowledge. Traditional modalities of judgment are partly explainable in terms of degrees of doubt and possibility. Negative and sublative consciousness, strictly speaking, is indicative of the searching intention of the knowing mind. In this search there is no *permanent* halting place. The positive outcome of negative doubt in the form of valid knowledge is itself open to doubt. "Doubt represents a mode of transition to negating annulment, which . . . can . . . appear as an enduring state [of knowledge]."[36] Sustained phenomenological reflection makes it plain that what appears *enduring* knowledge at one stage turns out to be *passing* or even fleeting at another stage. The complete sense of an object is never grasped as a *totum simul*. It appears *and* is constituted, disclosed *and* discovered, gradually by relentless intentional consciousness.

The line of Husserl's argument, paradoxically enough, reminds me both of Quine and Śaṁkara. The reasoning of Quine's argument in what he calls *semantic ascent* purports to show that it is not only possible but also desirable to do away with the distinction between object language and metalanguage. Ascent is possible *within* language itself without resorting to any metalanguage. The usual distinction drawn between "talking of objects *in* words" and "talking *about* object-mean*ing*-words in words" is misleading. Higher-level theories, physical, mathematical, and philosophical, can all be constructed out of words' mean*ing,* maybe a bit indefinitely, of middle-sized objects. From meager input of stimulus meaning massive output of theories and abstract entities can well be extracted. One need not think, as, for example, Husserl, Russell, and Frege do, that a *philosophy* of science or a philosophy of mathematics has a privileged or foundational claim over other disciplines regarding its permanent validity. Every system of knowledge, philosophical or non-philosophical, is open to question and revision.[37]

Śaṁkara's negative methodology, *neti mārga,* is marked by the importance it attaches to the concepts of doubt, negation, and sublation. For example, the perceptual knowledge of snake-*in*-rope and that of snake (without any trace of rope in it) are in a sense, that is, locus-wise, identical but undoubtedly differ in their (specific) cognitive content. Although the perception of a snake causes fear, in the percipient, no such thing happens when the rope is perceived as a rope. The knowledge of a rope in its articulate form negates the doubt whether the snakelike object is really snake. Additionally, it is expressive of the affirmative judgment that what has been perceived is indeed snake. Interestingly enough, Śaṁkara points out, this doubt-removing empirical judgment, despite its seemingly "enduring" character, ultimately turns out to be transient or sublative when the knowledge of ultimate reality, *Brahman,* dawns upon the perceiver.[38]

The crux of Śaṁkara's argument is this. Neither in the object of illusion nor in that of empirical knowledge is the ultimate reality completely manifested. This haunted or intentional consciousness of incomplete manifestation explains the knowing self's *initial* recognition and *subsequent* negation of different grades of objects, perceptual, empirical, and even relatively transcendental. In a serious sense Śaṁkara is empirical realist; for he accords graded ontological status to the negatable objects. But in another sense he is an uncompromising transcendental idealist. Because, according to him, from the ultimate (*Brahman*) point of view, graded or regional ontologies simply do not exist or make no sense whatsoever. Within that *Brahman* no difference, no relation and no other empirical category has any place. This ultimate unity knows no plurality, duality, still less dualism, within it. It is absolute unity and identity without difference, and, strictly speaking, unthinkable and inexpressible. One can speak about what it is *not:* none can speak about what it is. One who knows it becomes it. Knowing and being are identical at the highest level. True knowledge is knowledge by identity. Differential knowledge is empirical and therefore cancelable. The latter is due to objective *māyā* and subjective *avidyā,* suppression of truth and projection of falsity.

My main interest in referring to Śaṁkara's view in the phenomenological context is this. According to him, the basic nature of valid knowledge is freedom from contradiction, *abādhitatva.* If regional ontological entities, together with their epistemic corelates, appear coherent, one may take the resulting form of knowledge as valid. But the reason why most of our perceptual and inferential forms of knowledge are not accorded highest epistemic status is that at some stage or other, at some area or other, they prove contradictory or incoherent. The highest form of epistemic unity is achieved when the highest reality, the largest self in the knower, is itself realized. In that self-realization there is no trace of negation, imperfection, or contradiction. This highest form of self-realization is the highest form of ideal unity. In this uniquely realized ideality the distinction between the knowing subject and the known and the knowable objects is completely absent.

It is what is called *getting the got.* From the empirical point of view, in this knowledge all actual as well as possible objects are perfectly realized. From the transcendental point of view, all actual and possible objects are eternally present in and before the highest reality and there can be nothing new, no process in it. Why it is so is an enigma, an unanswerable question, perhaps a question without any sense from the highest standpoint. In the language of Vedānta what is momentary (*prātibhāsika*) point-instant, what is empirically durable (*vyvahārika*), scientific objects, for instance, and what is transcendental-spiritual (*pāramārthika*) are all in a sense attained and retained in an undifferentiated form within the absolute reality or knowledge. When Śaṁkara affirms it, as stated earlier, he does not deny the preabsolutistic existence of different forms of, different grades of, objects of knowledge and their knowing correlates. One who says this is also logically obliged to admit that lower-order objects are not only self-explanatory but also their fuller meanings are not disclosed without a back-up principle of the transcendental self. To put it differently, the latter, the supreme explanatory principle, prefigures in scientific as well as prescientific grades of knowledge. To put the same matter from the other end, prescientific and scientific forms of knowledge anticipate and intend, though inarticulately, the highest unifying and explanatory principle. Under the circumstances, one would be perhaps justified in attributing a skeptical character to the prescientific and scientific modes of knowledge.

It is well known that time and again Śaṁkara has been accused of being a crypto-Buddhist, *pracchanna Baudha*. Close analysis of some of his basic theses lends credence to this interpretation. Bhāskara, a contemporary of Śaṁkara, finds something wrong, a trace of Buddhist skepticism, in the latter's halfhearted recognition of the empirical (*vyavahārika*) world and assertion to the effect that at the realization of the highest and permanent truth (*pāramārthika satya* or *tattva*) what is empirical and temporal gets dissolved. The ontological status of the world of science and common sense is thus downgraded and that of transcendence glorified.[39]

Second, Rāmānuja also criticises Śaṁkara for harboring sympathy for the Buddhist mode of thinking. To understand the Absolute exclusively in terms of knowledge (*jñanamātra*), denying, by implication, the importance of such basic notions as action and agency, is criticized as an endeavor to avoid the truth. The Buddhist's emphasis on the notion of *nirvāṇa* seems to have unduly influenced Śaṁkara and perhaps that accounts for his underestimating the role of action and devotion.

Third, another consideration adduced to support the Buddhist influence on Śaṁkara is found in their common adherence to two types of truth, absolute (*paramārtha*) and conventional (*saṁvṛti*). In the empirical form of knowledge, *aparāvidyā*, only conventional truth is available. Only in and through a higher form of knowledge (*parāvidyā*) can absolute truth be grasped. It seems that the crypto-Buddhism of Śaṁkara has been derived from his predecessor and celebrated thinker, Gauḍapāda.[40] It is to be noted that what Gauḍapāda calls conventional truth is, to Śaṁkara, practical truth. If the world of sense and science is only practically or conventionally true, it is clear that Śaṁkara is not prepared to accord it the highest ontological status. And there he sounds unmistakably pro-Buddhist; that is, somewhat skeptical about lower-level reality because of its cancelable truth appearance.

Finally, it has been pointed out that the concepts of nescience (*avidyā*), name (*nāma*), and form (*rūpa*) are analogously used by Mahāyāna Buddhism and Śaṁkara's Vedānta.[41] Without *avidyā* the empirical world remains inexplicable, with the dawning of true knowledge it disappears and, therefore, deserves no explanation. In his top-down system, Śaṁkara does not feel obliged to explain why nescience, name, and form, for example, are conventionally effective at all in the world of science. He seems to have lightly dismissed this question by reiterating his argument that from the highest transcendental point of view the world of sense and science, of name and form, is eternally nonexistent or negated (*trikālabādhita*). It never was, nor is, nor will be. Śaṁkara's denial of the puzzling mosaic of the worldly manifold seems to be both historically and conceptually traceable to Buddhism and associated skepticism. Ultimately he "denies" what he himself initially affirmed. How do reality, high or low, affirmation (*sat/asti*), and denial (*asat/nāsti*) go together. Does it not mean trying to speak the unspeakable?

A strikingly parallel line of transcendental argument is discernible in the works of Husserl. Let me put it very briefly. Husserl thinks that he is working in the tradition of Descartes, taking it far beyond him and making good use of the phenomenological insights of psychological epistemologists like Berkeley and Hume. His starting point is naturalistic, at least that is how he views it. As one among many other persons, situated in the world of space and time and belonging to a particular culture, he starts reflecting upon his own experience. His in-depth reflection makes it gradually transparent to him that the intentional character of his experience is inductive, anticipative, and self-transcendental. Even at the perceptual level knowledge shows its extraperceptual or potential capacity (*Mitwissen*). The internal horizon of perception suggests (*vordeutet*) that it, because of its intentional character, is destined to be self-expansive.

The first-level outcome of Husserl's phenomenological reflection is what he calls *descriptive psychology*. This is different from naturalistic or associational psychology because of its internal structure and growth potentiality. Strictly speaking, this structure is unsupported or open ended both ways, inwardly and outwardly. From the outward point of view, what is given to sense experience is not discrete, causally determined, and passively received. From the inward point of view, the embodied ego is like a bottomless pit, knows no boundary, and, on the contrary, on reflection, discloses its expansive horizons, comprising more and more things and beings. Husserl takes immense pains to convince his critics that his descriptive psychology, "starting point" of *transcendental-phenomenological idealism,* is not vitiated by solipsism. It is not idealism in any ordinary sense. For it does recognize the reality, though bracketed, of the scientific or natural world. But what it additionally claims to have done is to elicit from the natural givens many gradually transcendental meanings.

> Our phenomenological idealism does not deny the positive existence of the real (*realen*) world and of Nature—in the first place as though it held it to be an illusion. Its sole task and service is to clarify the meaning of this world, the precise sense in which everyone accepts it . . . as really existing (*wirklich seiende*). *That* it exists—given as it is as a universe out there (*dasseiendes*) is an experience that is continuous, and held persistently together through a thread of widespread unanimity—that is quite indubitable.[42]

Second, Husserl wants to show how this collective indubitable knowledge is achieved. In transcendental consciousness one exceeds ones individual psychological exclusivity and gets into the life of fellow human beings and contents of their consciousness. From psychological subjectivity one moves into transcendental subjectivity. Time and again one returns to oneself to start anew, to rediscover the world of science that is already there. It is a sort of radical beginning intended to find the foundation of what is already available. In fact, Husserl's enterprise bespeaks of his dissatisfaction with what doubt is about, what is given to him in natural experience and by positive sciences. In fact, contrary to the Quinean advice, he is in search of a "first philosophy." But it is a unique type of foundational philosophy, being itself, strictly speaking, unfounded; that is, open ended, marked by everexpanding or endless horizons.

The second level of the foundational structure has been called the *eidetic structure*. Existents are unified and shown to be meaningful in terms of eidos or essence. In fact, the objects that are unified are essentially unifiable as correlates of intentional acts and also because of their own objective generality. Humankind is credited with having spontaneous insights into these essences. Knowledge of essences is independent of knowledge of facts. Essences are not necessarily apprehended as objective. Descriptive use of essences for unification of objective facts do not make them objective. They well out of intuitive spontaneity. Husserl finds essences as morphologically exact and definite. Therefore, in their eidetic structures the worlds of objects, different regions of ontology, acquire and exhibit new meanings. The second-level structure of knowledge, despite its definiteness and exactitude, is not absolutely conclusive.

Husserl perceives a serious difficulty at this level of his inquiry. Transformation of everything objective by the transcendental method of bracketing (*epoche*) into something subjective looks paradoxical. Therefore, at the third level he wants to resolve the paradox by showing that right from the beginning what has been taken to be objective was essentially intersubjective. Unrelated and discontinuous objects, things, and beings in the world, out there in space and time, are meaningless entities or, strictly speaking, not intelligible at all. He rejects the idea that human beings are mere *subjects for the world;* equally unacceptable to him as the concept of human beings as *objects in the world.* Also he rejects the notion of God as the supreme principle of transcendental subjective unification. He proposes to solve the paradox by pointing out the immense potentiality, unifying and constitutive potentiality of every "I," every knowing subject. A truly radical beginning must take one back to one's innermost depth structures of subjectivity. There lies the principle and power of the constitution of intersubjectivity—the constitution of "all of us" "in" me. But the underlying "I" of this "me" is not a "functioning ego-pole" nor is it to be constituted as some mysterious entity lodged in my body. Strictly speaking, no intermediary or even final structure of meanings or ideal unities constituted transcendentally by us can be regarded as human. In a rather serious sense they are claimed to be impersonal, intersubjective, that is, objective, and culture neutral. The point basically interesting to me is that, in Husserl's admission, even the "final structures" are not really final. Our transcendental subjectivity does indeed try to find a final ground for itself but it never comes to succeed completely in this task.[43]

One observes a persistent ambivalence in Husserl's unfounded foundationalism. At times he speaks of the possibility of apprehending what is absolutely certain.[44] Again he concedes that none of our justifications can be "genuinely conclusive . . . new horizons can open up and cause the need for a renewed justification to arise."[45]

Once the ideal unity of God or its theoretical surrogate is taken off and it is conceded that the horizon of knowledge is everexpansive, its actual availability and the resulting cognitive certainty can hardly be achieved, except in a nonserious psychological sense. For, as I mentioned earlier, he wants to defend realism and idealism simultaneously. On the one hand, he gives the impression that he is realist and, on the other, he defends transcendental idealism. For example, in *Ideas* he observes:

> All doubting and rejecting of the data of the natural world leaves standing the *general thesis of the natural standpoint*. "The" world is as fact-world always there; at the most it is at odd points "other" than I supposed, this or that under such names as "illusion," "hallucination," and the like, must be struck *out of it* . . . but the "it" remains ever, in the sense of the general thesis, a world that has its being out there.[46]

Having said this Husserl speaks of the necessity of "radical alteration" of the naturalist thesis. This alteration is brought about by showing the "enduring presence" of the natural world "out there" as "a character which can function essentially as the ground of support for an explicit (predictive) existential judgement which is an agreement with the character it is grounded upon." This mode of argument is typically Cartesian, purported to show that what is being grounded (natural) and what is providing the ground (judgemental activities) can both be shown as consistent in and by the constitutive principles of transcendental subjectivity, higher-level powers of consciousness. In this transcendental act of radically altering *natural existence* into *ideal existence* one is *not* absolutely free. One's consciousness can only "disconnect" or "bracket" the natural from the judgmental, but one is not free to deny altogether the existence of the former. Then why this alteration? Why this judgmental transformation? Husserl's answer is: "We do so only to be clearer about the meaning of the bracketed, to see clearly and convincingly its relation with other objects, bracketed and unbracketed, actuals and possibles."

A similar line of argument, a comparable sort of bottom-up and top-down movement between realism and idealism is evident in most of his other writings. In *Crisis* he observes: "In advance there is the world, ever pre-given and undoubted in 'ontic certainty' and 'self-verification'."[47] But for getting to the clearer meaning of this "ontic certainty" and "self-verification" the world needs to be further analyzed, systematically and intentionally. This analysis first relates "the world" to "transcendental subjectivity as objectified in mankind." But to make one's "natural world" cognitively and apodictically sharable by all alike internal-intentional reflection is absolutely called for. This alone, Husserl affirms, can lift natural science to the level of a *philosophia perennis*.[48]

Time and again Husserl asserts that the objective world does not lose its existence and existential sense when it is considered transcendentally and yet he hastens to add that the all-sidedness of this objective world cannot be grasped except in and through its radical alteration by transcendental and subjective consciousness.[49] His repeated return to this point is to be understood as indicative of his dissatisfaction with the objective natural world. It is a cognitive dissatisfaction expressive of the incompleteness of this type of knowledge. It leads him to a transcendental-subjective investigation in which the objective is assimilated under the subjective and then rises above the polarity between subject and object.

This is one of the reasons why critics and commentators are divided in their assessment of Husserl's position. Whereas some, like Ameriks, are of the view that Husserl is a realist, many others, like Findlay and Morrison, think that his position is basically idealistic. Still others like Holmes maintain that his transcendental phenomenology is neutral between realism and idealism.[50]

I myself have tried to show elsewhere that there are undoubtedly prorealistic elements in Husserl.[51] No discerning Husserlian scholar can deny that his account of the natural world, the region of empirical objects, *taken by itself*, is realistic. But, like Śaṁkara's, Descartes's, and Kant's, Husserl's main interest lies in trying to *justify* our knowledge of the natural world, including all its objects, particulars as well as universals. Consequently, the basic aim of his methodology and epistemology is to show how naturalistic knowledge is to be validated. This aim itself betrays a sense of uncertainty with or skepticism about what we call the (natural) *scientific mode of knowledge*. This sets him in search of a "first philosophy" and a transcendental method intended to undo the miseffects allegedly associated with scientific naturalism. Without self-validating self-knowledge, he is sure, the positive science of the world cannot be recovered from its skeptical associations and vindicated apodictically. It is not without point that he asserts quite clearly, "Positive science is a science lost in the world." And to get it back he exhorts us to follow the Delphic motto, Know thyself![52] Husserl is bound to remind one here of the Vedantic principle of self-realization as embodying the highest form of knowledge.

Admittedly, at this transcendental level the distinction ordinarily drawn between realism and idealism, between the internal and the external, between the mental and the physical, makes no clear sense. Yet if the basic realism-idealism dispute is sought to be settled by answering some questions such as, Is there a world of material objects and mental beings that exists external to and independent of some knowing consciousness? The Husserlian response, it seems to me, would be No. And, therefore, one who is faithful to his texts has to admit that primarily he is an idealist. The basic motive of his transcendental idealism is to get away from cognitive uncertainty or skepticism.

Provisionally assuming that idealism fails to contain skepticism, I propose in the next chapter of our study to turn to a strong form of realism defended by Moore. I choose Moore because of his radical criticism of and opposition to skepticism.

7

For and Against Skepticism: Moore and Wittgenstein

1. Moore's Criticism of the British Empirical Idealism and the Alleged Self-Refuting Character of Solipsism and Skepticism

To understand Moore's opposition to idealism one should first look into his criticism of psychological empiricism, which came down from Locke to Mill. By his own admission, Locke was enquiring into the origin, certainty, and extent of human knowledge, together with the grounds and degrees of belief, opinion, and assent. None of the British empiricists disputed the possibility of having *certain* knowledge at the practical level. The most important and common point of their views that was not at all acceptable to Moore was their psychological and introspective method. Therefore, although in Bradley's *Principles of Logic* Moore found, at least so he believed, a sustained and possible criticism of psychologism, he thought he had found a good theoretical means to refute idealism. But it is to be remembered that at that time, 1897, Moore, like Mill and Bradley, believed in the reality of mental ideas. He was yet to be familiar with the full implications of Frege's view (1884) on the subject. But soon he found a flaw and an inadequacy in Bradley's criticism of empiricism. He discovered that Bradley's own thought was infected by the same error as that of the British empiricists, who could hardly distinguish idea as symbol from idea as the symbolized. Bradley's distinction between the ideal content (predicate) and its ground, pure sentience or reality or a part of it (as subject of judgement) did not appear very clear and plausible to Moore.[1]

The second plank of Moore's opposition against idealism is to be found in his attack against the Berkeleyan principle, *esse est percipi*. In his celebrated paper "The Refutation of Idealism," Moore first tries to show that the principle rests on the failure to appreciate the distinction between "*esse*" and "*percipi.*" To think that essence of a thing consists in its being perceived is to miss the point that there may be unperceived objects, essential or existential. To assume that the meaning of *percipi* is a part of the meaning of *esse* is wrong. Equally wrong is to assume that *esse est percipi* is a necessary proposition. Therefore, Moore concludes that to hold that the proposition *esse est percipi* is necessarily true is nothing but to hold that "something else plus *percipi* entails *percipi.*" This is patently illogical, circular, and therefore, rejectable.[2]

In the second phase (after 1910) of his criticism of idealism Moore relies mainly on common sense for his refutation of idealism. He does not go to the extent of maintaining that Berkeley's position is meaningless or necessarily false. Where Berkeley goes wrong, according to Moore, is to misconceive the relation between physical fact and mental fact. To put the issue in his own words: "I hold . . . that there is no good reason to suppose either (A) that *every* physical fact is logically dependent upon some mental fact or (B) that *every* physical fact is *causally* dependent upon some mental fact" (emphasis in original).[3]

As regards Berkeley's defence of the existence of external facts as continuously perceived by God, Moore's position is plain. To postulate God's existence for proving the existence of material objects like tables and chairs is contingent upon our ability to explain the veracious nature of Berkeley as a person. What he maintains is that the truth of such commonly believed and shared propositions has a sort of foundational status in any intelligible description of the world.

The other argument invoked by Moore against idealism takes the form of a critique of the theory of internal relations. Both in his "Defence of Commonsense" of 1925 and his "Proof of an External World" of 1939 Moore's attack on idealism substantially rests on the criticism of this theory. What Moore tries to show, mainly against Bradley, is that relation is not internal to *relata*. It is not a property belonging to either of them, or both of them taken together. Bradley's difficulty with external relations seems to consist in his misconception that reality of relation is to be understood in a way quite different from how we understand the meaning of the terms. Here, too, Moore mainly relies on common sense, although he takes pains to present it in a very refined logical form.

Refutation of idealism, Berkeleyan or Bradleyan, does not by itself vindicate realism. Besides, one has to bear in mind that Moore is not in favor of naive or direct realism. Russell's argument against it is well known. "Naive realism, i.e. the doctrine that things are what they seem, leads to physics, and physics, if true, shows that naive realism is false. Therefore naive realism, if true, is false; therefore it is false."[4]

Moore's commonsense realism, though in some respects similar to direct realism, should not be taken as naive. For, like Russell, Moore admits that the physicist's analysis of what a chair is, what its physical (atomic) constituents are, is correct. But he is convinced that the physicist's correct *analytic* view does not oblige him to give up his commonsense realism. The former is different from the latter only because of its theoretic-analytic picture. At one stage or other this picture has to prove itself consistent with the commonsense view of the same thing. In and through logical analysis, meaning analysis or reduction, one and the same thing may be differently represented. But representational difference does not mean denial of the substantive identity of the real thing. To many this analysis of the complementary relation between the commonsense view and the scientific view of things appears reasonable and acceptable. However, some others differ on this issue: they are not prepared to agree that "table" and "atomic constituents of table" are referentially identical. Noncongruence of physical analysis and meaning analysis give rise to a sort of skepticism.

To contain skepticism Moore feels justified in upholding the basic validity of commonsense truisms. The truisms he makes use of to refute metaphysical idealism, nominalism, and skepticism, briefly speaking, are (1) material objects and mind do exist; (2) *material object* and *mind* are class terms and the reality of what they stand for does not consist in their being perceived; (3) space and time are real; (4) objects included in (1), (2), and (3) are related in different ways as relations are real; and (5) the preceding truisms are known to be true by all of us.

The foundational status that Moore wants to bestow on these commonsense truisms is not as solid as they appear. It is not with such *words* as *material object, mind,* and *existence* that philosophers like Moore quarrel. Their main concern, despite their *theoretical* commitment to common sense, is to analyze their meanings, how they are to be taken for the purpose of vindication or rejection of idealism or skepticism. The reality of space and time is also not an undisputed thesis. It has been understood and explained in different, at times contradictory, ways as is evident from the views of Nāgārjuna and Śaṁkara to Bradley and Einstein. The larger issue of the nature of *reality* need not be gone into because of its extremely controversial character. Further, we have just noted how relations are oppositely understood by the internalists like Bradley, Bosanquet, and the externalists like Russell and Moore. Finally, the last truism of Moore's realistic foundation, (5), as we know, invites immediate rejection of the skeptic, who is likely to question not only the truth claims of (1), (2), (3), and (4), but also, perhaps more so, of the additional claim that this so-called truisms are *universally* known to be true.

Let one not assume that as the defender of commonsense realism Moore is not aware of this possible and apparently grave criticism. His reaction is like this. First, criticism from different quarters—subjective idealism, objective idealism, nominalism, and, particularly, skepticism—is bound to be different. Besides, each of these schools formulates its view differently. Therefore, the question of meeting of different formulations of different forms of criticism can neither be seriously taken nor answered without presupposing a specific (maybe bracketed) point of view. That this sort of criticism may be raised against the target views of Moore does not in any way establish the *privileged* claim of Moore's commonsense realism. Another important point to be reiterated here is that we all know that the position that Moore's (1), (2), (3), and (4) are true is open to contest by skeptics, both of the moderate and radical varieties. The moderate skeptic, like others, is *committed* to his own conceptual scheme. But the radical skeptics of the Buddhist variety, for example, need not be *committed* to any conceptual system whatsoever. They refute, at least that is what they claim, other's schemes making use of the latter's arguments, standards, definitions, and so on, and they have no peculiar positions of their own to defend nor any arguments to advance.

In view of this observation how can one make good sense of the foundational propositions of Moore's commonsense realism? I think what Moore has in mind in support of his position is this. Whatever may be one's theoretical stand, at some stage or other one has to accept the above propositions seriously. His own task is to show that what we take as true as a matter of common sense can be defended even through careful logical analysis. Analysis and explication do not necessarily dissolve the problems of philosophy. In this respect, Moore's view is different from both those of Wittgenstein and Quine. In a way it may be seen as closer to Popper's on the subject. For common sense, like science, also has its explanatory role, however imprecise. Needless to add, Moore would reject Popper's criterion of falsifiability to draw the line of demarcation between the two. On the contrary, he would try to show how scientific analysis, rightly understood, stands close to and proves consistent with common sense truisms. From another point of view, that is, from that of ontological relativity, Quine also may be credited to have a sympathetic view toward the conceptual system of commonsense realism. But what he is likely to object to is its lack of simplicity or ontological austerity. To establish commonsense realism it is not enough to criticize and reject Berkeley's idealism.

Positively speaking, a realist has to establish the existence of external world, meeting all possible objections against this thesis and also to show that our knowledge of the external world is true. To achieve this end the strategy the realist follows is called, or may be called, *analysis*. Interestingly enough, this strategic *analysis,* carefully followed and scanned, can hardly be regarded as a part of what we ordinarily call *common sense*. Strictly speaking, Moore's "common sense" is clearly uncommon. It is full of professional acumen and subtle logical analysis.

Moore argues to show that both solipsism and radical skepticism are self-refuting or self-contradictory. To say that we do not know for certain the existence of material things or other selves, besides (or including) oneself, is inconsistent with the actual fact that we know from the history of philosophy several such philosophical views. Moore's point may perhaps be easily met by pointing out that our claim to have known other selves' solipsistic-skeptical position is itself subject to the blemishes of solipsism and skepticism. In support of this position it may be pointed out that the solipsist or the skeptic need not subscribe to Moore's type of argument purported to refute solipsism and skepticism.[5] In brief, Moore tries to prove the existence of external world by using a sort of *Modus Tollens* form of argument. It may be put thus: if the skeptic is right, I do not know that the external world exists; but I do know that it exists, and therefore the skeptic is wrong. That Moore's argument is not conclusive and that it is rebuttable may be shown by a counterargument of this form: if the skeptic's position is true, then I do not know the existence of the external world; but the skeptic's position is true, and therefore I do know the existence of the external world. If the premises of Moore's argument are taken to be true, it is difficult to find fault with the premises of this counterargument. In response to this stance Moore reverts back to the last proposition, (5), of his commonsense truism; namely, it is not enough that the premises of the argument are *true,* what in addition is required is the *knowledge of truth*. Moore's view is that this requirement cannot be satisfied by the skeptic. But radical skeptics may always take the position that their counterargument purports only to demonstrate the vulnerability of the opponent's, in this case Moore's, argument and not to justify their own. The skeptics' argument, as mentioned earlier is like purgative medicine; it goes out together with what it purges. Let me make it plain. Moore himself concedes that it is difficult to disprove the basic point of the skeptic. Somewhat in the Cartesian vein he admits: "I cannot see my way to deny that it is logically possible that all the . . . experiences I am having now should be dream-images. And if this is logically possible, and if further the . . . experiences I am having now were the only experiences I am having, I do not see how I could possibly know for certain that I am not dreaming."[6]

However, this *logical* concession does not lead Moore to abandon his commonsense truisms. When the principles of logic are found to be inconsistent with logical truisms, the former have to be subordinated to the latter, or, alternatively, their mutual consistency has to be analyzed and shown. In other words, the *logic* of the skeptic does not shake the commonsense realist's belief in the existence of the external world. To believe that the external world exists is one thing, but to make one rationally certain about this belief is quite a different thing.

2. Different Forms of Realism and Moore's Proof of External Realism

One point keeps on disturbing us: is there anything in us that can make us certain about our knowledge of an external world. A further question may be raised regarding the character of that something which claims to have the capacity to give us certain knowledge about an external world. Is it psychological or epistemological? Another question to be taken into account here is, Which of these two sciences, psychology and epistemology, is more respectable or comprehensive? According to some thinkers like Quine, psychology is more comprehensive and dependable than epistemology as ordinarily understood. Others like Popper take epistemology more seriously and are afraid of psychology because of its psychologistic undertone, among other things. It is clear that, although the former are proreductionist, the latter protranscendentalist. The common point between the two is to recognize science as a going and growing concern and without any finalistic aim. In different ways both are realists. Quine's realism, as we have seen, is very liberal, but Popper's realism is relatively conservative. Neither of the two proposes to offer a proof of an external world to vindicate his commitment to realism. One might say that both of them are internal realists but not in

Putnam's sense.[7] As distinguished from these moderate views against skepticism and for realism, I recall Moore's famous "Proof of an External World."[8]

The immediate provocation for Moore's proof is said to have come from Kant's observations that the existence of the things *outside* us has never been proved. Kant certainly has given us numerous *considerations* in support of his empirical realism. He has also told us why transcendental realism cannot be proved. The ground of proof cannot itself be proved. But certainly it is an open question whether Kant's *considerations* for empirical realism really amount to *proving* the existence of external things. It is clear that Moore is not convinced by the Kantian arguments. That is why he undertakes the responsibility of convincingly proving the existence of an external world.

Skepticism may be viewed from two ends, internal and external. That the words *internal* and *external* are notoriously vague is a well-known fact and has been pointed out by many philosophers. The charge of vagueness, perhaps also that of ambiguity, may be raised against another comparable pair of words, *inside* and *outside*. Physically speaking, these distinctions are drawn in terms of space. But the concept of space has not been uniformly interpreted by the physicists. According to some of them, space and time are *absolute* and independent of our "place" in them. Even those who like Einstein speak of *relativity*, do not say that it has anything to do with our perception or conception of space and time. Relativity is relativity of relation *between* space and time or, one might say, *in* space and time. But the contrary views are also available. Some philosopher-scientists like Leibniz have construed them as "internal" forms of consciousness.

The whole distinction between what is internal and what is external appears very problematic if we look at the issue from the commonsense point of view. Somewhat (but not quite) perplexing is the Kantian distinction between the form of outer sense (space) and the form of inner sense (time). Forms of sensibility as Kant understands them are not "external" to us in any ordinary sense. If spatial distinction or determination is "located" within sensibility or consciousness, people like Moore strongly feel, we step into the trap of idealism, whatever might be its color or shade, and we retreat from realism. So when Moore seeks to prove the existence of an external world the sort of distinction between the "external" and the "internal" that he has in mind is extremely close to common sense. The behaviorist's denial of what we ordinarily call *mental consciousness* or the physicalist's readiness to regard the irritations of human nerve endings as physical appear quite contrary to common sense and, therefore, unacceptable. Both behaviorism and physicalism are *theoretical* positions, more or less remote from our practical life.

Moore's proof of an external world is apparently very simple. He holds up his two hands and making a gesture with the right hand says "here is one hand," and making another gesture with the left hand says "here is another." Having made the gestures (expressible in those two sentences) he thinks that he has proved that two human hands exist. To him this proof of the existence of hands is a proof of existence of an external world. Moore's assumption is transparent: hands are parts of an external world. Proving the existence of *his* hands Moore claims to have provided a "perfectly rigorous" proof of the existence of external things. He finds it difficult to think of a better or more rigorous proof of anything. Questions are settled and doubts removed by this sort of "absolutely conclusive" proofs.[9]

Unless we are in a very theoretical or philosophical vein, it is unlikely for us to feel induced to dispute the validity or at least attractiveness of Moore's proof. But persons of common sense may point out that Moore's very undertaking to provide a proof of the existence of an external world is highly philosophical. Persons of common sense are unlikely to take this responsibility upon themselves. To them it is such a simple and clear thing that it requires no explanation or proof. Having themselves undertaken the philosophical task it is not quite appropriate for them to appeal to common sense and, in addition, to criticize the people who propose to dispute such proof for their alleged unnecessary theoretical or philosophical zeal. What they miss, perhaps unintentionally, is their own philosophical (not commonsense) commitments.

Against Moore's proof several objections may be raised. But before raising the objections I want to raise a question about the *way* he tries to prove the truth of his thesis. The way in question clearly suggests that he has some *philosophical* skeptic in his mind whom he intends to convince and silence by using the *logical* method of deduction. He believes he has met all the requirements of valid deductive proof. First, it is, he thinks, not vitiated by *petitio principii;* that is, the conclusion "two human hands exist" does not repeat the premises, "here is a human hand" and "here is another human hand." The former is separate from the latter; the truth of the latter is independent of the truth values of the premises. Even if the premises are false, the conclusion can be true. Second, the premises are known to be true. Moore fails to understand how one could possibly be uncertain of one's own hand. Finally, given the truth of the two premises and the knowledge of them, the conclusion is validly drawn from the premises, when they are taken together. The logical neatness of the structure is no proof of his claim that his premises are true and that he knows them to be true. Because he is up against the philosophical skeptic, the latter is perfectly at liberty to say: "Though Moore swears by common sense he is relying on an uncommon philosophical sense and his knowledge claim about his hands is easily questionable."

Let us try to construct an argument that can show the untenability of Moore's knowledge claim about his own (or even another's) hands. I can well think of "my" hand. I can also think of a contingency, say, an accident or an unavoidable surgery resulting in the severance of "my" hand from my body. It is also conceivable that the surgery may succeed in saving my hand but in a form so badly mutilated that even I may fail to recognize it as *my* own hand. This argument about one hand may easily be extended to another hand. Given this situation, Moore's conclusion, "two human hands exist," is not of much force in proving the existence of an external world. If we recall the contingency of an accident or surgery just stated, the premises of the Moore's proclaimed valid argument cannot be given any privileged position for proving the existence of an external world. If there is no *known* way of showing that a particular hand, or two particular hands, are mine, Moore's premises fail to satisfy the second requirement of a valid argument. Unless Moore knows that a particular hand is his and that he has the right to be certain about it, he cannot draw the conclusion that he comes to draw from his premises. On the other hand, if he knows that "his" hand is really his and that he has the right to be certain about it, the conclusion of his argument adds precious little content to his so-called proof of an external world.

Ontologically speaking, Moore's hand or for that matter my hand is at a par with the tables, chairs, and other objects that I perceive around me. Moore seems to have gone wrong on several counts in his proof of an external world designed to refute the skeptic. First, as already stated, Moore has no right to be certain about the existence of his hand as a proof of an external world. His awareness that a particular hand is his is (merely) perceptual. Like other perceptual judgments, it may well be wrong; the philosophical skeptic at least has the right to raise this point against him. Second, if Moore shifts his ground of proof from perceptual judgment to common sense, he does not improve his position in any way. For in that case he does not need any proof at all. One could also ask, If common sense is so "solid," where is the need for a proof of an external world? For whom does Moore need the proof? Obviously, he already knows that his hand is not only a part of an external world but also a proof of it. For the skeptic however, this proof carries no conviction; in actual life the skeptic does not feel the need of a proof to convince him of the reality of the world he lives in; in his philosophical moods the skeptic does not see how he could conclude the reality of an object from the bare fact that he happens to perceive it. Moore's proof at best is question-begging and at worst superfluous.

3. The Variety of Uses of the Term *Know* and the Resulting Difficulty of Removing "Doubt"

Unless a reasonably clear sense is made out of what doubt is, Moore's antiskeptical argument in the form of a proof of an external world does not cut much ice. The word *doubt* has been used in more than one sense. It is, in truth, multiguous as multiguous, as the other word *know* so frequently in use in philosophical writings. On the ambiguity of *know* innumerable views have been expressed by philosophers. The word *know* has been used in more or less different senses in the following sentences.

1. Churchill knows Mahatma Gandhi.
2. The devil knows how to drive a racing car.
3. I know what toothache feels like.
4. Fear is unknown to God's nature.
5. I know my enemy's mind about my promise to help him.
6. I know that there are three sofa sets in my drawing-room on the other side of that wall.
7. I know what is in the safety deposit box of that liar.
8. I know that this hand is mine.

If these different senses are borne in mind, to say "I know that *p*" is very misleading. For, unless we know what stands for *p,* it is of no use to try to speak something in the form "I know that *p*."

We encounter a comparable predicament with the diverse uses of the word *doubt*. Doubt may mean so many things: questioning, suspicion, disbelief, unbelief, distrust, mistrust, incredibility, untenability, unreliability, misgiving, and self-doubt. The list of "the synonyms" can easily be extended. But the question is, Are they really synonyms?

Once we remember the various senses of the word *know* and those of *doubt,* the whole *theoretical* picture of skepticism becomes extremely complex. The sentence in which *know* figures may be doubted, not necessarily negated, in very many ways. Even the extent of complexity indicated by the preceding remarks centering round the ambiguity of *know* and that of *doubt* is not sufficient to give us a clear idea of the whole issue. In case the use of a word within a short sentence does not provide a sufficiently clear context, we do need, in fact we make implicit use of, a much larger and more complex context, which is indicated by some such expressions as *form of life* and *language game*. The expression *form of life* has been in use for a long time. In the recent past, it has been extensively and often very perceptively exploited, among others, by Dilthey and Wittgenstein. The concept of *language game* has been specially coined, refined and used by Wittgenstein. Both in *form of life* and *language game* things and beings, their positions, dispositions, and behavior, are related very complexly. Unless this complexity is borne in mind, the full or clear meaning of these linguistic expressions cannot be grasped. Even clarity knows no terminal point. It is always a matter of degree.

Dilthey's and Wittgenstein's views make us aware, of course in different ways, of the enormous complexity of the problems of skepticism and those of relativism and their interrelationship. To illustrate briefly the points just mentioned. What exactly is to be understood if the content of the sentence 1, Churchill's knowing of Mahatma Gandhi, is said to be dubitable. It may seem an easy question to answer. One might say that there is no "*Mahatma* (noble-souled) Gandhi, there is only Gandhi." In fact there are many persons with the family name Gandhi. Therefore there may be doubt about which one (Gandhi) of them we are talking. The *existence* itself of Gandhi or Gandhis may not be doubted but Churchill's knowledge, rather his claim to it, may well be doubted. Somebody may point out that what I call *knowledge* of Gandhi is merely an *opinion* about him. Somebody else may add that the very possibility of knowledge being dubitable, the question of Churchill's knowing Mahatma Gandhi or a table makes no difference. Thus it is found the question of who knows and what is "known" is irrelevant. This does not bring our perplexity regarding the content of sentence 1 to an end.

Further questions may be raised regarding what *doubting knowledge* means or is like. Can *knowledge* be *doubted?* How in those cases where knowledge is *defined* in terms of *truth* can we meaningfully speak of "doubting knowledge" or "knowledge doubted"? If by *knowledge* we mean, as most of us do, "scientific knowledge," apart from other forms of knowledge, it is very difficult to imagine that it is not open to question and correction. The forms of knowledge that are questionable and corrigible are of course dubitable.

But the question is, What do we achieve or gain by doubting knowledge? Do we improve or precisify it? Assuming that we do, the question arises, Why and how can doubt improve our knowledge, make it more precise, and render it more error-free than it would be if it were not subjected to doubt? Is doubt itself a form of knowledge? Or is it based on knowledge? Or are doubt and knowledge coordinate species? Coordinate species of what, of which genus? Or, should we take them, *a la* Descartes, as two modes of the same consciousness? Is it for this reason that *cogito ergo sum* and *dubito ergo sum* are often taken at par to serve the antiskeptical purpose?

The logic of doubting underlying the different cases listed earlier (sentences 1 to 7) is bewilderingly variformed. For example, the logic of doubting the devil's knowledge of how to drive a racing car (2) is quite different from the logic of my knowledge of what toothache feels like (3). To pronounce credibly on one's ability to drive a car we are required to see that person in operation, but not necessarily so. An expert in the matter may form an idea of the driver's ability by asking some appropriate questions. But who has seen the devil or could put questions to him?

In the case of knowing what toothache feels like I myself may not have an opportunity to feel or perceive it at the moment. I may have to depend on my memory, but memory may fail me. With the passage of time it may get faint, unreliably faint. It is also possible to think of a person who never has had a toothache and, consequently, no memory of it. Even then it may be suggested that one can "justifiably" use the expression *toothache,* provided one knows how and when to use it or is initiated into the rules of the language game relating to *toothache*. In that case one's own feeling of *toothache* is not a necessary condition for one's successful use of the word *toothache*. Had toothache been one's peculiarly private feeling it could not be used *communicatively;* that is, successfully. From the cases of the repeated use of the expression we are obliged to believe that though from some very complex reasons we think that *toothache* refers to one's private feeling yet in a none-too-metaphorical sense it is "transferable"; that is, communicable. If "I" is not traceable, as Hume argues, to speak of "my toothache" makes no (private-psychological) sense. The same argument may be extended to "your," "his," or "anyone's" feeling or knowledge. Exploiting this Humean insight Wittgenstein prefers the locution "feeling" to "my feeling" or "your feeling." Nonprivate or public knowledge is a collateral consequence of this view.

By using the language game–theoretic strategy Wittgenstein tries to point out two things: the immensely diverse uses of the word *know,* and also the varying degrees of success in removing doubt. In effect, following Wittgenstein, one cannot reach a *general* solution of the problem of the skeptic. Because, rightly understood, the problem of the skeptic is not identical in all cases. Sometimes our doubt is rooted in the inadequacy or absence of perceptual information or obstruction (6 and 7). Sometimes it is due to the unreliability of the inference on the basis of which we claim knowledge (4 and 5). But close scrutiny reveals that even these psychological or logical expressions like *perceptual* and *logical* are somewhat misleading. For, the seemingly same word *know* is put, *almost* imperceptibly, to different uses. Those who do so and others who "understand" it are not ordinarily *conscious* of the variety of uses, for the "bases" or "deep structures" of ordinary language are not *equally* available to all its users. Though hidden or often unconsciously assumed, these "bases" impart intelligibility and stability to our discourse. Though "internal" to (ordinary) language game, these "structures" are not arbitrary or privately introducible.

But to say that "we are not often conscious of the variety of uses" makes sense if only it is presupposed or assumed that there is at least one "available" way of being conscious about it. This is not to suggest that there is an essential or unique use of *know* that provides us the *core* meaning of the word. It only means that, given the context of the use of the word, there is *a* possible way of deciding or agreeing whether the word used has been proper or not. Once that possibility is open to *us,* it is also possible for *us* (the same or at least similar *us*) to come to a conclusion regarding the propriety of the use of the word in question, in this case *know.*

4. Social Contract Theory of Meaning and Wittgenstein's Critique of Moore's Antiskepticism

One might say that the Wittgensteinian strategy of combating the skeptic is bound to remind one of what may be called a *social contract theory of meaning*. But the legitimacy of the contract itself, together with its rules, may be questions. In other words, the suggested strategy of combating and containing skepticism, one might say, is itself open to the charge of skepticism. Neither the *social* character of the contract nor the sanctity attached to it is a good enough premium to ensure its acceptance by all in perpetuity. Even the parties to the contract may back out. If this line of argument is pressed, the game-theoretic strategy of antiskepticism may be declared vitiated by the fallacy of infinite regress. We can always take a socially contractual stand or enter into an agreement and yet may be questioned. Being questioned, we may fall back to another socially contractual stand or agreement and again we may be questioned. And so on. But the *practical* question is, Are our social contracts questioned in this way? The most consoling response may be No. But does this consolation amount to refutation of skepticism? Or, is it a mere respite against skepticism for the time being? Or, one might say, as Kripke does, that Wittgenstein's (game-theoretic) solution of skepticism is itself "skeptical" as Hume's is and not "straightforward" as the Cartesian solution is. Even if it is assumed for the sake of argument that Kripke is right, it only means that Wittgenstein meets one form of skepticism by another and that if we stop somewhere it is temporary and *practical.* But, then the charge of infinite regress remains, despite the "unsurveyability" of the infinity.

Rightly understood, the old question reappears here. How do we take the word *skepticism?* It may be taken in a theoretical vein. It may be taken in a practical vein. Neither the practical nor the theoretical is uniform. Even "the practical" may be construed in very many ways. The same is the case with "the theoretical". So, within the language game(s) there is no significant distinction between "the internal" and "the external" questions. It is not that only *within* the language game–theoretic context are we obliged to do away with the distinction between "the external" and "the internal," between "the theoretical" and "the practical." It is always an "internal" question in the sense that it is *relative,* always relative, to the concerned form of life or discourse. But simply because of the relativity of the meanings of the expressions, Wittgenstein insists, they are *not* to be taken as *arbitrary.* Here his position needs social contractual defense. We must remember that this defense itself is not beyond question or doubt.

A close study of Wittgenstein's *On Certainty* shows that he first rejects Moore's *absolute* distinction between what is external and what is internal and then assimilates Moore's epistemological question into his philosophy of language, the concerned form of life. Wittgenstein tries to impress upon us an important point that no clear sense could be ascribed to *doubt* unless the language game in which it figures is somehow mastered. "If we are not certain of any fact, we cannot be certain of the meaning of our words either."[10] For example, unless I know for certain that it is *my* hand, it is impossible to prove the existence of an external world by using it as a (true) premise. I have already argued earlier of a possible situation in which one may fail to identify "my hand" as really mine. The orientation of my argument is primarily ontico-epistemological. The same point has been differently presented by Wittgenstein.[11] The gist of his argument is that one can always be doubtful about whether one's hand is really one's. This doubt or uncertainty is not only about one's hand but also about one's words, about all other related things.

Yet, in fairness to Wittgenstein, it has to be admitted that he who raises all these doubts also shows a *practical* way out of them. It is in the practical use of language, in and through our discourse of life, that we discover the intended meaning of our expressions and also find out where we go wrong and how we do so. In other words, there are certain rules in each of our language games. Our social contracts (of meanings) are not without rules. Contracts work essentially because of their ruleness. When we fail to follow the rules it is not only that the concerned contract does not work but also that there remains nothing like a contract that could possibly be recognized as such and acted upon. Rules are not merely "regulative" of the possible moves "within" the game but also "constitutive of it."

When it is said that rules are both regulative and constitutive of games what is meant is this. Rules are not taken as regulations having validity of their own outside the game in question. They do not stand apart or outside the game. From this one must not think that a particular game could be what it is without the rules within it. A game *without* rules is no game. Rules without or outside the game are not rules either. To speak of rules that are not rules of some game makes no sense.

But even within the rule-governed games moves are available that are not uniquely determined by any rule. The same rule or set of rules may be followed differently, strongly or weakly. For example, a football player may kick the ball, strongly or weakly, in the sky or grazing the ground and in various other ways. None of these *ways* of kicking is either *specifically* allowed or disallowed by some rule. Rules are like boundary conditions, violation of which renders the moves unintelligible as part of the game in question. If some player dribbles with the ball outside the marked ground, it is not a part of the game, although, observationally speaking, it appears to be so.

To say that within the rules different moves are possible is to suggest that the "same" word or expression may be differently interpreted. Availability of different interpretations of some particular sign or string of signs shows that it has no meaning uniquely its own. The sign + may be read in different ways: plus, the sign of addition, the sign of the Red Cross, and so on. What the sign + *exactly* means depends upon how it is used, where and before whom. In effect, the meaning of a word can be understood only in the context of the entire form of life.

Even this way of putting the matter, it seems, is not likely to be endorsed by Wittgenstein. Because one should not get the impression that meaning is ascertained as a *result* of application of some rules embedded in a form of life external to the use or application of the concerned signs.

Meanings are not (statements of) truth conditions. It is clear that the later Wittgenstein implicitly rejects the classical realist view defended by Frege and found in the *Tractatus*. Instead, in *Philosophical Investigations* and other works he is primarily interested in finding out the conditions under which certain form(s) of words may be appropriately asserted or denied. Apparently his interest shifts from *truth conditions* to *assertability conditions*. What, in addition, he has in mind is to find out the role and use in our lives of our practice of asserting or denying the forms of words under the concerned circumstances. Assertability conditions have often been called as *justification conditions*. But within the liberal fold of Wittgenstein's rule-governed language game no justification could be interpreted or taken as conclusive or unique. Assertability or justification conditions can only indicate the legitimacy of certain uses of language or linguistic behavior.

It is not only inconceivable to have a language game without rules. It is equally inconceivable to have a language game that is not played by *several* players or a community. Neither Moore nor Wittgenstein, for example, can legislate or constitute a game and play it *alone*. If the legendary Robinson Crusoe is condemned on the alleged grounds that he on his own can neither have a language game nor the ability necessary to play it, it rests on a misconception. Unless it is assumed that Robinson Crusoe has been *exiled from* or has lost his *access to* a language game of which he was (or at least could be) a player, to speak of his inability to have a language of his own and to use it does not make any sense. Whether Robinson Crusoe has a language and can use it can be meaningfully decided only if he is brought back under the appropriate conditions in which he may be in a position to assert or deny something. If he has no access to the necessary assertability conditions, to speak of his inability to have a language of his own is not illuminating in the least.

The plausibility of Wittgenstein's refutation of Moore's antiskeptical argument is to be assessed against his (language) game-theoretic strategy. Under what circumstances by showing up his two hands can a person (say, Moore) *prove* the existence of an external world? Before we try to answer the question we must remember, as stated earlier, that the rules of playing the game (let us call it the Game of Proving the World by Hands, in brief GPWH) must be available to the players, to the persons concerned. The skeptic who raises the question regarding the existence of an external world is required to grasp such concepts as *external world, hand,* and *proof* and the rules of their use or application. Similar requirement has to be satisfied by the antiskeptic who undertakes the responsibility of refuting the skeptic's position and others who would be in a position to take part in the game GPWH according to its rules. In other words, the skeptic, the antiskeptic, and the other participants in the game must be able to play and follow or understand the game. The players of the game form a community. Theirs is *a* form of life. They have a sort of agreement between themselves. There must not be any Robinson Crusoe-type of person in the community. If the mentioned conditions are fulfilled, raising of a hand by somebody (with a particular gesture) may be interpreted as the proof of an external world.

5. The Infirmities of Moore's Proof of an External Realism or World

But does it really happen? This conditional form of proof seems to be flawed on several grounds. First, it is not enough for a person, Moore or somebody else, merely to raise a hand with a particular gesture and persuade others to accept it as the premise of a "proving argument." The event must be known to the concerned persons as intended for proving an external world. Otherwise, the act of raising the hand by itself is no "proving move," proving an external world doubted by the skeptic. The act of raising the hand should be like a commonly readable sign of the *claim* that it is a proof of the existence of an external world. The *aim* of the specific GPWH game must be known to the concerned persons. Otherwise, the outcome of the game, refutation or failure of refutation of the purported proof, would not be intelligible to the players and the watchers of the game. Rightly understood, the players, the watchers and their acts are complexly interwoven in the game.

Second, within the game in question, GPWH, which conditions

and what combination of them constitute a proof cannot be (at least not easily) shown to the satisfaction of all concerned. What particular move of the hand and what specific gesture would be a proving move can hardly be uniquely identified. The conditions of proof may be satisfied or remain unsatisfied in very many ways. It is almost impossible to indicate the *extent* of satisfaction of the conditions that would be acceptable by the concerned community as proof. Even if one has grasped the concept of proof, whatever that might be, how and when to use it cannot be *unanimously* decided upon. The "agreement" in this respect is bound to be partial. Can partial agreement be called agreement at all? This question and the related difficulties mentioned here are indicative of the "skeptical solution" of the problem raised by the skeptic.

Third, the weakness of the skeptical solution has to be viewed in the context of the difficulties associated with the concepts of *criteria* and *checkability* used by Wittgenstein. It is not at all easy to be assured of the satisfaction of the conditions necessary for proving the *aim* of the game GPWH. One of the players of the game may feel that both the necessary and the sufficient conditions of proof have been satisfied. Another may feel that only the necessary conditions have been satisfied. But the *responses* of different players are unlikely to be uniform. Given this constraint, it is indeed difficult, if not impossible, to prove what the antiskeptic wants to prove. The criteria that may be used for removing the "private" character of feeling satisfied (in respect to the conditions of proving) cannot be easily spelled out. Consequently, the proclaimed "public" character of the satisfied conditions cannot be conclusively established. Similar considerations may be raised against the *checkability* requirement. It is not easy to spell out which sort of check (and at which stage) can remove all doubts, answer all questions. In vain one searches for a "straight" antiskeptical argument from Wittgenstein against the skeptic's problem.

Still another hidden aspect of GPWH merits our attention. It has been raised and critically commented upon by Kripke. It seems that Kripke is of the view that unless we recognize a minimal *uniformity of human nature,* the responses of the players of any language game remain an inexplicable mystery. The very fact that a game can be recognized and played by several persons according to rules is indicative of a much neglected truth, a cluster of truths. Both availability and usability of rules by us, the players of a particular game, cannot be accounted for until and unless it is assumed (1) that the concerned persons have their own identities and (2) that the identities of different persons have similarity or uniformity.

The critique of metaphysics underlying Hume's and the earlier Wittgenstein's denial of the human self or subject (as a substantive entity) raises many problems that it cannot justifiably claim to have succeeded in solving. Even if it is assumed that Wittgenstein's argument on private language (or privacy of any other form) has been successful, the larger question remains unanswered, How do we show the public character of a particular form of life? If as individuals we are required to be *self*less or *soul*less, devoid of privacy, so that our interrelationship, the game we play, the rules by which we are said to be interwoven, can be regarded as objective and public, what is the necessity of recognizing a particular form of life as really public? Is *form* or *life* by definition public? *Life* of what? If the privacy of the *individual* life could be the last refuge of the skeptic, what compelling reason is there to think that the privacy or exclusivity of a (particular form of social or) public life cannot perpetuate the same "mischief"? What is the fundamental difference between individual privacy and societal privacy? If egology could be sociologized, what prevents one from egologizing sociology? How can the two forms of life, public and private, be shown to be totally different? Is their difference to be understood in terms of *exclusivity?*

Even in our *silence,* are we not related to others? Is not silence at times communicative? Is it not a sort of withdrawal from the plenum or continuum of life? If language is a social institution, it is because human beings share the same nature. When Wittgenstein himself doubts the possibility of "lion's laughter," he implicitly recognizes the necessity of some human beings, beings like ourselves, to make language possible. The reasons for which lions cannot have a language and those for which we can need to be looked into.

In brief, we cannot meaningfully deny the peculiarities of our nature, whether we characterize it as "metaphysical and elusive" or "public and linguistically available." even if we cannot have a "straight" solution of the skeptic's problem, we need a coherent and intelligible account of our own selves, of talks about our own selves. *Self*less things do not exhibit any interest in talks about themselves. Must we banish "mental states" and their vocabulary from our discourse? Can we successfully do so? If so, at what price?

6. Ways of Understanding Wittgenstein's Prorelativistic Antiskepticism

To understand Wittgenstein's prorelativistic antiskepticism, one is advised to bear in mind the distinction between the *instrumental* use of language and the *noninstrumental* character of language. A language may be *used* or left *idle*. A book of chess or cooking is *useful*. The language used in this sort of book is put to an instrumental use. In this case language may be said to have an aim. But this aim cannot be ascribed to a language *as a whole*. Language *qua* language does not teach us either to play chess or to cook foods. To say that we need language (written or uttered or behavioral) is not to admit that language has an aim (or several aims) of its own.

With one and the same language many games may be played. In language games many things are present besides language. Human beings, nonlinguistic social institutions, nonhuman living creatures, and even inanimate things are interwoven in a language game. A novel or a cinema may be taken as a piece of language game. How many "things" appear in it? Mastery of the rules of the game is the minimal requirement of playing it.

Wittgenstein, as we know, was not interested in epistemology in any ordinary sense. When he addresses himself to this problem he does so mainly as a philosopher of language or, more specifically speaking, as a theorist of meaning. Left to himself, he would perhaps like to introduce himself as a careful recorder and reporter of the *practices* of language. In an important sense his philosophical activities exhibit a distinct practical character. However, this is not to deny that what he says on the practice of language does not throw up a theory in its wake on the subject.

When, on the request of philosophers like Norman Malcolm, he agreed, toward the end of his life, to write down his response to Moore's "A Defence of Common Sense" and "Proof of an External World," he undertook a larger responsibility which comprised, among other things, his view on others' antiskepticism and the special logical status of some "empirical propositions."

It is clear from Wittgenstein's account that he was not prepared to travel as far as Descartes, readily agreeing to doubt the existence of ordinary things like chairs, tables, and human bodies. Like Moore, he saw that the "empirical propositions" asserting these things have a special status of their own. However, he took pains to show that the skeptic's ways of putting across the nature of doubt, knowledge and certainty are faulty. Nor was he satisfied with Moore's ways of refuting the skeptic by proving the existence of an external world and upholding common sense.

Given Wittgenstein's view of language *as a whole*, the question of doubting it does not make any sense. The language of doubt has to be a part of the concerned language as a systematic totality. Only within the total whole could the words and expressions regarding doubt be meaningful. To assert or deny anything about a language as a whole is devoid of sense. The conditions under which assertion or denial of a language in its totality could be given meaning are just not available to any one of us. Language games can be meaningfully played only within the *whole* of language. The latter provides a sort of foundation, a set of ground rules, as it were, for playing all the possible meaningful games.

As a corollary of the preceding view Wittgenstein is unable to endorse universal skepticism (which apparently he ascribes to Descartes). One cannot meaningfully speak of being deceived by an all-powerful evil genius, unless one is clear at least of what is meant by *deceive*. If we go on saying "our perception deceives us," "our memory deceives us," "our inference deceives us," and so on, we fail to convey any coherent meaning to the speaker and the hearer of these expressions unless we all, including the hearer, are reasonably clear about the meaning of *deceive*. The language game of deception must hinge on something that itself is not deceptive.

Wittgenstein repeatedly returns to this line of argument in his *On Certainty*. "If you are not certain of any fact, you cannot be certain of the meaning of your words."[12] Unless we are certain of what *deceive* means, our use of it will not serve any purpose. "If this deceives me, what does 'deceive' mean any more?"[13] Unless a nondeceiving sense of reliance is available to us, the talk of deception makes no headway.

Wittgenstein, somewhat like Descartes, speaks of the necessity of some "reliable" ground or nondeceptive foundation that would lend meaning to the talk of deception. Descartes could have, epistemologically speaking, stopped at the level of self-existence as the most reliable or trustworthy ground of doubt as well as knowledge. But, as we know, he retreated from self-existence to the existence of a nondeceiving God. A similar line of argument is in Berkeley as well. When Wittgenstein says that "a language-game is only possible if one trusts something"[14] he is exploiting a Cartesian-Berkeleyan insight. Although Descartes and Berkeley wanted a "straight" solution of skepticism, Wittgenstein's reference to indubitable direct knowledge of something, "taking hold of my towel without having any doubt," for example, does not aim at achieving that end. In fact, the sense of *sureness* attached to taking hold of my towel is not of knowing. At least Wittgenstein is not prepared to use the word *know* in this context. By implication, it is being conceded that one could be sure of something without knowing it.

When Moore says, "this is my hand," he has no doubt about his utterance and its meaning. But there is nothing compulsive about his meaning. To utter "this is my hand" is not a meaningful proof of the existence of an external world. One could say as well "this is hand" and claim that "this (utterance) is a proof of the existence of an external world." Certainty is not always a proof of knowledge. Unless what is uttered or written can be *tested* or *checked* against trustworthy facts, it cannot be accorded a special status of knowledge. But once we remember that even the trustworthy facts against which our knowledge claims are checked or tested are open to more than one interpretation, different uses, the supposed special status of the reliable or trustworthy facts is substantially lowered.

7. Where Wittgenstein Stands Close to Moore on the Point of Antiskepticism

Wittgenstein is prepared to recognize the special status of the propositions that Moore uses as the premises of his argument for proving the existence of an external world. But he takes pains to show that *that* special status is not enough to prove what Moore wants to. One could be sure or certain about one's own name, about the brain under the skull, about the towel one is holding in one's hand and so on. And yet one could be wrong in each case. No belief, no knowledge claim is accepted in isolation. Each as a part of a large system, surveyably large, is accepted. But in testing a part of the system may be disowned. By disowning a part of a whole we do not reject the latter. After all our belief system is not like an axiomatic system. The logic, inductive or deductive, of accepting or rejecting certain beliefs cannot be *absolutely* framed, shown, or used. The logic of proving or disproving is not as sacrosanct as it is often thought to be.

Wittgenstein rules out the possibility of universal doubt. Every doubt must have its foundation. Universal doubt, if any, cannot be shown to have a foundation. "A doubt that doubted everything would not be a doubt."[15] In other words, to the talk of universal doubt no coherent meaning could be attributed.

However, Wittgenstein's profoundationalism is to be taken with a pinch of salt. For, according to him, nothing is foundational in isolation. We are *taught* that something is foundational.[16] To be taught or instructed about what is foundational, we are required to be told of the concerned *context*. When we are certain we are so only within the bounds of the concerned language game and on specific grounds.[17] If Ludwig Wittgenstein does not doubt that he is L.W., there is a specific reason for it and that reason is available within a language game in which L.W. figures and can mean something that is indubitable. The logic of why Ludwig Wittgenstein does not doubt that he is L.W. can hardly be described. He advises us to "look at the practice of language" so that we can "see it."[18] Strictly speaking, the logic of seeing or showing is not itself visible. Kenny has perhaps rightly suggested that foundational logic is the logic of showing and not of seeing.[19] It has to be gathered from the use of language.

The last recorded view of Wittgenstein on the matter seems to be reminiscent of the view of his younger days. Once upon a time he used to speak of logic as a practice that "takes care of itself" and needs no justification, inductive or deductive. Somewhat in the same vein he speaks of the foundation of doubt in the last days of his life. And apparently he himself was aware of it. But it seems to me that the logic of *Note Book* and *Tractatus* that takes care of itself is different from that of language game developed by Wittgenstein in the later part of his life. "The earlier logic" is *simple* and paradigmatically shown by the form of the elementary proposition. "The later logic" is complex, practical, and available only within the concerned language game. The foundationalism of the *Tractatus* and that of the *Philosophical Investigations* and *On Certainty* are considerably different. The former may be described as founded foundationalism and the latter problematic or conditional foundationalism.

The main difference between the two phases of Wittgenstein's philosophical career may be indicated in another way. Although he was thinking over the themes of the *Tractatus* he was more or less reacting to the influential philosophers of the time, notably Russell. His ontology was Russellian but he was trying to get out of the logic of Russell, particularly the theory of types and the theory of description. Positively speaking, he was trying to develop a logic of his own, true to his atomistic ontology. Russell's dilution of his earlier doctrine of denotation in terms of the theory of description apparently disturbed Wittgenstein. If the denotation of the denotative expression is said to be logically paraphrasable in such a manner that its realistic ontological commitment appears very thin, if not nil, Wittgenstein apprehended that the high tide of nominalism was bound to prove unstoppable. The theory of description was to him a disturbing prelude to nominalism.

Though he took note of Russell's notion of "knowledge by acquaintance," this form of knowledge seemed to him primarily postulational and psychological. Its *logic* was not manifest. The logical empiricist in Wittgenstein was not prepared to accept any reality that does not manifestly disclose its logic. This craving for manifest logical disclosure of reality was obviously a proof of his strong realism. Wittgenstein was notably worried over the problem posed by empiricism and its close ally, associationist psychology. James's "stream of consciousness" was certainly an improvement on the association psychology but he did not know how to capture it. How could the passing thought be the thinker? The logical apparatus at his disposal and the ghost of Hume did not allow him to capture consciousness, self or subject. As the acquaintantial base of description does not lend realistic legitimacy to description, so the talk of consciousness, the language in which we talk about consciousness, does not become convincingly clear and distinct simply because there is something like consciousness. Wittgenstein did not deny consciousness on the grounds that it was not there. He refrained from talking about it because he felt that he did not have the language necessary for the purpose; that is, either to affirm it or to deny it.

He was not in favor of tampering with the life of the ordinary language. The thing in the ordinary language that disturbed him most was its nondisclosing and nonmanifest grammatical surface. He was looking only for that disclosed and manifest grammar of language that does not need any "external" certification regarding the truth of what it says. In other words, Wittgenstein was looking for an indubitable form of language in which what is captured or contained is unmistakably and manifestly articulated. If this *form* of language is really available, he thought, the problem, rather the pseudo-problem, of skepticism, would stand dissolved, thereby rendering every effort to solve it superfluous or uncalled for. The *Tractarian* logic aims at dissolving the pseudo-problem of skepticism.

But the basic question raised in the *Tractatus* could not be answered within its own confines. Is the form of language that unmistakably certifies the truth of what it says humanly available? Where is that form, the elementary form, of that language to be found? If it is available and if it is successfully usable, why do we come across the problem (or even the pseudo-problem) of skepticism? If truth is manifestly found in the form of elementary proposition, how can we possibly go wrong in our bid to capture or picture of what is the case?

Does the *form of language*, logical or ordinary, have a life of its own? How does it essentially differ from the *other forms of life?* Does (or can) the logical form successfully tell us about its identity? Can we identify it in isolation, outside the forms of our own life? How can we talk about it? How do we distinguish between the right form and the wrong form (of language)? How do we draw the line of distinction between the surface form and the hidden form, the surface grammar and the hidden grammar, of language? Can one manifest language be mapped on another (equally manifest) language? How do we distinguish between the form of "the clothing" and the form of "the clothed"? If every piece of language is in place, why do we talk at all of "right use" and "wrong use" of language?

8. The Point of Continuity between Two Phases of Wittgenstein's Approach to Skepticism

All these questions seem to suggest one basic thing: the meaning of the forms of language have to be gathered from within the forms of the life we live. The latter are manifestly complex; and that being the case, the former cannot be very simple either. Artificial over-simplification, unaided by interpretation, creates confusion and does not provide clarification. Even the questions of simplicity or complexity cannot be satisfactorily settled in an abstract or isolated manner. What is theoretically simple may appear practically complex. And often the converse appears to be the case. For example, a simple equation may represent a very complex state of affairs. On the contrary, a complex whole, a flower for example, a multipetaled and multicolored flower, may look very simple. Much depends upon how we view it, how we live it, how we are disposed toward it, and so on. The question of simplicity and complexity cannot be decided in an "abstract" and "external" manner. Its answer is to be sought within the form of life where it arises.

It is true that "the logical language" defended by Wittgenstein in the *Tractatus* is for the third person, that is, for public acceptability, and that the forms of natural language supported by him in his later works are understandably different. But the difference is not as radical as it is often made out to be. The sort of objectivity sought to be ensured by the picture theory and verificationism of the *Tractatus* is at least partially retrieved and retained by the rules of language game and the denial of the privacy of language. In effect, the rejection of the private language argument is intended to defend the *public* character of language, natural or logical.

Wittgenstein's discussions of skepticism in the *Tractatus* and his later writings, particularly *On Certainty,* exhibits substantial continuity of his thought. In both the phases of his thought he views the issue of skepticism from within the concerned forms of language. In this respect, the difference of his approach from that of Moore is notable. In the *Tractatus* he states that the issue of skepticism cannot be meaningfully put into words in the form of a question and, cannot, therefore, be answered either.

> When the answer cannot be put into words, neither can the question be put into words. . . . If a question can be framed at all, it is also *possible* to answer it.(6.5). Scepticism is *not* irrefutable, but obviously nonsensical, it tries to raise doubts where no question can be asked. For doubt can exist only where a question exists, question only where an answer exists, and an answer only where something can be said.[20]

It is clear that, to Wittgenstein, doubt is not an ordinary mental state. It is a propositional attitude. Doubt, according to the Tractatarian Wittgenstein, lacks a determinate sense. Wittgenstein, under the influence of Frege, used to maintain at that time that an expression having no determinate sense is meaningless. It is well-known that later Wittgenstein changed his view on the subject. The reason why Wittgenstein could not find any meaning of *doubt* is that there was nothing standing for it and available in language. Lack of sense makes it impossible for one to express it in proposition.

Perhaps it would not be correct to think that Wittgenstein was opposed to the sort of skepticism associated with scientific hypothesis and improvable theories. Perhaps he had in mind the classical skeptics like Pyrrho. Questionability of the commonsense belief was not of interest to him. What he tried to attack is abstract philosophical skepticism. Philosophical skeptics of the classical tradition tries to show that their *philosophical* opponents, logically combated and cornered, are obliged to fall prey either to the fallacy of circularity or that of infinite regress. Their last resort in despair is to reiterate their position without giving any reason. In other words, the target of the Tractatarian critique of skepticism is the philosophical skepticism that claims to have shown the impossibility of philosophy as a theoretical, discursive, or analytical discipline. But *within* the framework of the *Tractatus,* within its language, neither the basic thesis of classical skepticism nor its successful critique could be framed. Consequently, it would not be wrong to maintain that the earlier view of Wittgenstein on skepticism is like his view on the mystical.[21] Neither the contention of the mystic nor that of the skeptic could be put across in a meaningful or publicly sharable manner.

Certainly in his later works like *On Certainty,* as we have seen, Wittgenstein's criticism of skepticism has taken a different route. It has been developed in response to his criticism of Moore's way of refuting skepticism by proving the certainty of commonsense truisms. Here he tries to show that Moore's purported refutation turns out to be inefficacious. Wittgenstein's later approach to the issue is to be understood against the background of his changed view. While he was working on *On Certainty* he had already given up the picture theory of proposition, atomic facts, eternal objects, logical space, name (in the strong sense), and so forth. In brief, the rigid framework for the purpose of capturing an a priori structure of the world was abandoned. His later approach was marked by the richness and complexity of natural (linguistic) expressions used in the context of the ordinary forms of life. But one important thesis of the *Tractatus* was retained in his later philosophy: certain elements of language, logical constants, *and, or, not,* though meaningless without anything to stand for them (in the world), are functionally or combinotorially extremely significant. His fundamental idea that logical constants are not *representatives* (4.0312) is successfully exploited in Wittgenstein's critical examination of Moore's refutation of skepticism. In his later works the issue of *doubt* is discussed in the multiple contexts of ordinary languages.

My earlier analysis of Wittgenstein's response to Moore highlighted two aspects, one negative and the other positive. In the negative aspect I highlighted the contexts, as formulated by Wittgenstein, in which the weakness of Moore's proof of an external world becomes evident. In the positive aspect I tried to show that Wittgenstein's formulation and analysis of skepticism, being less ambitious and more specific than Moore's, has succeeded in putting the issue of doubt in clearer perspective. Moore's defense of common sense, though clear, is not philosophically sound. Nor does his proof of an external world appear to be successful. But one point that Wittgenstein takes from Moore is important; namely, *doubt* makes no sense without presupposing something certain or at least less dubitable. But this is an insight that is not peculiar to Moore; one can easily trace it back to Descartes, or Berkeley, or even to Hume, the philosopher of common sense.

Having gone through Moore's so-called refutation of idealism, one cannot think that the Irish Bishop was not careful enough to recognize the "hard reality" of the material world or that he was trying to deny our commonsense beliefs. When referring to an "external material object," the common person does not really mean the existence of an *external material object* out there in space but rather something like "an idea" produced [in him or her] independently of [his or her] will." Berkeley anticipates the objection that "by [one's principles] all that is real and substantial in nature [would be] banished out of the world, and instead thereof a chimerical scheme of *ideas* [would] take place." In response, he asserts that all the objects of commonsense beliefs like the sun, the moon, the stars, houses, rivers, mountains, trees, stones, and our own bodies are real and substantial. All these "remain as secure as ever, and as real as ever." This is in brief what Berkeley claims to have shown clearly in his work.[22]

To doubly assure us of his opposition to skepticism he affirms: "That what I see, hear and feel both exist . . . we are not for having any man turn sceptic and disbelieve his senses; on the contrary, we give them all the stress and assurance imaginable; nor are there any principles more opposite to Scepticism than those we have laid down."[23] This strategy is very common in philosophy. Empirical idealists are not required to contradict the commonsense truths or doubt the beliefs of the person in the street. What they do is to give their interpretation of what the people say and believe in their "prereflective" and "preanalytic" moments. The alchemy of the philosopher's "reflection" and "analysis" changes the picture "radically."

Hume's stance on the issue of skepticism regarding ordinary people's beliefs is striking like Berkeley's. Questioned whether he is "really one of those sceptics, who hold that all is uncertain," Hume predictably responds, "This question is entirely superfluous . . . neither I nor any other person, was ever sincerely and constantly of that opinion."[24] Referring to the question of the external world Hume, like Berkeley, sounds even more emphatic. He draws a distinction between the "*causes* which induce us to believe in the existence of body" and the existance of body itself. Unless the latter is believed the question of causal efficacy of what may lead us to believe in the existence of external world makes no sense at all. To Hume causal arguments carry more weight than probabilistic arguments providing "a superior kind of evidence." "One would appear ridiculous, who would say that 'tis only probable that the sun will rise tomorrow or that all men must dye."[25]

As pointed out, the idealist can, without offending the commonsense realism, unbracket all the bracketed or preanalytic concepts of common sense. This is precisely the Cartesian strategy vigorously revived and pursued by Husserl and his followers. And, therefore, well up to a point commonsense realism, subjective idealism, and even transcendental idealism may work together in their justification of the existence of external objects of commonsense beliefs. But, ultimately, as we know, they part their ways in theory. I repeat "in theory." That in practice most of us, including philosophers and scientists, substantially agree is rarely disputed.

8

Examination of Some Views on Skepticism

1. Different Forms of Language and Their Strength and Weakness, and How Well-Founded Are the Foundations of Mathematics as Language?

Where should we (or do we) gather the meaning of the forms of language? Are we to gather them from within the forms of the lives we live? The latter are manifestly complex; and that being the case, the former cannot be very simple either. Artificial oversimplification, unaided by interpretation, creates confusion and provides no clarification. Even the questions of simplicity or complexity cannot be satisfactorily settled in an abstract or isolated manner. What is theoretically simple may appear practically complex. And often the converse appears to be the case. For example, a simple equation may represent a very complex state of affairs. On the contrary, a complex whole, a flower, for example, multipetaled and multicolored, may look very simple. Much depends on how we view it, how we live it, how we are disposed toward it, and so on. The question of simplicity and complexity cannot be decided in an "abstract" and "external" manner. Its answer is to be sought within the form of life where it arises.

In this context one has to understand the intended meaning of Neurath's boat analogy and Quine's extensive use of it. A boat, language, and science, even if damaged or doubted, are all going concerns, though admittedly of different sorts. An unrepaired (or even a sunken) boat does not cease to be a boat. Even an unspoken or dead language remains "archaeologically" meaningful. An obsolete paradigm of science, as a chapter of history of science, remains interesting and instructive. However, this is not to suggest that all concerns go equally well and always.

Language, as we have noted earlier, has its different forms, logical and ordinary. Sometimes we draw a distinction between written and spoken language; within each of these forms further distinctions may be easily drawn. There are different *styles* of writing. The same language may be uttered *phonetically* in different manners. *Logic* is no omnibus term. Distinction is drawn between standard and nonstandard logic, between two-valued, three-valued, and many-valued logic. Other well-known distinctions of logic are indicated by such terms as *deductive, inductive,* and *probabilistic.*

Though we often speak of the difference between common sense and science, science and philosophy, and philosophy and common sense, we are never quite sure whether these differences rest on inviolable distinctions. One among many available descriptions of philosophy, given by Quine, is that it is neither more nor less than science. A slight variant of this description is that philosophy is self-reflective science. A more circumspect form of this view is that philosophy is physics conscious of itself. Evidently the last definition draws a refined distinction between science as a whole and physics as a form or a part of that whole.

There is, perhaps, nothing grossly wrong in saying that philosophy is refined, conceptually better organized, common sense. However, organization entails pruning and precisifying. In its better-organized form science certainly looks more elegant. But in the process of organizing itself the commonsense view of the world loses much of its interesting ruggedness and realistic flavor. The reorganized world of common sense, found in science, is bound to undergo certain changes, adding certain new features and shedding some of its old features. This story is repeated with understandable variation while we see the transformation of the world of science into that of philosophy. We might say that the same world is found in common sense, science, and philosophy. But because of the different ways of organization and conceptualization we speak of different worlds: viz., the commonsense world, the scientific world, and the philosophical world. It may be pointed out here that however diverse might be the ways of our conceptualization and organization of the world (assuming for the time being that we are dealing with the same world at each level) the one thing that we are always concerned with is language. In and through the same language we conceptualize and organize. When we say this we are not forgetful of the different possible distinctions drawn within the language. We can draw a picture of the world in ordinary language; we can do so in mathematical language; and, again, we can repeat it, in a more generalized form, in philosophical language. But this *linguistic* method of "unifying" our different approaches to the understanding of the world need not be taken as the last word.

When I say this I have in mind mathematics as language. Conventional wisdom advises us to accept mathematical truths as foundational. But the initial difficulty that one feels in accepting this advice is the availability of at least three different "foundations" of mathematics: logicistic, formalistic, and intuitionistic. For the convenience of discussion I propose to refer briefly only to logicism and intuitionism.

Toward the end of the last century the question of the foundation of mathematics started receiving attention from mathematicians, logicians, and philosophers, mainly because of the discovery of the paradoxes of set theory. Afraid of the possible presence of similar paradoxes in other branches of classical mathematics, they addressed themselves to the issue of foundational consistency. Mathematical truths, they felt, must be consistent.

Though Leibniz and Dedekind had vaguely anticipated the program of logicism, it was left to Frege to develop logicistic thesis. According to him, laws of mathematics, analytic in character, are implicit in the eary principles of logic and embody a priori truths. Algebra, analysis, and even geometry are deducible from the laws of number.

Russell, who was independently working on the foundation of geometry and mathematics, also came to the same conclusion that the fundamental laws of mathematics could be deduced from logic. While the final results of his work was in press, Russell, it is well known, discovered a paradox in the published results of Frege's work. This disturbing discovery put Frege's foundational consistency claim to serious jeopardy. It was now left to Russell and Whitehead to develop logicism in a way free from paradox or inconsistency, providing a perfected mathematics leaving no room for doubt in it. Russell was pretty sure that the principles of logic were truths and therefore consistent. However, this optimism was not shared by his collaborator and teacher, Whitehead, who observed as early as 1907: "There can be no formal proofs of the logical premises themselves."[1]

A distinct streak of Platonism was evident in Russell's thought at that time: the principles of logic and the objects of mathematical knowledge exist independently of any mind; mathematics offers truths about the physical world. His view was more radical than Frege's. The physical-mathematical truths appeared so unquestionable to Russell that he dismissed every skepticism about them. "Of such scepticism mathematics is a perpetual reproof; for its edifice of truths stand unshakable and inexpungable to all the weapon of doubting cynicism."[2] The axioms of *Principia Mathematicia* (*PM*), giving grounds for deduction of theorems of logic, are well known. One of the important theorems is the principle of reductio ad absurdum, which states that if the assumption of *p* implies that *p* is false, then *p* is false. In *PM* the notion of the propositional function is defined intentionally. For example, the propositional function, "*X* is good," stands for the set of all good objects and does not state the names of the members of the class, a requirement of extentional definition. Second, *PM* introducсs the theory of types to avoid the paradoxes of self-reference. To put the matter in terms of sets: one cannot speak of a set belonging to itself. If one says *a* belongs to *b, b* must be of higher type than *a*. Extending the point one might say that a propositional function cannot have as one of its arguments anything defined in terms of function itself.

But the theory of types, on scrutiny, was found to be extremely complex. To avoid the complexities the axiom of reducibility was introduced. The axiom says that any proposition of higher type is equivalent to one of first-order. Two other important axioms introduced by *PM* pertain to choice and existence of infinite classes. The axiom of reducibility has been criticized by, among others, Poincaré, Weyl, and Ramsey. Russell defended it when it was first criticized but the later Russell became diffident about its tenability. Also criticized was the axiom of infinity, central to the structure of arithmetic. Besides, the question was raised whether this axiom is the axiom of logic at all. The axiom of choice also turned out to be extremely controversial. The form of the axiom is, "Given a class of mutually exclusive classes, none of which is null, there exists a class composed precisely of one element from each class and of no other element." Though this axiom was found necessary to reduce parts of classical mathematics to logic, Russell and Whitehead themselves did not feel convinced whether it could be considered as logical truth.

With the basic three axioms of logicism, reducibility, infinity, and choice, under fire, the entire system seemed to be shaky. Russell himself was not sure how to show clearly the derivability of mathematics from logic. Russell started retreating but did not readily give up. However, his critics were unrelenting. Poincare, for example, thought that it was marked by sterile manipulation of logical symbolism and could hardly touch the concrete and complex nature. Second, Weyl wondered that, if mathematics is purely formal discipline and its theorems follow from the laws of thought, how could it be of any use in dealing with the different sorts of natural phenomena, geometry of space, acoustics, electromagnetics, and mechanics. Hempel also expressed doubt regarding the reducibility of the nonarithmetical parts of mathematics, such as geometry, topology, and abstract algebra to logic. But interestingly enough, Quine's contrary view to the effect that mathematics *is* reducible to logic and that geometry, topology, and abstract algebra also could be accommodated within the general structure of logic does not appear very plausible to Russell himself. Notwithstanding Quine's and Church's different types of defense of logicism, Russell felt skeptical about it. He does not conceal his disappointment with the elaborate but negative outcome of his main work, for his aim was to "discover" "solid foundations" of mathematics.

> But as the work proceeded, I was continually reminded of the fable about the elephant and the tortoise. Having constructed an elephant upon which the mathematical world could rest, I found the elephant tottering and proceeded to construct the tortoise to keep the elephant from falling. But the tortoise was no more secure than the elephant, and after some twenty years of very arduous toil, I came to the conclusion that there was nothing more that I could do in the way of making mathematical knowledge indubitable.[3]

If the foundation of logicism proves shaky, the fortune of the opposite approach, of intuitionism, does not appear very promising either. The intuitionists have tried to establish the truth of mathematics on the basis of some nonsensous powers of human mind. The names of Descartes, Kant, and Pascal are closely associated with the origin of intutionism. But Kronecker may be regarded as the immediate forerunner of intuitionism as we find it today. His well-known epigram is, "God made the integers; all the rest is the work of man." Others who contributed to the growth of intuitionism are Borel, Lebesgue, Poincaré, and Baire, each in different ways critics of logicism. Brouwer ultimately brought together the ideas of these mathematical luminaries into a reasonable unified system. Like the founders of logicism, he also was very confident that at long last he had established the foundation of mathematics.

According to Brouwer, mathematics is a human activity or mental construction. But these activities are not to be construed as empirical. Mathematical concepts, the integers, for example, are born out of immediate certainties of the mind. Distinct events in a time sequence are intuited by the mind. The successive natural numbers are formed by repetition. Mathematical thinking composes self-evident truths and does not depend on logical implications. Intuition itself determines the correctness and acceptability of mathematical ideas. Besides the natural numbers, Brouwer found that addition, multiplication, and mathematical induction are intuitively clear, distinct, and certain.

It is to be pointed out here that Brouwer draws a fundamental distinction between mathematics and language. To him, mathematics is fully autonomous and independent of language. Language is only of communicative or instrumental value. Mathematical truths may give rise to words and symbols in one's mind but the latter have nothing to do with the truth of the former. The intuitionist is of the view that logic of the logicist is a sort of language and cannot provide access to mathematical truths. Perfection of logical forms does not in any way add to mathematical truths. Logic depends on mathematics. Mathematics does not depend on logic. Logic as language is less certain than our intuitive mathematical concepts. This is not to deny the intuitive availability and acceptability of some logical principles and procedures that can be successfully used to assert new theorems from the existing ones. The logical principle that is freely and frequently used by the intuitionists is the law of excluded middle. It asserts that from the *history* of reasoning regarding finite sets we come to know that every meaningful statement is either true or false. Thus obtained and accepted, it was then taken as an independent a priori principle and was unjustifiably applied to infinite sets.

The history and psychology of mathematical reasoning make no sense without language. Without linguistic "devices" like signs and symbols, how can the contents of mathematical knowledge possibly be retained, repeated, and expanded? Though we call it a *device,* language is like a mother's womb in which mathematical knowledge, like other forms of it, is conceived, grows, and then is delivered; that is, made publicly sharable. Not only abstract language but also concrete images, visual or tactile, are of immense help in creative mathematical thinking.[4]

Brouwer's understanding of history suggests that classical logic was abstracted from the mathematics of finite sets. Forgetting this restricted origin, some logicists and formalists put logic above and prior to mathematics and unjustifiably extended it to the mathematics of infinite sets. This mistake gave rise to antinomies in set theory. It has been pertinently observed by Weyl that for an all-knowing God the principle of excluded middle may be valid because he can survey the infinite sequence of natural numbers and all at once, but it is not available to *human* logic. The denial of the law of excluded middle in the field of infinite sets opens up a new possibility, a third state of affairs; viz., admission of propositions that are neither provable nor disprovable. Intuitionism rejects not only infinite sets but also a large part of analysis, including pure existence proofs.

To save mathematics and its foundation from the so-called anarchy let loose by intuitionism, Hilbert asserted that, unlike Kronecker, he did not need any God. Nor, unlike Poincaré, was he in need of mathematical induction. His rejection of Brouwer's primal intuition was firm and categorical. Equally firm was his rejection of the Russell-Whitehead axioms of infinity, reducibility, or completeness. To them the proof of consistency was not available. For humankind cannot forsee all possible implications of the said axioms. From his words it is clear that Hilbert felt pretty sure that his proof theory would be able to solve and settle for good the problems of consistency and completeness.

But, as we all now recall, Gödel shattered the dreams of all those foundationalists, logicists, set-theorists, and formalists who claimed to have established foundation even of intuitionism. Weyl, apparently in a sad mood, commented that mathematics, due to God's existence, is undoubtedly consistent but it is due to the devil that this consistency is not provable. The basic point of Gödel's paper, "On Formally Undecidable Propositions of *Principia Mathematica* and Related Systems," was that the consistency of any mathematical system, comprising even the arithmetic of whole numbers, cannot be established by the logical principles accepted by different foundational schools. Equally devastating proved the outcome of Gödel's incompleteness theorem. It states that any formal theory, *T*, adequate to embrace the theory of whole numbers, is incomplete, if consistent. As a result of Gödel's finding, the Russell-Whitehead system, Hilbert's axiomatization of number theory, and the Zermelo-Fraenkel system were all found to be seriously flawed. The price of consistency turned out to be incompleteness. Gödel's incompleteness theorem is in a way a denial of the law of excluded middle. To that extent it is a concession of intuitionism. But Hilbert, the most aggrieved party in the controversy, stood fast and unconvinced by Gödel's discovery. A member of the school, Gentzen, liberalized the methods of proof available in Hilbert's metamathematics by using transfinite induction and succeeded in proving the consistency of number theory and a part of analysis.

But after Gödel's revolutionary works mathematical foundationalism could never be the same again. Nearly a decade later he struck again. In 1940 appeared his *The Consistency of the Axiom of Choice and of the Generalised Continuum Hypothesis with the Ax-*

ioms of Set Theory. To start with, it seemed to strike a somewhat optimistic note, for it stated that because of the consistency of the axiom of choice and the continuum hypotheses the situation was somewhat reassuring. But later developments showed that Gödel's results did not rule out the possibility that either or both of the axioms of choice and the continuum hypotheses could be proved on the basis of the Zermelo-Fraenkel axioms.

In 1940s two important developments took place, centering around the problem of mathematical foundations. One could be traced to Wittgenstein who was apparently influenced by some of the Brouwerian insights. According to this view, mathematics is a particular form of language game with its own rules. Wittgenstein's *Remarks on the Foundation of Mathematics* highlights, among other things, that meaningful talk of abstract mathematical entities cannot be basically different from other sorts of talk available in natural language. Besides, different branches of mathematics cannot be provided a unified foundation without disregarding the important difference between the areas of discourse. A popular belief, somehow supported by mathematicians themselves, is that the objects of mathematical propositions are abstract, eternal, and highly idealized, whereas physical shapes and areas are "unshaped," "uneven," and "ill-organized." According to this view, mathematical propositions disregard the perceptual properties of real objects and provide us highly ideal pictures. This impression is strengthened by the geometrician's assertion that triangles, circles, and so on in themselves are perceptually unavailable and that their use value consists in mapping the physical things by superposing them on the latter.

Wittgenstein rejects this view. According to him, in geometry, for instance, we are not concerned with two types of objects, ideal and real or perceptual. Our concern is with the sentences or the propositions discussing different sorts of objects and their mutual relations. When one does mathematics one does not use mathematical propositions about number, set, continuity and so forth. The point that Wittgenstein is trying to make assumes special importance when, for example, we talk of infinity and continuity.[5] Talk of continuity or infinity is neither endlessly continuous nor literally infinite. The meanings of these concepts are to be gathered from the ways these are used. The talks to be meaningful must be somehow surveyable; that is, available within the *practical* limits of language.

Contrary to Brouwer, Wittgenstein asserts that the meanings of the propositions containing the concepts are not derived from anything like pre- or extralinguistic intuition. Their meanings are to be gathered from some well-established procedures, practices, or uses. *Justification* of mathematical propositions is not to be found in any intuition. The *reasons* for *doing* what mathematicians do within mathematics are not external to it. If, for example, following some procedures they get some results they are interested in, they think, they are justified in following that procedure. It is like a sort of successful measurement. But, as we know, there are different ways of measuring. For measuring liquids like water and milk and hard things like steel and gold, minute particles, and astronomical distances we do not follow the same measure. Yet we keep using the word *measurement*. Though the "same" word is used, its meaning is different in different language game–theoretic contexts. Clearly this line of Wittgenstein's argument in mathematics unmistakably shows his aversion toward any sort of essentialism or foundationalism.

Second, another criticism against foundationalism was voiced by those like Courant, Birkhoff, Synge, and Von Neumann, all of whom emphasized the desirability of maintaining close relations between mathematics and empirical sciences like physics and expressed their disappointment over the withdrawal of the mathematicians from the world of mundane reality. For example, Neumann observes that if a mathematical discipline distances itself from its empirical source, it gets into grave difficulty. To him, the right remedy seems to lie in the rejuvenating return of mathematics to its source; that is, "the re-injection of more or less directly empirical ideas."

This approach echoes Descartes's critical attitude toward abstract geometry. Descartes was opposed to the idea of *abstract* geometry for the sole purpose of "exercising the mind" and favored physical geometry "which has for its object the explanation of the phenomena of nature." A similar view is echoed by Quine when he says that mathematics should be viewed as a theoretical part of the natural sciences, "comprising truths or hypotheses which are to be vindicated less by the light of pure reason than by the intricate systematic contribution which they make to the organising of empirical data in the natural sciences."

Another radical critic of mathematical foundationalism is Imre Lakatos, basically a Popperian by training. According to him, mathematical conjectures are not altogether different from the physical ones, nor can their refutability, as it happens in empirical sciences, be ruled out. On the contrary, rational mathematicians, normatively speaking, should try to test their mathematical conjectures. What is more, according to Lakatos, "mathematical discovery" is a genuine, and *not* metaphorical, expression. Unexpected mathematical discovery refutes the accepted mathematical propositions. Of late, this approach to mathematics has been appreciated and followed in different ways by Crowe, Browder, and Kitcher.[6]

The sad but instructive outcome of the different forms of foundationalism in mathematics has been aptly summarized in a story referred to by Morris Kline:

> On the banks of the Rhine a beautiful castle has been standing for centuries. In the cellar of the castle, an intricate network of webbing had been constructed by industrious spiders who lived there. One day a wind sprang up and destroyed the web. Frantically the spiders worked to repair the damage. They thought it was their webbing that was holding up the castle.[7]

The two main lessons of the story are (1) damage repair or "monster barring" is not what is expected of a creative mathematical thinker, and (2) the history of mathematics, though differently interpreted, has its enduring value.

2. On the Nature of Language and Human Knowledge of the World: Some Indian and Western Theories

Attempts have been made to think of our relation with the world dispensing with the idea of language. Instead of looking at the world through language we may go down further and try to see how language itself has come into being and been able to perform the sort of work it is doing now. In other words, our linguistic competence itself may be the object of our investigation. In this connection, extensive use of the ideas of physiology and psychology is unavoidable. But, then, it may be pointed out that the concepts to which we are obliged to turn for understanding our psychological competence or linguistic competence are themselves embedded in some language or other. Of course, in an obvious (not trivial) sense all sorts of our understanding the world, our relation with the world, are embedded in language. This does not prevent us from following other ways of our understanding the world, our relation with the world, or our place in it.

The tendency found in some thinkers to draw a line of distinction between prelinguistic intuitive certainty and intuitive certainty as an epistemological concept seems to rest on a misconception regarding the nature of language. Whatever is available in language is taken to be somewhat discursive, psychological, and therefore, not universalizable. Besides, language is recognized as a regional cultural institution. Therefore, unless it could be shown that this or that language, though regional or relative in character, has a universal foundation and that what can be formulated or presented in one language could be made available as well in other languages without substantive loss of meaning or certainty of content, the universal character of language as such cannot be proved nor can it be taken as foundationally true.

In India some language philosophers of the monistic persuasion like Bhartṛhari, Sureśvara and Maṇḍana Miśra developed a theory according to which meaningful language, *sphota,* or Master Sound, *sabdavṛtti,* or power of language, is ultimate reality. It is *universal* in nature but expressible in *specific* forms, primary and subsidiary. Sometimes this view is called *sphotavāda.* Roughly speaking, it asserts that meanings, which are universal and unchanging in their essence, are gradually disclosed because of their inherent power. All sorts of ignorance (*avidyā*) disappears when one realizes the true meanings of such great statements (*mahāvākya*) like "I am Brahman" and "That thou art." Monistic hermeneutics is said to be the key to freedom from doubt. Sometimes it is called *śabdādvaitavāda,* semantic monism. In Sanskrit, *śabda* means both *sound* and *word.* What is, physically speaking, sound is also a unit of meaning. Units may be small or big, letters, words, or sentences. Though the details of the theory propounded by different thinkers are different in some important respects, their essential unity, viz., the universal nature of language and its units, is of importance to us. If this view could be defended, linguistic arguments against the view of foundational unity of human knowledge could be substantially salvaged.[8]

A comparable line of argument is found in the Cartesian linguistics. Descartes's philosophy of language, shaped in the model of his innatist epistemology, until recently, was suspect to many professional linguists. Fortunately, historical researchers have succeeded in establishing the view that the Cartesian insights against skepticism and relativism had indeed been taken up and worked out by some of his followers in the seventeenth and eighteenth centuries. Chomsky's attribution of credit to Descartes in this context is well-founded, both conceptually and historically. The basic point of the Descartes-Chomsky thesis is that the difference and relativism of the surface features of language must not make us oblivious of their deep-rooted syntactic unity. The concept of universal syntax embedded in this view is purported to save our diverse syntactic rules from narrow regionalism. Subsequent investigations and criticisms brought to light the inadequacy and the limited character of a pure theory of syntactical meaning. Rules of combination of words or expressions by themselves fail to disclose the full meaning of human expressions. For the purpose of correctly determining the meaning of words and their combinations we have to know, besides the syntactical rules, the contexts in which they are used, the ways they are used, and also the force and tonality of their use. In other words, the syntactic approach needs to be supplemented by semantic, phonological, and sociological details. This composite theory of language, highlighting the universal features of different languages, without disregarding its individual peculiarities, is intended to minimize the problems of relativism, underdeterminacy of translation, and, above all, skepticism.

Making good use of this linguistic approach to the theory of knowledge, some people believe that they have succeeded in containing skepticism and that the nature of language, rightly understood, does not in any way promote skepticism. But language alone is not to be deemed the only devil for keeping our knowledge of the world uncertain. The nature of the knower, "I," for instance, has also to be deeply looked into. The "I" as affiliated to God, and not on its own, is the foundation of universal knowledge. God provides the necessary connection between intramonadic and intermonadic knowledge.

Some thinkers address themselves to the question of why we take as certain the world as we experience it. Have we good reason to be sure that the world as presented to our senses is the world as it really it is? Descartes, for example, tried to answer this question. The question or the problematic form of it represents the skeptical aspect of his thought. When he thinks that he has found a reliable answer to the question he returns to the world as he finds it as a man of common sense. And, then, as a philosopher he reflects on it and reflection yields the conclusion that this world consists of two sorts of reality, matter and consciousness, but both are expressions of one supreme reality, God. It is not that there is no linguistic presupposition in the Cartesian formulation of skepticism and its refutation. It is there, but it is primarily assumed and not explicated. What is explicated, delineated in detail, is the dubitability of the world as available in and through senses. One may say that it is conceptual exploration. It may be described also as experiential mapping.

3. Is Knowledge Certain? Hume's Question and Kant's Answer; Quine's Question

There are some thinkers who start from the assumption that we have certain knowledge of the world, at least of the empirical world. Thereafter they raise the question of how we, human beings, have been able to acquire this sort of definitive knowledge of the world. Kant's basic question, How are synthetic judgments possible a priori? is of this nature; that is, clearly precritical, if not dogmatic. He starts from the unquestioned assumption that synthetic judgments a priori are possible. Thereafter he confines himself to the question *how* they come to be possible. He does not go to the very fundamental question, whether in fact they are possible. By not raising this question he refuses to discuss the basic issues of skepticism with the skeptic. On the contrary, he tries to tell the skeptic why one should not start from the "wrong" end. To him, this skeptical end is the wrong end. But the point of interest to note in Kant is that in the course of answering his basic question he places before us a set of concepts and categories, which, to start with, is not evident to us. In and through transcendental reflection on what is given he tries to show that the seemingly meager given is loaded or impregnated with many hidden promises and potentialities. His "blind" given, rightly understood, is not really blind. It has in it both eye to see and the light to show us the transcendental way up. Kant's transcendental argument, developed step by step, through imagination, understanding, and apperception, discloses the hidden "ocular structure," meaningfulness of the given. In a way Kant shows that, though the full and definite meaningfulness of the given is not disclosed to us at the initial stage, its promise is implicitly held out from there. Kant's antiskeptical strategy is both preventive and promotive. It prevents initial (or prejudgmental) skepticism and promotes the certitude of knowledge, scaling the higher transcendental reaches. Because his starting point is necessary and a priori judgments of science and mathematics, he is not required to encounter the cognitive potentiality, if any, of the "blind" given, the stuff of knowledge.

Kant's method of showing our place in the world is primarily conceptual and categorical, not linguistic. Our bodies, minds, and souls and their relations with the natural world are conceived by him not through the idioms of everyday language. On the contrary, he finds that these idioms are embedded in and expressive of a unified set of concepts and categories that is gradually disclosed to us through reflection. The structure of our understanding is such that every possible object of experience is assimilable under it.

Given this constitutive structure of understanding, he cannot think of any object of it that can possibly disrupt or damage it. The conditions that make the object of knowledge possible cannot be undone by the latter. Kant cannot think of any cleavage between conceptual scheme and content.

Even what is not assimilable within the conceptual scheme of Kant does not threaten the latter from "outside." Strictly speaking, what fails "outside" the scheme is not *quite* alien to it. The ideas of Reason—God, Freedom, and the Immortality of the soul—though (in a way) external to the scheme, provide it with ideals to aim at. Viewed thus, even things-in-themselves are *normatively* available within the scheme. The scheme of our understanding has in it some ideals of synthesis that are not born out of the empirical given. This construal of the relation between "the empirical" and "the ideal" leads us to a dualistic interpretation of Kant or, more broadly speaking, to a scheme-content dualism.

It is clear that the denial of the scheme-content dualism is intended to combat skepticism. To *justify* a particular scientific paradigm of knowledge, in this case Newtonian mechanics, Kant painstakingly constructs a categorial framework that, as stated earlier, can never possibly encounter a falsifying evidence from without. The understanding that constructs theories of science also constructs the evidential findings purported to support the former *unilaterally.* Antievidence *within* the framework is not available and from *without* not admissible. And, therefore, the question of "discovering" falsifying evidence to test genuinely scientific theories makes no sense *within* the Kantian framework. This explains, at least partly, the sweeping Kantian dictum: "understanding which makes knowledge possible also makes nature possible."

In brief, naturalistic skepticism of the Humean origin is sought to be contained by a unique, at any rate claimed to be unique, epistemic-constructionist framework. Through a series of ascending syntheses and under and overarching apperceptive unifying principle, Kant builds up his theory of scientific knowledge that is never *open* to "disturbing" or falsifying evidence. This immunity against all possible errors or the claim of finality is due to the very constructionist strategy that has been followed in developing this theory by Kant. He had been working under the impression that Humean skepticism was due to his failure to draw the important distinction between "the transcendental" and "the empirical," "the ideal" and "the actual." If the former terms of the conceptual pair are denied and reduced to the latter terms, he felt, together with the scheme-content dualism, reductionism and skepticism stage a come back.

Therefore, he was in favor of retaining the dualism to bar the monster of skepticism.

It is not quite clear whether one should use the word *dualism* or *duality* in this context. If we say that there is a *dualism* between what is *in* the scheme and what is *external* to it, it is not easy to define the relation between the two. If what is *external* is really *alien, alien* in the strong sense, to what is *internal* to the scheme, we do not know how we can possibly conceive of it and speak of it meaningfully, it is partly for this reason that we encounter a systematic problem in defining the relation between "the empirical" and "the transcendental," between "the actual" and "the real." The problem may be put in a dilemmatic form. If "the transcendental" is said to be available within "the empirical," the former ceases to be transcendental and becomes hardly distinguishable from the latter. And in that case we are landed in a sort of naturalistic *reductionism* or, in brief, *naturalism*. Naturalism and empiricism go well together. But, as we know, this alliance smacks of *skepticism*. From Hume to Quine one can easily trace this proskeptical naturalistic trend.

Though this naturalism is often characterized as proskeptical, its strength must not be underestimated. As we have noted earlier, Hume's skepticism, based on naturalism, is of a peculiar kind. It recognizes the world in its ordinary, natural, or unbracketed sense. Neither in physics nor in history does he defend skepticism of the classical type. On the contrary, he justifies induction quite seriously and because of its consistent practical efficacy. This line of his argument can be traced in the writings of many other British philosophers like Reid, Mill, and even Strawson. It would be rather trite to think that naturalism as such is destined to degenerate into skepticism.

A somewhat similar and plausible defense could be offered also on behalf of Quine. Despite his theses of indeterminacy of meaning, ontological relativism, and antifoundationalism, his theory of knowledge is claimed to be reasonably sound both on conceptual and doctrinal sides. In his conceptual studies Quine is engaged in clarifying concepts by defining them, defining the obscurer ones in terms of the clearer ones and the less obvious laws in terms of the more obvious ones. This strategy, according to him, is certainty maximizing and, to him, the ideal starting point is the self-evident truths as found in the definitions of the systems like *Principia Mathematica*. Analysis of "number" in terms of sets was aimed to provide "foundation" for mathematics and somatological analysis in terms of nerve irritation was meant to provide foundation for our knowledge of the physical world. This explication of truth in terms of semantic concepts of meaning or reference represents an enterprise to provide the conceptual foundation of analytic philosophy. But to what extent it has succeeded is a separate question.

In a sense Quine is justified in asserting that from Hume to the present day empiricists have not added much to the doctrinal side. But for his departure from Hume's naturalism Kant perhaps could not make his antiskeptical claims plausible and respectable. But it is to be noted that not only from Hume to Kant but also from Kant to Quine a consistent trend of constructionism is clearly evident. Each one offers us a method of construction, construction of the world we are called upon to deal with in science. For the sake of simplicity one may say that Hume's method is psychological, Kant's epistemological, and Quine's logical. But the implied distinction need not be taken as very hard and fast. Where the Hume-Quine approach is significantly different from the Kantian one is this.

The former does not rely upon or make excessive use of metaphysical presuppositions. Hume outrightly rejects metaphysics at the theoretical level. What he leaves unsaid about it, its *practical status,* is carefully attended to and explicated by Quine. Instead of consigning metaphysics to flames he shows how his constructionist strategy can substantially save a large part of Hume's metaphysical beliefs and bring them in harmony with the other planks of the Neurath boat. The reason is partly to be sought in the Quinean reduction of semantics to metaphysics. The ontological questions of object that figure within any theory cannot be decided from the outside. That is how the Quinean can do away with the traditional distinction between what is claimed to be metaphysics and what belongs to the realm of semantics. In his scheme of thought there is no metaphysical absolute: God, self, or even a rabbit. This way of undoing the meaning-metaphysic distinction also helps him to highlight the inadequacy of the classical realism-idealism dispute.[9]

4. Quine's Question Extended from Epistemology to Ethics: Some Indian Views Recalled in This Context

If from epistemology we move to ethics, seek to do away with the distinction between the actual and the ideal, and try to understand the latter in terms of the former, we end up with a sort of moral skepticism. It is no wonder that in the area of knowledge, where we are obliged to infer the (not given) general from the (given) particulars, our inference, inductive inference, is sought to be justified by Hume in terms of common sense, practice, customs, and so forth.

Similarly, in the area of morals the naturalist Hume tries to defend human ideals in terms of their conformity to tradition, custom, convention, and so forth. In brief, Hume maintains that whatever is obtained from experience has to be justified within and in terms of experience. There is no way out of experience. *New* experience has to be justified by *old* experience. Nothing born of experience can be rationally legitimized by something that is not itself experiential. "The transcendental" is an empty speculative term. The talk of justifying "the empirical" by "the transcendental," of "the actual" by "the ideal," is symptomatic of the metaphysical forgetfulness of the common parentage of the both. "The transcendental," like "the empirical," is born of experience. "The ideal," like "the actual," is born of experience. Still the main reason that we often are misled to believe in their essential distinction is our speculative propensity, customary attraction toward the flight of imagination, a sort of antinaturalism. Because of this propensity many of us think that the ills and evils of skepticism are rooted in (epistemological) empiricism or (ontological) naturalism or both, and the only way to get rid of these infirmities of our human nature is to pin our faith in some firm or unshakable transcendental principle or being. It is this faith, *not* experience, that explains our uncritical or precritical commitment to "transcendental principles" or rigorous knowledge, the everelusive God above us and the immortal soul within us.

In view of what has been said it is not at all surprising that Quine finds no progress beyond Hume in the "doctrinal side" of empiricism. Even the conceptual refinement that has been achieved by philosophers and scientists since the days of Hume does not appear to be sufficiently impressive, despite Quine's view to the contrary. It is no wonder that Quine finds that "the Humean predicament is the human predicament."[10] Hume found that all forms of knowledge, common sense and scientific, and ideals, ethical and aesthetic, are *ultimately* traceable to sense impressions, directly known. When Quine says that all forms of ontology are extensions of somatology, he is not saying something quite different from what Hume said, using different idioms. By breaking the unnatural Humean fork, the untenable analytic-synthetic distinction, an old empiricist "dogma," he thinks, he has only "naturalized" naturalism and extended its frontier, bringing even pure logic and mathematics within it. From tables and chairs to numbers and sets all sail together in the Neurath-Quine boat. And the boat is claimed to be natural.

The Quinean way of unifying the talks of objects, from physical and physiological to logico-mathematical, is semantic; that is, linguistic as distinguished from metaphysical. When mystics or metaphysicians claim to have a definite vision or view of their objects, they unconsciously slide back to their preferred background of language and wrongly expect others to share it and thus to leave behind doubt about their knowledge. It is true that metaphysical beliefs, rather the objects of those beliefs, are not self-expressive or self-justifying. Perhaps it is equally true that our meaning-determining rules are not arbitrarily chosen or legislated. An element of naturalism, practical advisability, and the question of simplicity are also involved.

The commonsense view of the world has a unity. The observational-experimental view of science also exhibits a unity of its own. The metaphysician's world, though seemingly outlandish, has some very known features in it. All unitary views have between them many overlapping areas. This is not to deny their differences in many important respects. The reason is not far to seek. One's preferred language is metaphysical, containing in it, for example, *God, external world,* and *invisible souls* or *monads*. Interestingly enough, all these objects can be, in fact have been, explicated by both commonsense language and scientific language. In the process of explication every language eliminates some of the aspects of the "commonly recognized" objects. In some cases even well-recognized objects of one language are eliminated by the process of explication followed in another language. The question of *existence* of this or that object cannot be decided *absolutely;* that is, independent of the framework of this or that language. Whatever language is chosen, ultimately it is open to modification by the experience of the language user, the language using community. Therefore, the question of change, revision, repair, and so on, can never be ruled out.

The time-honored distinction drawn between reason and experience, naturalism and antinaturalism, realism and skepticism, theoretical reason and practical reason makes little or no sense in the context of Indian philosophy. The knowing mind is never credited with the presence and function of a priori forms or categories. What make validation of knowledge (*pramā*) possible are called *pramāṇas,* instruments of knowledge: perception, inference, analogy, verbal testimony, scriptural authority, postulation. Understandably, all *pramāṇas* are not recognized or given equal weight by different schools of philosophy. In the present context, the view of the Nyāya-Vaiśeṣika, logical realism, is interesting. From the Western point of view, its theory of knowledge would undoubtedly appear proempirical and its ontology, interestingly enough, strictly realistic. No trace of the Humean predicament is to be found in it, neither in its theory of inference nor in its theory of ethics. Even in its theory of meaning, in the determination of the relation between words (*pada* or *śabda*) and their meanings (*padārtha*), the logical realist avoids skepticism by resorting to a sort of transcendental-conventional argument, anticipating the Berkeleyan approach to the subject. The conventional relation between word and object, though confirmed in human experience, is due to God and, therefore, free from the vagaries of human volition and legislation. The talk of ontological relativity makes little sense to Nyaiyāikas. To them, the ultimate ontological categories, substance, quality, relation, action, universal, and so on, do not owe their origin to any human perception or conception. On the contrary, our perceptions and conceptions, including the judgmental ones, correct or erroneous, are explainable in terms of (1) the said ontological categories and (2) the recognized *pramāṇas* (perception, inference, and so forth). Even skeptical phenomena are claimed to be intelligible because of the stable and universal ontological furniture of the world.

Two points may be briefly noted here. First, different schools, even rival ones, notwithstanding their commitment to different sorts of ontologies and *pramāṇas,* have their own subtle ways of recognizing what is transcendental. Buddhism, despite its accent on perception as the primary source of knowledge, recognizes a peculiar mode of knowing in terms of which what is *not* transient, Buddha's cosmic body's *dharmakāya,* for example, can be grasped (but not communicated). Nāgārjuna's inability to speak with certainty about what is the highest reality, *nirvāṇa* is also worth recalling in this connection. He expresses it in the form of a tetralemma: (1) it is neither positive (*bhāva*), nor (2) negative (*abhāva*), nor (3) the conjunction of both, nor (4) their disjunction. Similarly, but from a quite different point of view, the logical realist speaks of super-sensuous (*yogaja*) perception in which the limits of space and time, the threshold and height of normal perception, can be overcome. Neither ontological austerity nor ontological profusion, neither epistemological austerity nor epistemological profusion has any necessary connection with skepticism in Indian thought.

Second, the Western distinction between theoretical reason and practical reason seems to be somewhat irrelevant in the Indian context. Most of the Indian philosophers, belonging to different schools, are of the view that theoretical knowledge, though very important, is not an end in itself. Knowledge of what is real is to be cultivated mainly for its liberating influence. Knowledge removes the biases and prejudices, *saṁskāras,* of the human mind and helps human beings be good and right and, ultimately, free. This transcendental idealism is not only an ontological doctrine but also a practical or ethical one. The question of truth is not a matter to be decided exclusively in terms of abstract reasoning. The importance of the practical experience of the people, *lokavyvahāra,* is taken to be the end of all disputations. For example, what determines the definitive identity of water is not a formula, H_2O, but its capacity to quench thirst.

5. Is Skepticism Inherent in Theory or Practice? Can Naturalism or Reductionism Show a Way Out or Must One Accept the Third (Scheme-Content) Dogma of Empiricism?

As I have said earlier, the cluster of problems designated by *skepticism* may be approached from very many points of view, physiological or somatological, psychological, epistemological, linguistic, semantic, physical, or cosmological (with or without theology in it). These approaches can certainly be conceptually differentiated and discerned, but even after most perceptive differentiation and refined discernment it will be observed that many areas overlap, interpenetrate, and interact. This fact of common sense may be put in this way. Though, normally speaking, I live my life as an undifferentiated whole, it may be represented in different ways. As a man of theoretical pursuit I do certain things that, ordinarily speaking, I do not follow in my practical life. For example, as a philosopher of science I am a fallibilist, a skeptic of a sort, but perhaps I do not give that impression in the field of my day-to-day work. If I am pressed to answer the question whether I can validly draw a sharp line of distinction between my theoretical life and my practical life, perhaps I will be obliged to confess my inability and say some such thing as, "I really do not know where exactly and how should I draw that line." But having said that, even after that confession, I (and perhaps many other persons like me) continue to affirm and defend my theoretical views and practical decisions. This shows that even within the whole of our own lives we draw, rather we are obliged to draw, lines of distinctions, compartmentalizing our otherwise one "natural" life.

How natural is this "natural life"? The very fact that we are obliged to compartmentalize it, both theoretically and practically, shows at least one thing very clearly: even our self-encounter, self-experience, is not uniform, monotonous, and repetitive, at least not in the strict mechanical sense. Our abilities to draw a workably valid distinction between the "mechanical" sense and the "free" sense within our own lives show that we are *free* in a peculiar, almost inexplicable, way. The circumstances of my life do not affect me always in the same way. I am not environed identically over a long period of time. To put it from another side, my reactions to my environment undergo significant changes over the time. Sometimes I react instantly and impulsively; sometimes I react in a very studied manner. There are moments when I feel like reacting but hold back my reactions and feel tense within. It may be shown that there are various other shades of one's reactions to one's own environment. The picture becomes even more complex when we recall the endless feedback character of the relations between the elements of our environment and the conditions of our lives. Reflectively speaking, one can pertinently raise the question, Can we draw a valid line of distinction between (our) life and (our) environment? Do they not merge into an undifferentiated whole? If they are distinguishable, to whom do they appear so? The distinction that I draw between my life and my environment is perhaps not available to others, at least not in the way it is to me. Can others view me, experience me, in the way I do? Is the involuntary nervous tic of my body a part of my physiology or my environment? Does naturalism entail the denial of the peculiarity of the first-person ontology? Do we (I, you, he . . .) all sail together with tables, chairs, numbers, sets, and so on, in the same Neurath-Quine boat?

One way of dismissing all these questions is to advise us, as Carnap does, that we must draw a distinction between "the material mode of speech" and "the formal mode of speech," between "object language" and "metalanguage." Otherwise, we are warned, even our questions will fail to convey their intended meanings and, consequently, the answers will remain unintelligible and unconvincing. It is interesting to note that Quine ignores this advice of Carnap. For he finds that Carnap's advice marks a departure from naturalism. Without entering into the merits or otherwise of naturalism one can point out that underlying the distinction suggested by Carnap is a "natural" human tendency. The distinction between the said two modes of speech is not totally unnatural. To put it differently, it may be said that even within what is called *natural* are shades of difference. Reductionism, like truth, can never be achieved in an absolute or thoroughgoing manner.

If abstract entities like numbers and classes may sail together with concrete entities like tables and chairs in the same boat, one might say that Quine has done nothing wrong in booking knowledge as a fellow passenger on the boat. If knowledge is rightly understood, it is not something quite different from tables, chairs, numbers, and sets. By "right understanding" what he means here is "naturalized" understanding. In the name of naturalizing epistemology Quine tries to show that even the abstract forms of knowledge, logic and mathematics, for example, are rooted in and traceable to primitive somatological phenomena like the irritation of nerve endings. Because of the "unnatural" distance of numbers from the irritation of nerve endings we tend to forget the natural or somatological origin of mathematics. This is not to suggest that Quine is unmindful of the difference between numbers and their natural origin. The fact that disturbs us is not only the difference but also the distance between "natural home" and "natural origin." The natural home of numbers is traditionally known as mathematics. But when Quine shows us the long and the tortuous way from the irritation of nerve endings, the "natural home" of numbers, to their better known *local* address, mathematics, we can *theoretically* follow it. However, we are not *practically* inclined to traverse the *whole* way. To get the naturalized map of epistemology if we are required to study the whole thing, it is indeed a predicament, reminiscent of the Humean predicament. The *whole* in which the *local* addresses of somatology, mathematics and their connecting path are shown may well be correct or faithful. The point that disturbs us is that in the *reductive* whole the *local* differences between somatology and the higher reaches or abstract levels of epistemology are considerably ignored. If the localism of ordinary epistemology appears unduly restrictive, the holism of naturalized epistemology seems tortuously elaborate.

Are not many of us very critical of speculative metaphysics on the alleged ground of its essentialism, excessive reliance on the tenuous spiritual relation between the smallest things and the highest being? Do we really enhance the cause of antiskepticism by marshaling facts and arguments purported to show the doctrinal unity of "scientific" naturalism or "metaphysical" spiritualism? Are we not called upon by both logic and science to pay serious attention to graded difference, refined definition, and categorization of things and beings preferably on the basis of their local attributes?

Reductionism has both charms and dangers. Nature has no natural map of its own. There is nothing innate and immutable in it. The maps we draw of it present (or represent) it in different ways. The problems with demonstrative nouns and ostensive definitions are peculiar to philosophers (of language). The naturalistic method of learning and teaching what the simple words like *water, dog, tree* mean appears very nonproblematic, to start with. But when we ask some such questions as, Is frozen water water? Is the tail of a dog,

dog or doggish? and Is the branch of a tree a tree? we realize that it is not easy to ascertain the meaning even of simple words mentioned earlier. Ostensive or demonstrative identification and individuation are not as simple as we are often told by the parents and the primary school teachers. Both Wittgenstein and Quine spilled lot of ink to clarify the problem. Neither somatological satisfaction and confirmation (of Quine) nor game-theoretic rule compliance (of Wittgenstein) takes us very near to the solution of the meaning problem even in its most elementary stage. When we say this one might remind us that our discourses are primarily, almost exclusively, theoretical, and that in practice, we do ostend, identify, and individuate things by words or its gestural equivalents. At one stage or other of our discussion of meaning and knowledge we are forced to draw a line of demarcation between theory and practice, although this line is neither sharp nor permanent.

In this respect Quine echoes James and anticipates Davidson. Neither to James nor to Davidson is the scheme-content dualism or the sharp distinction between realism and skepticism acceptable. What glues together our beliefs and makes us accept them as true is not their individual "confrontation" with what is there in the world. From coherence theory of truth some philosophers with foundationalist leanings are reluctant to move to coherence theory of knowledge because of the nonavailability of this confrontational relation. In the absence of the objective truth conditions of knowledge, they feel, they are being pushed to a skeptical position. But Davidson thinks otherwise: "What brings truth and knowledge together is meaning" and coherence of beliefs works as "a test for judging that objective truth conditions are satisfied."[11]

He wants to have best of both the worlds, scheme and content, beliefs and things. On the one hand, he states, "truth is correspondence with the way things are" and to make his realistic bona fide clear, goes to the extent of emphasizing the "non-relativized, non-internal form of realism." At the same time, he denies (1) the explanatory role to "true," (2) the existence of relations "being made true" obtained between interlocked beliefs and the world, and (3) the availability of good points of debate between realism and antirealism. His anxiety to commit himself to the notion of correspondence is understandable because it is a recognized escape route from skepticism.

Equally understandable is Davidson's opposition to the idea of determining the empirical truth of our knowledge in terms of individual or isolated sentences. That is why he rejects Schlick's foundationalist epistemology and feels drawn towards Neurath's coherence theory. In fact the paradigm of Wittgenstein's picturesque elementary proposition deeply impressed many logical empiricists like Schlick, Neurath, and Popper. At the same time, they felt the problem of reconciling basic statements with higher-level theoretical statements, foundationalism with empiricism, and empiricism with realism. It did not prove easy for them to show how a network of statements, beliefs, and attitudes is related to the world. This question is bound to come up if the correspondence theory is seriously meant and pressed to do its minimal ontological duties. If individual beliefs, to be true, have to do something in the way things are in the world, the backward pull of coherentism should not be there at all. It is there to retain at least the traces of realism in epistemology. Things of the world cannot *causally* account for the beliefs that form the main *corpus* of our knowledge. If causal *justification* sounds naturalistic in the pejorative sense, the other horn of the dilemma is that every belief is embedded in other beliefs, forming a network of beliefs. So, on the one hand, Davidson's theory of knowledge has all the weakness and strength of methodological holism as well as those of methodological localism. In this respect his difference from Quine is worth noting. He himself confesses: "the foundations of knowledge must be subjective and objective at once, certain and yet open to question."[12]

It is not surprising that to retreive his position Davidson claims that "beliefs are not historically or causally arbitrary; even if our *reasons* for our beliefs are always other beliefs, the *causes* sometimes lie elsewhere."[13] If further restrictions are not put on the statements, the notions of "reasons" and "causes" may sound essentialistic. The search for justification proves futile in this case. To save the desperate situation, Davidson returns to Quine's methodological holism or Neurath' boat analogy. He says, language by its very nature is intersubjectively sharable. When we hear other peoples' words and sentences, we succeed in most cases in understanding them. Beliefs and sentences are true as a lot and known in that way. To say this is not however to justify our knowledge or belief but only to indicate our ways of understanding them. In order words, it is a simple statement of *practice* and not of much informative value. A theory of understanding of language by itself is not a surrogate of sound epistemology.

6. The Primacy of Practice over Theory, of Observational Language over Theoretical Language, and the Relevance of the Issue to Combat Skepticism

The primacy of theory is the verdict of life as we ordinarily live it. Not that theory is an outlandish phenomenon. Theory is very much a part of our practical lives. Their close interrelation can hardly be denied. What is affirmed is primacy of the say of theory in conceiving, formulating, and deciding the acceptability or otherwise of theory. Views, beliefs, and theories become problematic when they are found to be incompatible with experience. Incompatibility may be both practical and theoretical. Intertheoretical incompatibility is not unknown either. At times, incompatibility between high-order theories and low-order ones is consciously sought so that the former may be tested by the latter. It is instructive to note that for testing high-order theories relatively, low-order ones are found to be more in demand and use. Here one also finds the rationale, though disputed, of the foundational claim of observational language. Observational language stands closest to the practice of life. It is at this ground level that the beliefs and views of the members of the same speech community most credibly converge and cohere.

We have neither the time nor the intention to test the verdict of our direct experiences (available in observation language) in terms of high-order theories. Rather, we test them almost unconsciously against the experiences of our fellow human beings. For this purpose the fellowship is primarily determined in terms of social proximity and use of the same language. Identification and individuation of objects are found to be best available at the ground level of experience and language use. However, this is not intended to deny the very possibility of error and doubt even at this level. The point is that here the threat of skepticism seems to be least. When we are close to our experience and equally close to the members of the same speech community, our cognitive claim is extremely meager. This substantially explains our minimal vulnerability to skepticism at this point.

The problem of skepticism, though essentially human, seems to be best understandable in and through language and not traditional epistemology. In epistemology our two main concepts are *reason* and *experience* and their different *forms*. As we know, the concepts of reason, experience, form, and their cognate expressions are open to many interpretations. Interpretations differ both theoretically and socially. For example, in theory of language such questions as, What is rational? are not answered identically or unanimously. The empiricist says, "What is based on experience is rational." The rationalist's response is, "What follows from self-evident reason is rational." Apart from the fact that the assumed dichotomy between reason and experience is not universally recognized, the other point to note is that terms like *reason* and *experience* are often used in an *abstract* manner. Abstract theoretical discourse thrives on its theoretical and abstract characters. When these terms are related to practical discourse, much of their associated mistiness disappears or starts disappearing. At least one of the main reasons why of late theory of knowledge is being increasingly brought closer to language, ordinary or otherwise, is that it provides a better, practically better, frame of reference for discussing and understanding the traditional problems of the theory of knowledge in which "reason," "experience," and their like figure extensively. In *use* of language the sense of the words used is understandably better captured. Use is always perused or followed together with its context. The contextual texture of language has an indefinable concreteness that answers many questions and removes many doubts before they are even expressed.

Wittgenstein's preference for ordinary language is *not* to be related only to clarity. The question of clarity and that of what is being clarified are not quite separable. To think that theory and logic, mostly couched in extraordinary language, have nothing to do with ordinary language is a mistake. Theory may be formulated in a technical language but its life is rooted in an ordinary one. Mainly for this reason the tenability of even very abstract theories is decided at their ground-level empirical roots. And at that level the language used is ordinary, no matter whether it is verbalized or not. Mainly in this context one has to understand Wittgenstein's unwillingness to draw a sharp distinction between object language and metalanguage. The rules for construction of metalanguage lie hidden in ordinary language. They are culled from the latter.

In the same context one recalls Quine's ingenious effort to derive logical language and its apparatuses from ordinary language. Note here that instead of using the language of reason and experience of traditional epistemology he tries to get to the roots of reference along the routes of child psychology and ordinary language. Even there he prefers observational sentences to observation (as a form of experience) for the purpose of scrutinizing how we practically use language to refer, individuate, and categorize the objects of the world. Our acquisition or the mastery of language seems to be the outcome of an ongoing, unconscious and holistic enterprise, not piecemeal exercises. The logician understandably likes to see the sentences in segregation or isolation, though in life they are interlocked. Sentential interlocking is the symbolic, that is, linguistic, outcome of observational aggregation.[14]

Where and how are we to look for sentential interlocking and observational aggregation? Why do we raise this question in the context of understanding the problems of skepticism? First, we are told that the observational sentences are more reliable than observations because the former are said to be more concrete and easily perusable than the latter. Second, emphasis has been laid on the interlocked character and level of observational *sentences* and on their *observational* (experiential) basement. The assumption is sentences in isolation are abstract and less meaningful; and the same in (systematic) relation relatively more concrete and meaningful. Similarly, the basis of assumption is that observations and their traces in isolation are abstract and less meaningful, and in (coherent) relation they are more concrete and meaningful. Naturally, maybe wrongly, the question is raised, Is there any foundational basement where the units, psychological or linguistic, of knowledge are available in an irreducible form?

This question, one might allege, rests on a muddleheaded bucket theory of mind. We may reject the bucket theory. We may even reject the very theory of mind on the alleged metaphysical ground voiced by philosophers like Hume. But to explain the very possibility of knowledge, however weak might be its form, we are required to explain the aggregation of information, its preservation and symbolization. One (like Popper) may not believe in the bucket theory of mind, another (like Hume) may reject the metaphysical theory of it, but hardly can one altogether dispense with a theory necessary for explanation of the availability of language, its expressive, cognitive, and other capacities. What I suggest here is this. We cannot completely do away with something like mind or human nature, whether we construe it somatologically (Quine-Smart), dualistically (Popper-Eccles), or transcendentally (Kant-Cassirer). Our psychosomatic complex, whatever name we give to it, is indispensable for explaining (1) what is given to us over a long period of time, (2) its retention, (3) its conceptualized form or (4) otherwise, (6) its application, and (7) availability of rules for correctly applicating these and distinguishing the correct applications from the incorrect ones. It may initially appear that the "human mind" or "human nature" is being endowed with too many functions. If we look into the close interrelations between the functions just mentioned (and some others left out), it will be seen that the denial of these essential requirements of the very possibility of a theory of knowledge or, more fundamentally speaking, a theory of language does raise insuperable problems that make it difficult for us to explain the questions regarding the interlocking of observational sentences, aggregation of observations, and the like. Mainly for this basic reason thinkers as widely different as Nāgārjuna, Śaṁkara, Hume, Kant, and Wittgenstein have to refer repeatedly to some concept of mind or other, despite its allegedly elusive or fluxist or transcendental character.

Besides the mind, another notion intensively exploited to explain the possibility of knowledge and counter the criticisms of the skeptic is the *social* character of language. The disparaging reference to Robinson Crusoe, mentioned earlier, is very instructive in this context. Positively speaking, what is implied here is this. What we need is not merely a theory of mind or human nature but a *social* theory of it. The reason for this need merits careful investigation. Society, roughly speaking, is a network of interindividual relationships. It is more durable than the individuals who enter into and go out of the network. Often this network is called *structure*. Whatever name is given to this entity, it is more abstract than the individuals, whose dispositions and actions are understood in its terms. Individual human beings are perceptible; but the network or structure of interindividual relationships is available only to our understanding, not our perception. The society that we always, almost unconsciously, assume and to which we so frequently refer, in a sense, is invisible and theoretical.

Interestingly enough, humankind has often been defined in terms of this very theoretical and abstract society. The definition of the human being as "social animal" is widely acceptable. Equally acceptable definitions of the human are "rational animal" and "linguistic animal" (or "semiotic being"). Scrutiny reveals that rationality, sociality, and the sign-using (that is, linguistic) capacity are interlocked or go together. The one thing that these three human capacities have in common is *power of abstraction*. Without reason we cannot disengage ourselves from our sensory experiences. That we can rise from the level of sense perception to that of imagination, from the level of imagination to that of understanding or further is due to this abstractive capacity. The fact that humans do not identify themselves with their bodies and can view themselves as members of a family or enlarging social aggregates like tribe, class, and nation is understandable again in terms of their abstractive capacity. Human beings are not confined to or shut up in the immediacy of their somatological experience. The ends of their nerves do not mark the end of their world. The expanding ontology of society, in spite of its "invisibility," is in a way native to human nature. Similarly, it can be shown that the human referential capacity moves from the concrete empirically identifiable things, from "bits" to "blocks" and beyond, relatively abstract understandable things. Because of this abstractive capacity starting with meager empirical input we can reach out to the wide world through theory and practice, through thought, action and speculation.

7. Relative Strength and Weakness of Somatology, Psychology, and Sociology in Understanding and Containing Skepticism

Skepticism, understood thus, is not a pernicious view. It is a description of the human position in the world as well as the nature of his knowledge. Skepticism disturbs only the theorist in us and not the concrete living individual. As noted earlier, when we are close to our experience, we are not likely to be disturbed by our skeptical moods. When the *social* frame of our living is small, we are not likely to feel lost. Similarly, when we individuate or identify objects by ostension, we are not likely to be misunderstood. Misunderstanding, misidentification, and misinformation prove to be really disturbing, wide of the mark, when we are away from our immediate experience, far away from our bodily existence, or on a speculative excursion.

In the light of this diagnosis and description of skepticism, the natural remedy that commends itself to us is to remain close to our body and experience, bodily experience and avoid speculation and living in larger social contexts. The question is, Can we, human beings, accept and practice this commendation? The answer suggested by our social evolution and linguistic development is clearly in the negative. Social scientists tell us that in the natural, that is, presocial state of society, human life was nasty, brutish, and short, hardly social in the ordinary sense. Even those who think that the natural state of society was idyllic concede that it was inadequate for the realization of the higher human needs. There is a consensus that the fullness of human life cannot be attained by atomic individuals. The individual's urge for a rich social life accounts for the evolutionary enlargement of social aggregates from family, clan, and so forth to the confederation of tribes, nations, and so on.[15]

Ontologically speaking, we are not passively posited in the world. We are open to many influences and ourselves influence our environment, near and not so near. Whether our linguistic capacity is native or acquired may be disputed. But it is clear that, open to the influences of the world and given our psychosomatical traits, we increase, or are obliged to increase, our commerce with the world around us. This world expands. Collaterally, we are called upon to enrich our language to articulate our experience of the world, the world that informs us and shapes us and that, in turn, is informed and shaped by us. The human referential and communicative capacities grow with our increasing dialectical commerce with the world.

Because elements of skepticism are found ineliminable from *human knowledge,* we often come across a theoretical tendency to take knowledge as a sort of *superhuman phenomenon,* not natural in any way, not subject to the contingency of the social or practical life. When I said earlier that our psychological approach to knowledge does not turn out to be very promising, I did not suggest that it is pernicious or inherently defective and, therefore, to be abandoned. What, instead, I suggested was that, compared to psychology, sociology was a more developed discipline and that practically it stood closer to our system of beliefs.

In spite of its theoretical persuasiveness, behaviorism, physiological and linguistic, is not central to our system of beliefs, including those pertaining to knowledge. Though no-ownership theories of knowledge have been persuasively defended by, among others, Bolzano, Frege, Ryle, and Popper the belief persists that knowledge is essentially human and social. Because of its socially sharable, communicable, and preservable character it is often accorded a glorified, if not reified, ontological status. There is no doubt that knowledge is better available to us in its social aspect. The problems ordinarily encountered in the psychology of knowledge to first, second, and third persons is somewhat reduced or minimized by the sociology of knowledge. By referring to social or objective conditions of knowledge variations or differences in the understanding of knowledge may be reduced or minimized but not totally removed.

Often it is said, as we indirectly concede, maybe in a limited way, that knowledge as a phenomenon has an element of subjectivity in it. Even a naturalist or reductionist view of knowledge cannot deny this aspect of knowledge. Even if humankind is said to be a thing among things, when its knowledge is referred to, it cannot be leveled and construed as causal collocation. If knowledge had extended beyond the physical bounds of the human biological organism, we would not have encountered the Humean predicament. The Humean predicament is born out of transcendence. It is only because we can more or less transcend the bounds of our body and share with others what I do and what happens to me and what they do and what happens to them that we are what we are and in course of time can become something more than what we are. The world impinges on us but not only at our nerve endings. It can reach us, get deep in us, skipping and crossing over the organic boundaries. The more important aspect of our relation with the world is that we can reach the world without necessarily going through the nerve routes. This is a sign of *our* freedom that is denied to other *things* of the world.

Humankind is more or less free in the matter of receiving information not only about the world around us but also about what happens within us. Information does not force its way into our body-mind complex. There is something in us, freedom or a capacity of consciousness, whatever we call it, that enables us to screen and interpret, to attend or ignore, the informational inflow. Where the outer world, which sends information to us, ends and where the inner world, the body-mind complex, which receives and interprets the information, starts cannot be very sharply demarcated. The difference between "the inner" and "the outer," though practically undeniable, cannot be clearly indicated. It varies from case to case. When we speak of a thing-being dichotomy, we highlight the ontological difference between what we are and how we are shaped by the world around us. Strictly speaking, we are not passively and unilaterally shaped by the so-called outer world. We have in us the power or freedom to go out of ourselves and change the world that ordinarily keeps on influencing us.

The population of the world in which we live is mixed. It does not consist only of things that are causally explained in science. The account of the world we have in physics is neither exhaustive nor entirely satisfactory. Our world is populated by fellow human beings. Nonhuman beings, things, and vegetation are also there to impress upon us. When we ontologically divide the world into two simple and broad categories, thing and being, one might say, we are oversimplifying an otherwise complex picture. Perhaps we do so for the convenience of understanding.

Human knowledge because of its complex nature is studied from multiple points of view, somatological, psychological, sociological, ecological, and so on. Needless to say, other possible approaches must not be ruled out. The point to be borne in mind is that these approaches are not necessarily exclusive. When we speak of sociology of knowledge, we do not propose to deny the psychology of it. What we want to highlight here is the *primacy* of sociology, the public or objective character of knowledge. Similarly, when we try to assimilate epistemology under philosophy of language, we are not unmindful of the psychology of language acquisition. Our main interest here is to minimize the effect of the impression of primacy associated with knowledge as a mode of experience (of this or that person). When we are engaged in a linguistic (not necessarily ordinary language) analysis of knowledge, we are in a way already in the area of sociology of knowledge. Language is a social institution. Its life consists primarily in its being used by human beings for the purposes of expression, communication, preservation, and so forth.

More than sociology, what lends knowledge an objective and larger perspective is ecology. In the ecology of knowledge we study not only the sociological or cultural conditions of knowledge but also the natural (including the biological) ones. The ecological approach has two forms, formal and informal. The formal view is information-theoretic and deeply influenced by the concepts and techniques of communication engineering.[16] The notions of human freedom and creativity are somewhat berated in this approach. In the informal ecological approach to knowledge both the natural and the cultural conditions of knowledge are given due recognition. Nor are the roles of human freedom and creativity underplayed.

When we speak of this different possible approach to knowledge, the point to be mainly noted is our understandable anxiety to keep knowledge above the board of skepticism. The implicit assumption is that knowledge *qua* knowledge, unless we take it in a proto-Platonic way, *a la* Bolzano and Frege, should be viewed within the "objective" scaffolding of semiology (philosophy of language), sociology, or ecology.[17] Otherwise, we are reminded, knowledge as a human phenomenon is dangerously vulnerable to the blemishes of skepticism. The traditional place of skepticism in the theory of knowledge lies between the extremes of Platonism, on the one hand, and fluxist psychologism, on the other.

8. How Is Skepticism in its Nonpejorative Sense Native to Human Nature?

Must skepticism be taken in a pejorative sense? Unless one insists on the *certainty* requirement or irrevisable truth (value) of knowledge, I find no compelling reason for taking a disdainful attitude toward any piece or system of knowledge. Unfortunately, our attitude toward skepticism is shaped rather prereflectively in most cases. Our intuitive notion of skepticism is closely associated with a sense of uncertainty. Our theoretical views, especially the precritical one, on the nature of knowledge convey the impression that uncertainty attending it adversely affects its dignity. But that it is not a very critical, reflective, and considered view becomes clear when we recall some other basic conditions that the theories of scientific knowledge are called upon to satisfy. For example, a theory of science is not regarded scientific in that strict sense unless it is verifiable or falsifiable; that is, somehow testable. Collateral to this *testability* condition or requirement is another: scientific theory must have *predictive* power. That is, it must be able to say something more, preferably much more, than that on the basis of which it is formed. Analyses of the said two conditions show—let us forget for the time being some other conditions that corroborate the same point—that dignity of scientific knowledge consists in its *openness* to testing by *other* findings, pieces of knowledges. Verifiers and falsifiers of knowledge are in no way irrelevant to it. For scientific knowledge is *not* self-evident.

The very necessity of *other* observation statements or relatively low-order theoretical statements for testing a particular theory arises mainly because of the openness or *incompleteness* of the latter. Notwithstanding the politeness of terms like *openness* or *incompleteness,* the fact remains that scientific knowledge is *not* self-evident. More emphatically speaking, *other*-evidentiality, that is, its testability in terms of *other* statements (observational or theoretical), is the hallmark of scientific knowledge. Once we understand this characterization of knowledge and its implication we have no option but to concede that knowledge is open to doubt: it is fallible. If the possibility of doubt or fallibility is recognized in principle, skepticism cannot be disowned—not, at any rate, in practice?

The same question arises, of course in a slightly variable form, in the context of the *predictability* condition of knowledge. One of the motive forces behind prediction is certainly the desire to test knowledge or, more precisely speaking, its theoretical moorings. However, another aspect of it should not be lost sight of. Prediction involves induction, inductive extrapolation, going beyond the given. Even in deductive testing, rightly understood, a "going beyond" is involved; it is indicated by such terms as *contentual enrichment*. The truth of the deductive consequences of a theory purported to test it add to its truth content. When the potential falsifiers (or observation reports) of a theory actually fail to falsify it, they "inductively" add to its possible consequences. The underlying logical anatomy of prediction and deduction is identical. The logical identity of the test structures cannot ensure a priori the *unity* of theory and its predictive—inductive or deductive—consequences. If the "consequences" of a theory are really intended to test it, the former can hardly be a priori declared to be consequences of the latter. The "consequences" may or may not confirm or verify the concerned theory.

They may infirm or falsify it. This *open* contingency, confirmation or infirmation, prevents us from substantiating the *unity* claim of theory and its *deemed* consequences. The unity of knowledge, like truth, is an ideal that is approximated. The metaphysical structure of the unity between theory and its supposed test consequences is largely (inductively) presuppositional, not experientially or phenomenologically available. But we keep on searching for the *possible* (ideal) unity in and through experience without being sure that there is *necessary* structure underneath, metaphysically sustaining the former. The resulting uncertainty or probabilistic character of prediction is a roundabout admission of the skeptical nature of our knowledge.

In a way the whole issue can be re-formulated as basically a human predicament, not exactly in the Humean sense. The reaches of the human senses are limited. About the very existence of supersensuous or paranormal capacities, said to be due to the yogic practices or God's grace, there are strong disputations by the critical-minded philosophers. Even those antiscientistic thinkers who concede paranormal (*aloukika*) modes of knowledge are not unanimous as regards their claim of public sharability. In a way the pejorative connotations of skepticism can be substantially dispensed with if we see it as the very expression of *human,* as distinguished from providential, character of knowledge. Elsewhere I have called it *anthropological rationalism.*[18]

Every form of skepticism, to my mind, is implicitly rooted in some concept of human nature or other. In a way Kant recognized this point very clearly when he asserted that what one can possibly know, do, and hope depends basically upon what one, that is, one's nature, is. Those who, like Śaṁkara, Leibniz, and Hegel, maintain that human beings are *essentially* gods and have potentially the capacity to know reality from the highest, transcendental, and, therefore, neutral point of view, do not encounter the problem of skepticism in the form it is discussed by Hume and Quine. Those who, like Descartes and Kant, think that the world of experience is purely causal and mechanical, encounter the problem in giving an unequivocal and straightforward antiskeptical epistemology. Situating humankind as they do in the causal mechanical framework, they show some hesitation in ascribing to it the cognitive capacity to know the real world in a definitive way. Therefore, they accord a lower ontological status to the world of science available to sense experience and testable by observation and experiment. At the same time, they resort to a transcendental argument purported to vindicate a higher form of knowledge, knowledge by identity or self-realization, for example, in which different forms of knowledge are said to be coherently available. As I will try to show, later on, in the next two chapters, even those who, like Hegel, do believe in the possibility of doubt-free knowledge have to offer an account why, at some stage or other, initial or intermediate, they have to encounter the shadow of skepsis.

Some of the reasons why most of the modern scientific and analytic philosophers are not deeply disturbed by the problem of skepticism may be noted in this way. First, they accept the current paradigm of science and confine their studies to bring out its philosophical, mainly methodological, implications and, in the process, explicate the nature of language and logic used by them. Second, they accept the current scientific theory of human nature, supported by the psychological and medical sciences. They accept this theory without much question. Third, the philosophers' disputes regarding the knowledge claim of the different branches of science are sought to be resolved by referring to the authority of the currently dominant paradigms of science. In a way most of our analytic philosophers, in fact, are philosophers of science. Knowingly or unknowingly they follow the Lockean underlaborer concept of philosophy. That partly explains why those who, like Husserl, Aurobindo, and Heidegger, prefer to philosophize in the so-called grand metaphysical style are not seriously taken or even carefully read. Even then questions of antifoundationalism, relativism, indeterminacy, and so on keep coming up. This is due to genuine diversity of human perception of what is genuinely scientific and what is scientistic. Finally, another issue that bothers the antimetaphysical philosophers is their anxiety to reach those of common sense. They wonder how the first-grade scientific specialists from whom they largely draw their inspiration succeed in reaching the common people in their quest to find out a suitable answer to this question. They want to strike a settlement between their philosophical views and the commonsense beliefs on the concerned issues. In a way philosophical anthropology, that is, studies in the nature of humankind and the sociology of knowledge, scientific and commonsensical, go a long way to define the scope and limits of the different forms of skepticism.

9. From Somatology to Sociology of Knowledge: from Experience beyond Experience

From slender empirical base we move, almost naturally, to varying degrees of generalized knowledge through inductive extrapolation or probabilistic expansion. In quest of knowledge we go beyond, we feel intentionally obliged to go beyond, the narrow base of what is obtained through sense experience. What is thus obtained has in it some transcendental cues. When we speak in the same breadth of the intentionality of experience and the transcendental cues of the given, we have to be clear about the relation of the two. Though our experiences, especially the immediate ones, are intimately related to, almost rooted in, our somatic existence, that is, body-mind complex, the limits of our body are not the limits of our embodied experiences. From the body we always go beyond the body. Our bodily organs perform the double duty of informing us of what happens in and around us and also of enabling us to try to get to the sources of the concerned information. Because many of our needs, their kinds and properties, are detected and identified in our bodily existence, the body is better equipped to know how to remove those needs. The traces of previous experience remain related, classified, and stored in the body. When we speak of the body's ability to reach out to the sources of needs felt in the body, this ability is nourished by the stored and sorted out information made available by experience.

In fact the human body works as a sort of antenna and transmitter rolled into one, two-in-one. Strictly speaking, the cues of experience are not exclusively "of experience," unless experience is construed in an abstractive exercise and as experience as such; that is, unrelated to its object and content. It is not that by exercising our extreme analytic acumen we cannot draw intelligible line of distinction between the "internal" content and the "external" object of experience. Both the content and the object of experience should be initially construed as bracketed because in reality they form a continuum. The main reason why the given (object) in the human context is said to be a myth is this. The given has not only an "external" address, it is almost immediately nourished, that is, interpreted, by internally available relevant classified information. Careful reflection on the given shows that its external address is not one. Numerous influences converge on it in a complex way. The internal aspect of this complex convergence is traceable with relative ease. But, rightly understood, experience is complexly related both subjectively and objectively.

The further we travel following the cues of experience, the broader horizons we see, the larger worlds we get into. This expanding world, in spite of its expansion, does not get totally detached from its empirical roots. Even when it is not empirically surveyable, we can somehow form an idea of it. This idea may well be criticized because of the partial invisibility of its fancied contours. This process of extrapolation and expansion is generally known as induction. The problem of induction is the problem of distance travel, distant from immediate experience. The invisibility of the fancied contours of the metaphysical world structure would not have bothered us if we were not already sold on the naturalistic idea that every fancied flight must have *an* empirical or factual anchorage.

We go beyond experience on the basis of our experience. The latter is not necessarily cognitive. The expansion of our world is not entirely due to the growth of knowledge or information explosion. Our will and its offshoots, expectation, need, hope, enlarge our world. They add wings to our thought closely tied to experience, and as a result of that the otherwise conservative thought tends to be daring and adventurous and tries to explore hitherto unknown distant horizons. The problem of induction, justification of inductive exploration, lies precisely here. To remain conservatively close to experience and have glimpses of the distant and larger world cannot be achieved simultaneously. If Hume's predicament (in relation to inductive justification) turns out to be essentially human in character, it is mainly because of our will to have the said two aims achieved simultaneously. We should be clear in our mind that slender empirical base of an expanding theoretical structure cannot have point-to-point, one-to-one, relation between its two ends, top and bottom. The problem is felt more acutely when on a narrow empirical base we try to construct, aided by imagination and speculation, a larger metaphysical world. This is another variation of the problem of induction.

The problem is sought to be resolved in various ways. One persistent suggestion is reductive; that is, to reduce the large theoretical world of science and even the larger metaphysical one into observational or direct empirical language. The problem is that, if we are brought back to the initially obtained empirical base and forced to commit ourselves only to observational language, we can neither construct higher level theories nor can we show their validity. If, on the other hand, we soar to the higher theoretical and metaphysical reaches, we are accused of being unfaithful to our experience. The worse accusation is that unfaithfulness to experience results in skepticism. Probabilism is also criticized as a sort of skepticism.

Reduction has other names and objects; viz., explication, simplification, generalization, logical parsimony, transcendental unification, ontological austerity. Broadly speaking, reductive program may be carried out in two different ways, linguistically and ontologically. Though some radical attempts, from opposite ends, have been made to show that, independent of language, ontology has no life of its own and that the human mind, *sans* language, has a direct access to (the structures of) reality. Heidegger and Wittgenstein, in some phases of their thinking, have defended the former thesis. One finds the latter one in Brouwer. But having gone through these two radical theses one gets the impression that the attempts to keep *reality* and *language* entirely apart are destined to end up in failure. Except in a very superficial sense one cannot be viewed as contained within the other. Neither "linguistic solipsism" nor "reality without language" make any coherent sense. One cannot be assimilated under the other.

The interesting question is how to understand the relations, simple and complex, between the two. There was a time when the issues related to reduction used to be discussed and clarified under the scopes of metaphysics and epistemology or psychology. It was customary in those days to discuss language under one of these (or both the) heads. In India right from the earliest times, as evident from the Vedas, the works of the grammarians and Patanjali, language has received close philosophical attention. If we are to believe Heidegger, the case was no different in the West. Only that aspect, until recently, did not receive its due attention.

Reduction program are initiated from two extreme ends and of two quite different forms. It may be of the top-down, that is, transcendental type or of bottom-up, that is, empirical, variety. The former structure is evident in Śaṁkara, Kant, and Hegel (under one interpretation) and the latter in Hume, Quine, and Goodman. First, psychologically speaking, the basic knowledge building block is sense impression. Second, ontologically speaking, the basic units of knowable objects are atoms. Third, linguistically speaking, the base of knowledge is word. Finally, semantically speaking, the basic unit of knowledge is sentence. It is not difficult to illustrate these four ideal types of reduction from the history of both Western and Indian philosophy. Mādhyamika Buddhist and Hume may be regarded as representatives of the first type. The second view may be ascribed to Democritus, the Vaiśeṣika, and the early Wittgenstein. The Naiyāyika and the causal referentialists like Kripke are proponents of the third view. And the fourth one are found in the works of Bhartṛhari, Frege, and the later Wittgenstein. One must not take these *ideal types* as historically accurate descriptions. Their value is only illustrative.

But, as we have already noted, neither reduction nor its possible outcomes, explication, elimination, unification, and simplification by themselves are enough to chase away skepticism. Every approach has its own share of problems. First, Nāgārjuna and Hume do not know how to firmly glue together the bits of transient sense experience. Second, Democritus, the Vaiśeṣika, and the earlier Wittgenstein could not give us a convincing accounts of how the atomic constituents of knowable states of affairs are configured—what exactly makes that configuration possible, durable, and soluble. Third, the history of original causal baptism, rigidly and insolubly relating names and nominata, is hardly more than a explanatory myth choreographed by realism and essentialism. To lend, as Nyāya does, God's sanction to the relation hardly improves the situation. Finally, sentential contextualism leaves its meant correlates indefinite and uncertain. Semantic holism of Bhartṛhari's *Vākyapadīya* fails to explain meaning speciation (of word, sentence) to our satisfaction. How does "universal meaning" individuate itself in smaller and graded contexts without loss of clarity or correctness of meaning?

One does not know how exactly to steer a middle course between the extremes of holistic reduction, metaphysical or linguistic, and empirical reduction of localism or minimal posits, ontological or semantic reduction. Must we extravagantly postulate, presuppose and turn a blind eye even to almost coercive sense experiences, or do we allow ourselves to get lost in the plenitude of pure experience without being able to make out its meaning? Whither to go?

10. Where Lies the Strength of the Repeatedly Refuted Skepticism?

In our Indian philosophical tradition the materialists (of the once influential Lokāyata-Cārvāka school) are said to have developed a strong antimetaphysical and an equally strong proempiricist theory of knowledge. According to them, perception is the only recognized source or means (*pramāṇa*) of knowledge. However, to accord recognition to perception (*pratyakṣa*) as source of knowledge (*pramāṇa*) does not mean that perceptual knowledge is beyond doubt. On the contrary, nothing is indubitable. Inferential knowledge is doubly dubitable. For *what* enables us to infer, that is, the ground of inference, is given to us in perception and, therefore, subject to doubt. Additionally, what is inferred (hypothesis or *pratijñā* or *pakṣa* lies even beyond the reach of perception.

To take the stock example, one can perceive only smoke on the yonder mountain but on the basis of it one goes beyond it and infers the existence of fire there. Smoke is *hetu* or reason or ground of inference and fire is probandum or *sādhya*. But the materialist, theoretically speaking, feels justified in being doubtful about the existence of a fire that is not perceptually available to him or her. As a matter of fact if the materialist, like Sextus and Hume, does not doubt the hypothesis, this does not validate it or removes its dubitability.

It is interesting to note that this stock smoke-fire example is not peculiar to the Indian philosophical tradition. It has also been referred to by the Greek skeptics like Sextus Empiricus. Some other typical cases and arguments have also been commonly used by the ancient Indian thinkers and the classical Greek philosophers. For example, the rope-snake illusion has been alluded to by some Greek thinkers like Carneades. Besides, the Buddhist logician's use of tetralemma in the context of the elusive nature of *nirvāṇa* has its Greek parallels. Whether the Greeks borrowed these ideas from the Indians or the converse is not a very interesting question for philosophy. Historically speaking, it is now generally agreed that from 300 B.C., if not earlier, there started a steady exchange of ideas between these two peoples through intermediaries as well as directly.

We find in the history both of Indian philosophy and Greek philosophy repeated and varied attempts to refute skepticism. This suggests unmistakably that this unorthodox trend of thought was influential enough to draw the critical attention of other thinkers opposed to skepticism or even fallibilism. Further, this also indicates that the simplistic or pedestrian, at time even vulgar, presentations of the materialist-skeptic position refuse to take, officially, due note of the strong points of this very old tradition. Those who tried to combat skepticism in India or Greece, it may be recalled, could not totally deny the presence of some sound elements of this time-tested school of thought. Even the wholehearted antiskeptics have to admit, willingly or unwillingly, that some common sense truisms have been highlighted in materialism. One must not judge it by its caricatured form—caricatured by its sworn opponents.

Even those who like Naiyāyikas are deeply opposed to the materialists cannot totally reject the latter's main epistemological thesis that, if we start from sense experience, the resulting forms of knowledge appear to be questionable. I would like to substantiate this point here by referring to two important concepts of the Nyāya-Vaiśeṣika school. I deliberately choose this school because it is more or less close to common sense, opposed to dizzy transcendentalism, and consistently professes realism. To my way of thinking, these three characteristics, taken together, are indicative of a thinker's willingness to meet the problems of skepticism on their own ground. In this connection I will briefly analyse the Nyāya structure of inference and the concept of causality within its limited context.

Strictly speaking, the Nyāya construal of inference is in a way basically causal and inductive. It is causal in the sense that there must be causal relation obtained between the probandum (fire in the mountain) and the ground of proof (visible smoke on the mountain). It is inductive in the sense that there is a passage from the known smoke to the (perceptually) unknown fire. That is what makes cognitive availability of fire *inferential*. The causal relation supposed to be existent between fire and smoke is the result of induction. This induction is based on two sets of evidences, positive (*anvayī*) and negative (*vyatireki*). In the example as provided by the Naiyāyikas one of the premises (*udāharaṇa* or instantiation) not only refers to the *presence* of fire in the kitchen but also its *absence* in the lake. This instantial premise seeks to confirm doubly, positively and negatively, the causal relation between the ground of inference (smoke) and what is to be inferred to probandum (fire or the mountain as qualified by fire). But for the Naiyāyikas an additional, fourth, premise (*upanaya*) is called for. Without it one might think that causal generalization by itself or in conjunction with only the first premise (*pratijñā* or hypothesis) and the second one (*hetu* or ground of inference) can warrant or justify the conclusion (*nigamana*). But that construal of inference seems unsatisfactory to the Naiyāyikas without invoking the fourth premise of *perceptual* application (*upanaya*) that explicitly states that *this* particular (*viśeṣa*) mountain, because of its smoke, is qualified by fire.

Even after having these spelled out or explained forms of the premises one can maintain that one has good reason to be doubtful about the validity of this inductively structured inference. First, skepticism is rooted in the proclaimed universal relation of pervasion (*vyāpti*) obtained between fire and smoke, or, more correctly speaking, between the fire-possessing character and the smoke-possessing character. How can one be sure that this generalization is *necessarily* true or exceptionless? To say that we have not yet seen any exception so far would beg the issue. Besides, the argument makes the generalization merely summative and its extrapolation very questionable.

Second, this smoke-fire pervasional relation is causally-inductively established and every case of its specific application is a case of extrapolation. Because it involves extrapolation beyond what is inductively available, the relational pervasion, the scope of questioning it cannot be foreclosed. The presumed ground of inference, that all hitherto observed cases of smoke have also been the cases of fire, by itself does not *entail* that the next observable case of smoke shall also be a case of fire. Even by mentioning two sets of evidence, positive (*anvayī*) and negative (*vyatireki*), one cannot change the basic logical weakness of the situation. The presumption of pervasional universality would still remain a priori and synthetic. The Naiyāyikas being opposed as they are to the concept of a synthetic a priori, how can they take this (*vyāpti*) relation as *necessary;* that is, noncontingent. If they take it otherwise—as contingent—the main plank of inference collapses.

Third, the case of instantiation has nothing more than a perceptual warrant behind it. Like all other cases of perception, this particular case of perception embodied in the fourth premise (of application) is subject to possible errors of perceptual judgment. At times this premise is said to be backed by *upamāna* (comparison or analogical) *pramāṇa.* Because its supposed strength depends both on right perception and right remembrance, and both are basically questionable, the situation does not improve by reference to comparison in addition to perception. Thus if one or more of the necessary premises of the inference remain contingent, the conclusion is open to doubt, at least in principle.

It may be true, as asserted by some Naiyāyikas, the sort of perception used in inference is rather peculiar in nature. Here through the perception of a *particular* (instantial) *sign* one gets access to, that is, comes to know of, the *universal* relation between two universal relata; viz., smoke-possessing character and fire-possessing character. To introdure this "peculiar" character of particular sign (*liṅga*) signifying universal relation, though ingenious, is not enough to stave off the possibility of perceptual skepticism. The solicited strength of universal relation (*vyāpti saṁbandha*) is not of much avail to salvage the situation.

Fourth, one may feel, in fact some writers have felt, the steps of the structure (*avayava*) of the Nyāya inference are complex and avoidably elaborate. For example, it has been alleged that the conclusion (*nigamana*), the fifth statement, "The mountain has fire in it," repeats the first premise (*pratijñā*) expressible by the "same" statement. Is it then not a case of mere repetition? Where is inductive augmentation of knowledge? The standard Nyāya response would be somewhat like this. "The mountain has fire in it" *as* hypothesis or probandum and "the same" *as* conclusion or proof are different in significance or intention. The latter has four warrants; that is, *pramāṇas,* in the forms of four premises behind it. Of the former the same cannot be said. Second, although the first premise *ascribes* fire to mountain, the conclusion *affirms* it. Apart from inference (*anumāna*) and perception (*pratyakṣa*), two other ways of knowing (*pramāṇas*) are also there, *śruti* (or verbal testimony) in the probandum and *upamāna* (or comparison) in the *upanaya* (or instantial application). So the conclusion or proof, rightly understood, is derived from the four premises (based on four different ways of knowledge) *taken together*.

Having heard the response the critic may still feel unconvinced. He points out that the epistemic strength of *comparison* depends very much upon that of perceptual tenability of what is being compared and with what it is being compared. Also the question of correct memory comes into the picture. Because both are under fire, as already mentioned, it is not of much logical significance to refer to them for the vindication of this inferential proof. So far as the probandum is concerned it is confessedly hypothetical (*śrutinirbhara*); therefore, its importance is purely provisional and contingent upon the (degree of) reliability of the "authority" in question.

Some of these criticisms are partially supported by ancient thinkers of the Śaṁkhya school who speak of the necessity of accepting additional five steps in the structure of inference. These steps pertain to (1) cognitive desire, (2) skepticism regarding the truth claim of the probandum, (3) the belief that the solution of the concerned problem is possibly available, (4) the aim of the investigation, and (5) the removal of doubt. The question of genuine necessity of these additional steps or premises has been rightly raised by some Naiyāyikas like Vātsyāyana and Udayana. But this very point has been raised also in the context of the inference structure containing five steps. Inference may be direct or indirect, prolix or precise, soritical or enthymematic. What matters most in ascertaining the *validity* of an inference is not the presence or absence of its spelled out form but the logical connections between the premises (explicit and implicit) and the conclusion. When the logical discourse is confined to the professionals, certain steps can be easily skipped or omitted because they are implicitly understood or silently shared. But in the context of *justification* or demonstration different steps need to be explicitly stated.

If one scans the suggested additional premises, at least one thing, already mentioned earlier, becomes very clear, viz., in the Indian tradition of philosophizing there is no hard and fast distinction between logic and epistemology, between formal logic and ontology. For example, some Naiyāyikas are of the view that what we call (in language) *logical steps* (of inference) are in fact ontological posits or segments of reality. In other words, logical connections are obtained not between statements or steps of inference but between states of affairs in reality. However, to insist on this distinction may give rise to avoidable confusion. After all the very fact that we need language shows that the relation between ontological entities or states of affairs cannot be meaningfully discussed, still less correctly ascertained and communicated without language.

Another basic thesis of Nyāya, highlighted by Candramati and Praśastapāda, for instance, states that all categories or reality are knowable (*jñeya*) and nameable (*abhidheya*). Therefore, to say that the structure of inference is exclusively or even primarily concerned with a *real* state of affairs and not with linguistic entities like statements is to ignore the very important role of language.

Besides, it goes against another important Nyāya concept of *liṅgaparāmarśa,* the relation between the probandum, on the one hand, and the object of inference pervading the ground of it, on the other. In brief, *liṅgaparāmarśa* obtains between a *particular* mountain and a *universal* relation between fire and smoke. To say this is to admit that a particular *liṅga* or sign can signify the universal relation making inference possible. By implication, this formulation of the concept performs two tasks at the same time; viz., it brings ontology back to its home, that is, language, and also it shows the ontological correlates of language.

This is an important strategy of the Naiyāyikas to refute the skeptic Buddhists' onslaught against his position. For the Buddhist logicians like Dignāga insist that inference is purely a judgmental affair and has nothing to do with ontological objects in the world. According to them, there is no ontology external to epistemology. If some blemishes are found there in knowledge it has to be removed by higher kind of knowledge like *prajñā* (intuition). To appeal to the so-called ontological states of affairs makes no sense to them.

The predicament of the Naiyāyikas over the scepticism about inferential knowledge centers around the nature of the relation of pervasion (*vyāpti*) obtained between the ground of inference and what is inferred. If it is insisted that the relata of this relation are particular (*viśeṣa*), then the generalized form of it is merely summative or restrictively enumerative. In that case inductive extrapolation beyond particulars (or even their summation) turns out to be hazardous; that is, skeptical. A consistent follower of this view has to admit ultimately that inference is always from particular to particular and that this inferential passage is made possible by certain rules of transformation that themselves need not be justified except in a weak sense; for example, as convention. This is the position one finds in the empiricist thinkers like Hume, Mill, and Ryle.

One grave difficulty of this position is that it cannot recognize any universal relation as *strictly* universal. To put it from the other end, according to this view, all universals are accidental and therefore contingent. In that case, the critic points out, the scope of plurality of causes is kept widely open. It is true that from the cause smoke we infer fire. But this is a factual truth. No logical necessity underlies and sustains it. One can always point out some cases of fire without smoke and also cases of smoke without fire. If this line of argument is accepted, inductive anticipation, prediction, and retrodiction turn out to be questionable. Second, it cannot rule out the possibility of a plurality of causes. The implication of admitting the plurality of causes is bound to be unacceptable to the realist Naiyāyikas, because it entails skepticism. Therefore, realists like Udayana try to show that causally efficacious relations obtain not between particular cases of smoke and particular cases of fire but between particular *types* of smoke and particular *types* of fire. Obviously this is a rearguard action to save the Nyāya realistic position against the Buddhist attack.

But this view, I suppose, has its own limitation. For the empiricist Naiyāyika without being inconsistent cannot claim that *all* the evidential conditions necessary for the production of effect are known, but has to concede at least in principle that besides the known conditions there may be some other hitherto unknown ones. In a way this makes the collocation of known conditions accidential and takes away, to some extent, its supportive strength needed to infer effect from cause and cause from effect. In view of this restrictive consideration it is difficult to endorse the strong stance taken by Udayana and other Naiyāyikas against the possible plurality of causes.

What one can advisedly do under the circumstances is to accept one of the following three lines: (1) causality is a law of psychological association, (2) it is a heuristic maxim, or (3) it is a mental construction or category used to organize and explain empirical phenomena. With Hume's name is associated (1). Ordinarily (2) is attributed to thinkers like Mill and James. In different ways (3) may be ascribed to Kant and the Buddhists like Dharmakīrti. For me the most instructive common point of all these lies in the *ultimate* inability to contain skepticism of some type or other. A Buddhist or a Kant says, "the causal world (or *saṁsāra*) is empirically real and reliably knowable but transcendentally ideal (or *śūnya*). The empiricists of the pragmatic persuation affirm nothing more than the *usefulness* of the causal maxim both in the fields of theory and practice. A candid Hume first concedes *theoretical indefensibility* of it and then, being careful as he is, hastens to add that it is *practically useful,* and this is primarily important.

The weakness of the doctrine of causal necessity becomes clear when we change the focus of our attention from the physical context to the human context, from event to action (*karma*). A necessitarian view puts both the Naiyāyika and the Buddhist into deep difficulty. Given that view, they cannot satisfactorily explain how human beings can possibly get out of the effects of their *karma* and attain liberation or *nirvāṇa*. In response to this very natural objection both Buddha himself and Gautama (the author of *Nyāyasūtras*) assert in different ways that the causes of human bondage, for example, ignorance, desire, imperfection, and evil habits and tendencies, can be known as well as removed. As heat can render a seed ineffective, similarly as the result of having genuine knowledge one attains liberation. Particular and successive *karmas* cannot be credited with the inexhaustible and permanent causal powers making freedom impossible. Several other considerations have also been offered to vindicate human freedom despite the causal or quasi-causal forces of habit, ignorance, and so forth. The very fact that one can know the causes of one's own ignorance and lack of freedom shows that one has the capacity to rise above them. The critical and sublative powers of consciousness make this rise possible.

In different ways all of them, critical thinkers of different persuasions, agree that a well-founded metaphysics or first philosophy of boundless scope is just not cognitively available. One may *believe* in it, or one may entertain a *faith* like that. But that does not make it rationally defensible. By highlighting this important point of agreement between rational thinkers one must not gloss over their equally important points of difference. For that is clearly against the basic spirit of critical inquiry.

To make the democracy of thought and action real and living we are first required to realize not only the futility but also the danger of such ideals as "universal agreement," "unanimity," "absolutely well-founded," and "apodictic certainty." Neither in science nor in philosophy can we actually realize these so-called ideals. On the contrary, the history of human knowledge is replete with contrary examples. The "same age" is differently understood and represented. The same view, despite its proclaimed rigorous proofs, is diversely interpreted. The lesson of all these developments is simple. Let us not deny ourselves the freedom of experience, expression, and interpretation.

11. How Is Experience Both Generative and Curative of Doubt?

It is very interesting to note that experience is both generative and curative of empiricism. Pure ideas, not nourished by or checked against experience, are static and empty. Those ideas, cognitively speaking, do not provide us any concrete sense of the world around us and our place in it. Nor can they work as motive forces of our action. Pure ideas are indeed empty.

We do not, rather cannot, live on the basis of empty ideas. Abstract scientific and mathematical theories are in a sense really empty. But, rightly understood, they are a skeletal anatomy, *figured* abstractions, of our concrete experience. Their fully fleshed-out figures are to be found in empirically applied theories and interpreted calculi, for example. Applicability and interpretability of abstract theories and formulas are indicative of the pregivenness of the world (of objects) in experience. Both "initially" and "terminally" ideas are environed and nourished by experience. Strictly speaking, human experience knows no clear-cut initial point or terminal point. A "boundless" world is prefigured in it. Therefore, in the context of human experience these expressions have to be taken with circumspection.

When it is said that experience is negative by its very nature, a very important point is made. Only in its skeletal form can experience be repetitive. Otherwise it is progressive or, occasionally, regressive. For example, mathematical formulas, unless applied or interpreted, are indifferent to experience. Looking at the anatomy of human experience many of us tend to forget that it is a contrived and remote representation of experience. Most of the contemporary scientific methodologists, deductivists as well as inductivists, are extremely inclined toward the formal abstract aspects of their approaches, apparently forgetting the phenomenological concreteness of the experience they are in effect called upon to deal with. This methodological one-sidedness of the scientific idealization of experience has drawn the pointed critical attention of Husserl. Referring to his view on the subject, Gadamer accuses him of this very one-sidedness, at least to a certain extent.[19] Husserl's effort to cure this illness, the one-sidedness of science, by tracing its empirical genesis seems very unpromising as motive force of our action. If scientific idealization, together with its attending abstractness, is to be blamed for its flight from historically tested, that is, socially dependable, experiences, Gadamer finds no good reason for praising the Husserlian proposal and attempt to vindicate scientific knowledge from methodological tyranny in terms of his "transcendental subjectivity," which claims to have the unilateral capacity to constitute the world(s) without consulting social experiences or the life world. No cognitive enterprise can effectively disregard its complex historical situatedness and yet be successful, relying only on methodological maxims, phenomenological or logico-empirical. Here it is interesting to note Gadamer's approving reference to Bacon to the effect that for reliable knowledge of Nature it is not enough to discard the "idols" or prejudices. What, in addition, is necessary is this: we must have both methodological rules of proceeding step by step from experience and socially reliable rules of interpreting our available experience. Simple enumeration will not do. Simple rule following will not do either. For valid knowledge we need a critical interpretation of experience.

9

Hegel and Heidegger on Skepticism: Some Problems

1. Two Aspects of Experience: Evidential and Explorative

Experience is both generative and curative of doubt. Doubt is due to the fact that the so-called Book of Nature, contrary to popular opinion, is not necessarily open. Often it is found to be very complex and closed. Underlying the complex phenomena lies their (more or less) simple law-governed structure. But even that structure is not manifest to the *human* eyes, to our *oculis rationis,* if any. By interpretation and reinterpretation of our experiences we have to disclose gradually their structure and the sub-structures nested within it. This process is endless. There is no rock-bottom structure. There is no going back to the so-called originary *ideal* science. By its very nature human experience is both forward-looking and backward-looking, retrospective and prospective, historical and explorative. Gadamer's criticism of Husserl on this point seems to be very insightful.[1] Neither phenomenological nor historical going back to the origin of science would help to free us from this defect of idealization. For experience *qua* experience, unless viewed as a part of a whole or a chapter in a historical tradition, appears to be of freakish character, meaningless.

The history of the scientific method shows that the pre-Baconian scientists were committed to remain close to experience so that they could claim certainty for their scientific theories. Consequently, most of them followed the method of simple enumeration, *enumeratio simplex*. It was left largely to Bacon to point out that starting from chance and stray experiences (or observation statements) we reach nowhere. Hasty generalization of everyday experiences yields no scientific theory. Bacon pleaded for *interpretatio naturae* in place of *enumeratio simplex.*

In our scientific exploration and progress, he argued, we are required to move step by step to tenable universals, simple forms of nature. The "particular experiences" that can lead us to universals are themselves implicit universals. Otherwise their "stray" character could not be undone and nothing universal could be derived therefrom. Logically speaking, it is impossible to obtain universals from pure particulars. This elementary truth was evidently known to the acute-minded Bacon. In recent years from Popper's negative methodology, marked by a pronounced anti-inductivism, one might get the impression that Bacon was unaware of the limitations of induction as a method of science.[2] To be fair to Bacon we must take note of his warning against the method of rash and premature anticipation on the basis of received opinion and prejudices. On this point we should closely look into what Bacon himself had to say. "Man, as the minister and interpreter of nature, does and understands as much as his observations on the order of nature, either with regard to things or the mind, permit him, and neither knows nor is capable of more."[3] Observations as such, that is, particular experiences, do not help us in understanding nature or mind. The *order,* which is already in nature and the mind (consisting of universals), has to be duly recognized. Otherwise, Bacon warns us, our generalization on the basis of chance experiences will give an incorrect idea of nature. Our interpretations of nature would be mistaken. That experience, unaided by interpretation, cannot be regarded as scientific was clear to Bacon. The subtlety and the complexity of nature are not instantly disclosed to us. The book of nature is not open: it has to be opened and read carefully. Unless the mind is purged of prejudices and remains conscious of the orderliness of nature, it will face difficulty getting into the secrets of nature. The very subtitle of Bacon's magnum opus, *Novum Organum,* "aphorisms concerning the interpretations of Nature and the kingdom of man," is instructive. The insight of Bacon opened up a new chapter in the history of scientific method. Opposed to the Aristotelian syllogistic logic, he laid primary emphasis on observational and experimental methods. But he was cautious enough to ask us not to indulge in rash and sweeping generalizations. For in that case we might wrongly be lured by metaphysical speculations, empty ideas, forgetting the ordered universals of the mind and nature.

Our understanding of nature is likely to prove *questionable,* if and when we fail to be aware of the prejudices that, almost unconsciously, seize us and influence our experiences and expectations.

Expectations or hopes, unregulated by experience, cannot free us from the influence of tradition on our mind. Uninterrogated or uncritical tradition, instead of bringing *new* aspects of nature to our attention, silently leads us along the path of least resistance. When our enquiry treads the beaten track and wants to remain always on the safe side, our knowledge is monotonously confirmed, not *critically* enriched, by experience. Unless experience proves more or less *corrective* of the already available experiences and their unity, its *confirmatory* character as such is not of much consequence to the growth or the enrichment of knowledge.

Experience that is merely repetitive and confirmatory can help us only in concept formation. But in cognitive quest what is more important is the correctness of the concept formed. This correctness can hardly be ascertained without applying the concept to actual experience. This question arises because experience has a duality within it. One aspect of experience is that it embodies some principle, a principle of expectation. This *subjective* aspect of experience is not the whole of experience. The subjective experience, marked by a sort of *expectation,* has something else as its object, the object that may fulfill the expectation in question. This indicates the *objective-negative* aspect of experience. Subjective experience needs objective experience to fulfill its negativity. Any and every object cannot remove this negation or fulfill this expectation.

The object that can do it is required to be determinate. The object that removes the negation of subjective experience not only removes the sense of need felt in experience but also enriches it, imparts a new character to it and makes it something new. In brief, experience has a *dialectic* within its own nature. When experience is construed devoid of this dialectic, it seems to be abstract, discrete, and empty. Without this dialectic we cannot explain how experiences are glued together, exhibit a universality, a growing universality, and do not fall apart. Unless we recognize this dialectical character of experience, experience ends up in the formation of concepts. And many of these concepts pertaining to a domain of experience prove to be inconsistent and occasionally even antinomous.

2. Hegel on the Dialectical Character of Experience: Self-Affirming and Self-Negating

Experience has two extreme phases, blind and antinomous. At one stage, it is interpreted as a blind manifold. At another stage, it appears to be self-inconsistent or antinomous. How can "the same" experience possibly lend itself to these two widely different interpretations? The answer seems to lie in a sort of transition from Kant's conception of experience to Hegel's. According to Hegel, experience has no life of its own. It belongs to someone's life, mine or yours. The question of blindness or antinomy arises only when experience is sought to be thought of in the abstract. Hegel construes experience as "skepticism in action."[4] The element of skepticism that emerges in experience is because of its negative-expectative character. In seeking to remove its experienced negation, experience goes out of itself. Because experience is necessarily human, it is we, whose experience we are talking of, that go out of the seeking circle of our experience and want to have the object sought after. The dialectic of experience, to start with, is the dialectic between the experience (subjective experience) and the experienced (objective experience). At the next stage, we discover that the dialectic is really obtained between the experiencer (the experiencing human agent) and the experienced (object). The object of the experiencing mind is "in-itself"; that is, has a status of its own. But this "in-itself" character of the object makes sense only in relation to the experiencing mind, a community of minds. If the in-itself character of the object is construed as something that has nothing to do with the experiencing mind, it becomes something like the Kantian thing-in-itself; that is, it loses its "community" and historical character. Only as the object of *our* experience can it answer our need, fulfill our expectation and impart a new identity to our experience. In this way our experience gains a dialectical and historical character.

The dialectical character of experience, as conceived by Hegel, shows two things. First, it does away with the static structure of experience as suggested by the categorical framework of Kant. Second, it highlights the importance of *negation* in the very nature of experience itself. But these two things make sense only under the presupposition that experience is of someone (or another) and addressed to something beyond its ken at any given moment. To speak rather impersonally of experience having negation within it and seeking transcendence beyond it makes no clear sense unless it is understood in its concrete human context. We are dissatisfied with our experience and want to transcend it, but not because of its negativity and inadequacy. In fact, the sense of negativity and inadequacy is of this or that person, the experiencer who is simultaneously conscious of what the experience is and what it is *not*. The positive aspect of experience is important because its presence makes one conscious of what one has and, through this consciousness, contributes to the realization of the goal sought after, the expected object. Without this positive base of experience it is not possible to be conscious of what *else* one needs to reach one's goal, to fulfill one's expectation. The sense of seeking a goal, going beyond the given experience, is essentially *skeptical*.

Negativity of experience is not only indicative of the skeptical mind of the experiencer or the knower but also of a promise held out to show the way to go and gain the necessary experience (of consciousness). The negative experience is not a roadblock or terminal point of a cognitive journey. It shows the inadequacy of available experience in attaining the ideal unity of knowledge and also suggests the possible experience that can compensate the need. So far as Hegel's dialectic of experience is concerned, the ideal unity cannot be completely attained by the finite human being. Through historical experience, humankind may, in fact does, progress further and further. But at no stage of our historical experience will we be able to be free from that skeptical sense that is born out of the gap between the positive experience and its negative goal. The goal aimed at is never completely realized. The wealth of experience certainly stands to humankind's credit. But the experienced person knows very well the imperfection of his or her experience. This awareness helps one to be free from dogmatism. But viewed from another standpoint, this rich experience makes one skeptical of the guarantee provided by that experience. The main value of experience consists in making us open to new experiences and in keeping us more or less free from a sense of dogmatism. But this skeptical disposition of the *historical* person opens up a new horizon to philosophers like Hegel. They think that at the human level there is no definitive way out of skepticism. Because of the finitude of human beings the ideal unity of the subjective experience and the objective experience cannot be fully or concretely realized. Hegel's philosophy of history ends up in a sort of skepticism and pessimism. Negative experiences, however instructive they might be, can push us forward only up to a point. But through the historical dialectic of experience we cannot reach the ideal land of complete certainty, the self-realization of philosophy, philosophy as the supreme science.

Philosophy of history shows both the meaning and the *inadequacy* of the meaning of history. In the realm of philosophy proper the job of the historian, whether the historian of science or social institution, is taken over by the philosopher. Philosophy here means self-knowledge, undifferentiated self-knowledge. At this level, the job of the finite person is taken over by the infinite God. In Hegel's philosophy of history, historical agents are essentially teleological instruments of God or Reason. In and through history human agents, interpreting their experiences, may certainly enlarge and enrich their world to a great extent. This process, taken to its logical end, tends to "melt" or finds its highest meaning in philosophy. History is the history of thought, the paradigm of thought is philosophy, the highest science, which knows no skepticism in it. The science without skepticism is neither historical nor human. The Hegelian solution of skepticism is superhuman and of little use to the sort of skeptical problem we are concerned with. But it provides us a deep insight into the nature of growing experience, human experience and its historico-dialectical meaning.

3. Heidegger on Hegel's Experience of Consciousness or Phenomenology of Spirit: The Limitation and Transcendence of Experience

We are told that for his book, *The Phenomenology of Spirit,* the title originally chosen by Hegel was "Science of the *Experience* of Consciousness." While the book was in press, another title was given to it, "Science of the Phenomenology of Spirit." Being very particular as he is about the precise use of words and their context Heidegger seeks justification of this interpretation of Hegel's philosophy (as science) in the titles of Hegel's main work. The said two titles of Hegel's work in question, read carefully, suggest that experience is essentially phenomenological disclosure of consciousness as Spirit.[5] But this Spirit is never available to the experiencer in its totality and entirety, providing a concrete and permanent, that is, non-historical, sense of certainty. Consciousness *presents* itself to the experiencer in different "shapes." Heidegger attaches special significance to the term *shape*. Spirit is available to us in and through experience in different shapes. It is unevenly layered. Negatively speaking, it is not phenomenal in the positivist, one-dimensional, sense. So, every presentation of consciousness or spirit in experience is marked both by negation and promise. Every experience is partly illuminating and satisfying and partly environed by doubt, a sort of drawing satisfaction-cum-doubt. Experience, on the one hand, puts us in doubt and the accompanying state of despair. The sense of despair would have condemned us to thoroughgoing skepticism, if experience of consciousness had been really phenomenal or one-dimensional. But Heidegger draws our attention to the original Greek meaning of *skepsis:* seeing, watching, and scrutinizing the presented beings of the Being.

The certainty-seeking duality of *skepsis* is evident in every experience. Every experience tells the experiencer what it is, how limited it is, and what cues it has for the experiencer to pursue in quest of the sought-after goal, certainty. The thinker or the experiencer is almost innately caught up in the crosscurrent of despair and hope, doubt and certainty. Experiences (of consciousness) come only to go away, appear in us only to disappear. This coming and going, certainty and doubt, define the very nature of consciousness for us. *Natural* experience gives us an insight into and yet keeps us away from *real* knowledge. The unity of the two that we seek keeps eluding us.

> Thus consciousness is in each instance a shape. *Skepsis* takes hold of consciousness itself, which develops into skepticism, and skepticism, in the appearance of phenomena, brings the shapes of consciousness forth and transforms one into another. Consciousness is consciousness in the mode of self-producing skepticism. Skepticism is the history of consciousness itself which is neither mere natural consciousness in itself nor mere real knowledge for itself, the first and foremost, in and for itself, the original unity of these two. . . . The history of consciousness brings forth consciousness in its appearance along with the image [of coming in and going away of phenomena]. . . . Skepticism in the process of accomplishing itself is historicity of history in the course of which consciousness works its way into the appearance of absolute knowledge.[6]

To Heidegger, skepticism is not a character only of knowledge as of this person or of that person; it has an impersonal character of its own. One might say that history is itself haunted as well as illumined by *skepsis*. Every age in its self-search and self-reflection perceives its imperfection and also its promised goal. In this respect, Heidegger is bound to remind us of Hegel. But once we recall the essential human underpinning of history, this impersonal image of *skepsis* is to be taken with circumspection. This qualification, however, does not apply to Hegel to the extent it does to Heidegger.

It is instructive that Heidegger's notion of skepticism has been developed, to a considerable extent, in and through his dialogue with Hegel. By way of *interpreting,* critically and creatively, Hegel's concept of experience, he shows that skepticism is an inalienable part of human nature, the nature of *Dasein* itself. Later we will see that long before his investigation into Hegel's concept of experience Heidegger was conscious of the inherence of skepticism in knowledge.

Somewhat following Hegel, Heidegger proposes to show how

skepsis is to be found in between the "natural consciousness" and "real knowledge." The former uses the term *consciousness* and *knowledge* interchangeably. But, rightly understood, consciousness clarifies knowledge and knowledge clarifies consciousness. They form a virtuous and developing dialectical circle. In Hegel's language *natural* consciousness shows itself to be *merely* the concept of knowledge or "unreal" knowledge. Unless the concept of knowledge is realized or unfolded, "real" knowledge is not available. But natural consciousness wrongly takes itself to be real knowledge. In this way it is alienated from its own nature or reality and thus gets plunged into falsity. Or, more appropriately speaking, natural consciousness's inability to grasp what it really is, its fall into unreal knowledge and yet its belief that it has attained truth, are not without significance. This movement of natural consciousness may be viewed as the way of doubt or despair. It is not "doubt" in the ordinary skeptical sense but, rather, a sort of departure from truth and yet a sort of passage to truth.

In natural consciousness one surrenders one's freedom to this or that authority, others' views, received opinions, and so on. Following this path, Hegel apprehends that one may have the vanity of this or that authoritative affiliation but no access to truth, honestly credible truth. In a way this approach of Hegel is reminiscent of the Cartesian method of doubt. Consciousness should follow its own detailed history and not another's authority. And only in this way might consciousness's own education take it to the level of science. Skepticism is important because through it one (as expression of Spirit) encompasses as well as interrogates the whole of phenomenal consciousness and ascertains its truth. Phenomenal knowledge, to start with, suggests that natural consciousness does not present us real knowledge. Consequently, one might form the impression that nature gets vaporized into the unreality of abstraction. But Hegel is earnest in his attempts to show that nature is not merely a concept, and that it is, in some way, real. The initial contrast between *natural knowledge,* knowledge achieved "without effort,"and *real knowledge,* knowledge of reality achieved as a result of seeking and striving, is posited not to be defended but sublated. The natural mode of consciousness is not present to us as object, as presence, with an accompanying sense of certainty. When the subject knows itself with certainty it is reality. When Descartes said that *ens verum* is the *ens certum,* he perhaps had precisely this in mind. The being of knowledge is present to the being of subject and this presentation is witnessed by Being as the measure or standard of the relation between the two, being-as-object and being-as-subject. The simultaneous presence of the two in and before Being is the point of sublation of the initial opposition between the two. Yet in a sense this point of sublation is a point on the path of knowledge, rather on the path for the quest of knowledge, and not a terminal point where the journey comes to a halt.

The duality of which we are speaking here may be viewed both in the "limited" human context (in the Heideggerian fashion) and in the "unlimited" Spiritual context (in the Hegelian fashion). To be cautious it is perhaps advisable here to bracket both the concepts, "limited" and "unlimited." So far as Hegel is concerned, he sees this dialectical development in the life of the Spirit. But to be fair to Hegel it should be mentioned that he does not contemplate the life of the Spirit in abstraction or isolation, disregarding the lives of the individual human beings in and through whom it actualizes or articulates itself. So far as Heidegger is concerned, the interplay between the thing and the being, the subject and the object, is construed essentially with the human paradigm in view. The changing identity of a thing can hardly be understood without its presence to being, and, conversely speaking, the openness of being is thing oriented. But even then his frequent reference to a distinction between beings and Being has led many commentators to ascribe to him an essentialist metaphysics that seems otherwise alien to his main existentialist trend of thought.

4. Life of *Skepsis* and the Ascending Shapes of Consciousness; Truth as Being-in-Itself and Knowledge as Being-for-Us; *Skepsis* as the Dialectic between the "Already" (Natural) and the "Not Yet" (Transcendental)

Both Hegel and Heidegger take note of and discern between the shapes of consciousness. A gradation of consciousness is between the life of the Spirit and the human life. The natural shape of consciousness is unstable and haunted by a sense of uncertainty. Its sense of uncertainty and its mode of instability are two sides of the same phenomenon. This phenomenon may be viewed both ontically and epistemically. Ontically speaking, Hegel would like to see the "demise" or end of this instability (of being) in the absoluteness of the Spirit, the supreme Reality. Heidegger, on his part, following mainly the footsteps of Descartes, returns to the human subject. But he realizes that this return does not mark the "demise" of *skepsis*. At no stage of its journey does natural consciousness appear to humankind or the Spirit as something that needs no further critical attention or scrutiny. Skepticism is there not only in the personal insight of an individual ego and its representation but also in the representation of the Spirit, unless and until the latter's self-encounter reaches the level of "unconditioned subjectness," the level where Being seizes the truth of beings with a sense of absolute certainty. In Hegel's system beings are objects, and when Being as reality attains absolute certainty, the former, beings-as-objects, are all actualized in it. This Hegelian construal of certainty, "demise" of *skepsis* and the end of histority do not appear to be acceptable to Heidegger. He has his own alternative construal.

Resolved not to commit himself to any transcendental reality that can present objects to human consciousness, Heidegger takes objects as human consciousness of a presence, the source of which is partly revealed (in the said presence itself) and partly concealed from the receiving human consciousness. Both the "giving" consciousness and the "receiving" consciousness are parts of the same consciousness. In a sense what is given as object to our consciousness is brought about by our consciousness. The former, the revealed, is not thrust upon us. In receiving it we form it. In its formation both its natural objectivity and our free subjectivity are conjointly at work. Also at work in this accepting-forming consciousness is *skepsis*. The *need* and the *will* of our receiving consciousness both guide and gaze at the arrival of the revealed object from the concealed area. *Skepsis* is symptomatic of what our consciousness lacks, that is, what it needs, and its deliberation at the suitability and otherwise of the shapes of the given.

Our natural consciousness is preoccupied with, almost lost in, the revealed objectivity of the unconcealed. It tends to forget or is somewhat blind to what Heidegger calls the *Being of beings*. This one-sided natural consciousness is both its history and its accompanying self-searching skepticism. Unless natural consciousness somehow succeeds in overcoming this overwhelming preoccupation with the given objectivity of the revealed, its skepticism proves merely repetitive (that is, barren), "absolute sophistry," and results in "empty nothingness." Here, Heidegger tries to impress upon us the inadequacy of natural consciousness in its one-sidedness, as a result of which its skepticism does not present us something new and needed by us. It seems that unless this natural consciousness can overcome its inertia, barrenness, and emptiness, it is condemned to die. Unless it succeeds in transcending its immediate existence, its end is inevitable. Interestingly enough, this "fall" or "death" seems to underlie the apparently death-conscious natural consciousness. Its surface indolence, "thoughtless indolence," is troubled by something underlying it, by something not yet revealed to it. The incompleteness of the shape of natural consciousness is disturbed and visited by something that was already there in a concealed form. When the indolent, if not dead, natural consciousness becomes a content of our consciousness, it gets back a life of its own, a needed impetus for progression.

Strictly speaking, our consciousness is neither totally lost in natural one-sidedness, stable objectivity, nor is it in the firm hold of real knowledge. Nor is there conjunction of two alien lifeless items, stable objectivity, and real knowledge. Our consciousness informs us rather of their unity. But this is a restless historical unity, marked by skeptical scrutiny and survey. In and through its shape of need consciousness is informed, however vaguely, of its goal. The vague awareness of the goal is not good enough ground for Hegel to admit that the goal itself is indefinite or indeterminate. In this view of Hegel one finds his commitment to a metaphysical view of consciousness of Reality that has little or nothing to do with the needs, strivings, and searchings of finite human beings and that has a prefixed goal of its own. To speak of *skepsis* as the center of tension of consciousness makes little sense unless this consciousness is emphatically described and recognized as basically human, freely human. In the context of natural consciousness Hegel himself gives the impression that skepticism is native to the nature of consciousness and has reference to human life. When he concedes that natural consciousness is not the heap of what is false, dirt, and in error, one feels inclined to accept that it does have a positive correlate. Fortunately, Hegel himself comes forward to affirm in a positive vein that natural consciousness is a seeking consciousness, not-yet-true consciousness, but engaged in seeking the truth.

The "I" is not barred from truth-seeking enterprises. Individual human beings are not isolated or barren "I's." Both as receiving and forming modes of consciousness they are interrelated and having a community of their own. Even at the level of natural consciousness this community of human beings is traceable and found to be engaged in a searching enterprise that aims at an awareness of the need to rise above the inadequacy of natural consciousness. Philosophy embodies this consciousness of transcending the dogmatism of phenomenal knowledge. Philosophy recognizes it as such and also overcomes it. In a way by its very nature philosophy as critique of the phenomenal knowledge gets into its nature, identifies its dogmatic limitation, and also discovers its promising cues. In this respect philosophy works as *skepsis* and accomplishes its work through detailed observation and through discovering the path of progression of natural knowledge.

Consciousness, for Hegel, is a divided, at least a seemingly divided, house. There are two aspects in it: consciousness distinguishes itself from what is for itself. To the latter it relates itself and yet differentiates itself from it. Consciousness thus turns out to be always consciousness of something, the latter is in the state of being known. The *being* of something *for a consciousness* is said to be *knowledge*. That which is presented to consciousness as content has a life of its own, and that life exists outside the relationship between itself and the consciousness for which it is a content. This side of the in-itself is called *truth* by Hegel. These two determinations of consciousness, knowledge and truth, are often differentiated as "being-for" and "being-in-itself." Heidegger bemoans Hegel's silence on the question of what exactly might be involved in these determinations. It was left to Heidegger to point out that "consciousness separates something from itself in such a way that it relates it to itself."[7] In consciousness distinctions are both drawn and overcome by consciousness itself. Here lies the *ambiguous* character and power of consciousness. For this reason our investigations into knowledge for its truth obliges us to move between truth as being-in-itself and knowledge as being-for-consciousness. Here we are put in a sort of dilemma. If we make truth of *knowledge* the object of our inquiry, we take, almost unconsciously, knowledge as being-in-itself, whereas in fact it continues to remain being-for-us. Thus we *invert* the positions of truth and knowledge and put ourselves in a sort of predicament. The truth that as being-in-itself was supposed to guide us in determining the truth of knowledge somehow becomes being-for-us. This is not the end of the story. As already indicated, the knowledge that as being-for-us has by process of inversion taken the place of being-in-itself, the original position of truth, puts forth, by implication, its claim to be its own standard (of truth). Does it mean that knowledge is a wanderer, wandering without knowing what is its standard of truth, or is it, without being conscious of its own identity and aim, seeking truth itself? If truth is there already within itself, what is the point in seeking it at all? Or, should we think that truth in its bid to be knowledge sets a new goal for itself, and similarly, knowledge does the same thing? Or, is it expressive of the very nature of consciousness in its self-differentiated form, as being-in-self *and* being-for-us.

The whole problem seems to be rooted in the fact that consciousness cannot go behind truth, being-in-itself, and ascertain what sustains it. The other aspect of the same story is that consciousness is equally unable to get behind knowledge, being-for-us, and ascertain what sustains it. These twin inabilities are present in and before consciousness and therefore their negations, the said inabilities, are not absolute but have positive implications. This is evident in Heidegger's remark: "in the nature of consciousness, knowledge and the object are split apart and yet can never part."[8] Natural consciousness reaches out to be an object as a particular being and then it goes to its knowledge of it as a being and remains there. This consciousness may be called *ontic* because it is derived from the Greek work *on*. This consciousness is a dynamic or historical being-toward. By thinking we, humans, impart new life, new impetus, to thought.

Thought is in our thinking itself. *On,* being-toward, is self-concealing. It is both present and the presence. The said ambiguity is self-concealing. To the efforts to reveal the meaning of the duality of what is present and the presence of being as thought, says Heidegger, we owe the beginning or history of Western metaphysics. Consciousness in its immediate representation of beings is ontic. Ontic consciousness is preontological. The mode of gathering together of beings into their beingness is signified by *ontological*. Before ontological truth in its differentiation from the ontic truth is available to consciousness they are in a way already together. But the togetherness, hidden togetherness, is in the way of disclosure or disconcealment in consciousness, in consciousness as knowledge. This knowledge-consciousness has to be distinguished from natural consciousness. The latter or the received opinion, being indolent, does not propose to get to the hidden. On the contrary, it hides itself behind the things that appear to it as true. It discourages *skepsis* to see the truth "behind" the truth. In its authentic moment *skepsis* may see the truth "before" the truth. If *skepsis* rightly performs its job, it can see the "nature of consciousness" as "the original unity of ontic and preontological representations." Consciousness succeeds in grasping its own truth.

Skepsis by its very nature is comparative, foresightful, and yet circumspective. But unfortunately, Hegel points out and Heidegger endorses, philosophy at times moves in rut, ceases to be skeptical and fails to be comparative, circumspective, and foresightful. Those are the moments of stagnation, regression, and backsliding in the life of philosophy. Without *skepsis* philosophy loses its necessary insight, hindsight, and foresight.

Unless consciousness, which is naturally natural, cannot rise above or go beyond its naturalness, it cannot discover its own standard of truth and give the same to itself. And without this standard it cannot assess and ascertain the progress of its journey toward and approximation to truth. It is particularly for this reason that consciousness as *skepsis* needs to be skeptical or self-critical. This need is not imposed from without, but is within the life of consciousness itself. In a sense it is natural to consciousness. In a sense it is beyond the naturalness of consciousness and therefore consciousness aims it as truth. Paradoxically enough, it has to be admitted that truth is "already" (naturally) in consciousness and "not yet" (transcendentally) present in consciousness. *Skepsis* is the dialectical movement between the "already" (natural) and the "not yet" (transcendental). Interestingly enough, the inside view of *skepsis* shows that "already" is animated by "not yet." And the latter presents itself in and as the former. If *skepsis* is the dialectical movement between the "already" and the "not yet," consciousness itself is *dialogue* between natural knowledge and real knowledge.

Resolved to deal with the *experience* of consciousness, the main subject matter of philosophy as science, Hegel looks into it rather penetratingly. Experience in the abstract fails to tell us of the nature of the object with which it is concerned. Experience as experience of some object proves informative or cognitive. The object of experience is situated in and surrounded by other (possible) objects. The object of experience draws in and presents new objects to experience. The new object shows the untruth of the old object. But, interestingly enough, truth of the new object is impossible without the untruth of the old object. Experience of *natural* consciousness, however, focuses its attention on its object and does not wander about and around what makes it possible. When natural consciousness brings to experience what sustains it, it ceases to be a mere representation and establishes itself as representation of Being. In other words, representation represents Being and in a way represents itself to itself. Representation thus shows itself to consciousness as an "inversion of consciousness itself." This inversion is "for us" and, at this stage, consciousness ceases to be natural. Inverted consciousness as the appearance of phenomena indicates something *else,* something beyond. Inverted consciousness works as *skepsis,* drawing our attention to other phenomena, other than those presented immediately to consciousness as experience. *Skepsis* performs two jobs: it gets hold of appearance or phenomena and, at the same time, reaches out to the absoluteness of the Absolute.

The inversion of consciousness takes us behind the back of natural consciousness. From this inversion we have repeated representations of natural consciousness and what lies beyond them and make them possible for us. The inversion inducts our experience into the process of progression with which science proceeds, dialectically disclosing Being to us. Being is not presented to us "automatically." *Our* inversion contributes to the presentation of Being to us. Even if we decide not to contribute, a contribution (in the form of inversion) takes place. Because we cannot totally enclose our consciousness, the Being of beings continues to be surveyed and anticipated by the *skepsis* that is inherent in our consciousness, Heidegger remarks: "the contribution of the inversion of consciousness consists in letting phenomena appear as such."[9]

The intriguing point to be noted here is that the phenomena that appear to us are always in the nature of steady invitation, anticipating and inviting other phenomena that are in a way pregiven in them. Phenomena invite no *stranger,* nothing *alien* to them. The invitees are kin and *willed.* Experience of consciousness makes apparent its *inadequacy.* The sense of inadequacy has *force* in it and assumes the form of *will.* When Hegel speaks of that experience as the movement that consciousness exercises on its self, Heidegger understands it as the "prevalence" of that force as will that wills the Absolute. Here the term *will* is not to be construed in its ordinary sense. The inversion of consciousness in experience to which we are obliged to contribute makes us increasingly conscious of the absence of the Absolute in and through its parousia.

> *[O]ur nature is itself a part of the parousia of the Absolute.* The inversion is the looking of skepsis into absoluteness. It inverts everything that appears in its appearance. . . . In the inversion, the presentation has the absoluteness of the Absolute before it and thus has the Absolute with itself. The inversion opens up and circumscribes the space of the historical formation of consciousness. In this manner it ensures the completeness and progress of the experience of consciousness.[10]

This progression is steady and skeptical. Because the Absolute in its historical formation is not absolutely present to us, not accomplished by or identical with us. In the course of the skeptical journey of the inversion of consciousness we are engaged in a *dialogue* between natural consciousness and the absolute knowledge. We cannot disengage ourselves from the dialogue. Because the inversion of consciousness is somehow informed of the absoluteness of the Absolute, it does have *skepsis* within it. The role of *skepsis* can be viewed from two ends. From the end of natural consciousness it works as a critical and propulsive force. From the end of absolute knowledge it works as an informative and drawing force. In either case its progressive and dialectical character remains undisputed. The dialectic movement of the experience of consciousness is opened up and sustained by the inversion. The same inversion also keeps the Absolute partly concealed from us. But, paradoxically enough, this is an *open concealment,* the concealment discovered by and disclosed in our experience. *Skepsis* forms the heart of the dialectic between the open and the concealed.

5. Experience as Engaged in Its Search for Truth is Naturally Skeptical

This *skepsis* has also its speculative flight capacity without which experience would have gathered itself completely in its natural consciousness. The secret of experience's capacity to take consciousness beyond the given (natural consciousness) is to be found in *skepsis*. It combines the dialectical-speculative powers of consciousness within itself. As said earlier, the skeptical powers are appropriative, critical, and foresightful. When Hegel changed the heading of his book from "Science of the Experience of Consciousness" to "Science of the Phenomenology of Spirit," he apparently did not grasp the non-phenomenalistic, dialectical, and speculative powers of *experience*. He was perhaps unduly obsessed by the unidimensional or phenomenalistic connotation of experience. In fact, Heidegger points out "the nature of experience is the nature of phenomenology." In brief, experience by its very nature is phenomenological. Disclosure is its very nature. But disclosure as such makes no sense unless there is something to be disclosed. Phenomenology is the skeptical experience of the Spirit and its parousia. The Spirit that is disclosed for skeptical experience of phenomenology is its subject, not its object (lying ahead of it) to be realized. The Spirit that is experienced is never available in its fullness or absoluteness. That explains the inherent skeptical nature of experience, inadequacy, and the searching character of it.

Natural consciousness by self-inversion moves toward its own home. This movement is a sort of appropriation. The concerned experience is the representation of the Absolute. The return movement of natural consciousness through its representation aims at releasing within itself the knowledge in which it is already there.

When the inversion of consciousness is achieved by us, Spirit appears to us, turns toward us. That shows natural consciousness is within us and not without us. It cannot be introduced where it already is. But this homecoming of natural consciousness must not make us think that it has left behind its abode among beings. The immediacy that marks the presentations of the objects of natural consciousness carries a sense-certainty within it. To start with, this certainty seems to be the Absolute. But the truth of certainty is different from the sense of it. The truth of certainty appears gradually through the experience of consciousness. Truth is mediated through experience, experience of the objects of natural consciousness. Objective differentiation of natural consciousness is gathered together in our knowledge, *our* knowledge as articulation of Spirit, the presence of Spirit in us. Truth is the whole of objective differentiation of natural consciousness as present before Spirit through our consciousness. But our consciousness by its very nature cannot have the whole as speculatively affirmed. In and through experience, this whole of natural consciousness in Spirit is steadily and gradually released. To speak of the truth of the whole in the abstract makes no sense to Hegel. Only in and through the experience of that whole, the bracketed whole, is truth gradually gleaned.

The gleaning may be shown both formally-dialectically and historically. But, to Hegel, these two ways are two aspects of the same process. The process is initiated and propelled by an inner necessity, a will, which is in the very nature of reality. Skepticism is born of (our) consciousness's will to get to truth. This will is native to (our) consciousness. The term *our* is bracketed here in order to indicate that our will to get to truth is still an aim to be realized, not yet actually realized. Consciousness is still in need of a *concrete* sense of self-fulfilment. *Skepsis* is lodged in our consciousness, manifests itself in the form of will and need. Because consciousness, through self-differentiation, not as such, not as an undifferentiated whole, wills to have itself concretized (as Spirit), it is followed, inspired, and sustained by *skepsis*. *Skepsis* works here both as critical capacity and as confirming capacity. The aspiration of *skepsis,* like that of consciousness itself, is to reach out to truth. In a way, skepticism is truth *seeking,* preparatory to and part of truth itself. In the whole of truth, in absolute self-knowledge, *skepsis* is *not* to be found. The end of skepticism does not mean refutation of skepticism. The expression, *refutation of skepticism,* makes no sense either to Hegel or Heidegger. For, this sort of formulation of skepticism might suggest that *skepsis* or doubt is the absolute antithesis or negation of truth, whereas, in fact, it is a *reflective* survey of, insight into, aspiration of consciousness for truth and also its "fulfilment." Contrary to popular belief, in its "fulfilment," skepticism does not really *die* or suffer *destruction*. It relives in our searching and reflective consciousness.

> A sceptic can no more be refuted than the Being of truth can be "proved." And if any sceptic of the kind who denies the truth, factically *is,* he does *not* even *need* to be refuted. In so far as he

> *is,* and has understood himself in this Being, he has obliterated Dasein in the desperation of suicide; and in doing so, he has also obliterated truth. Because Dasein, for its own part, cannot first be subjected to proof, the necessity of truth cannot be proved either. It has no more been demonstrated that there ever has "been" an "actual" sceptic (though this is what has at bottom been believed in the refutations of scepticism, in spite of what these undertake to do) than it has been demonstrated that there are any "external truths". But perhaps such sceptics have been more frequent than one would innocently like to have true when one tries to bowl over "scepticism" by formal dialectics.[11]

Heidegger's conception of skepticism is an inseparable part of his conception of truth (of knowledge). Because he defines truth basically in terms of humankind's Being-in-the-world and rejects the view of truth as the locus of judgmental assertion and also the view that truth is correspondence or agreement, he is committed to the finite or incomplete nature of truth. In his own language, truth is the gradual uncovering of humankind's Being-in-the-world. In a sense what is primarily true is the individual human being, the partially uncovered Dasein. Given this partial or incomplete character of truth, Heidegger finds himself unable to accept Kant's notion of firm truth defined in terms of the agreement of knowledge with its object. Also he rejects the neo-Kantian epistemology that defines truth as an expression of "methodologically retarded naive realism." Truth is to be understood neither as the subject-object relation nor as the real (judgment)-ideal(content) relation. For this sort of formulation presupposes a sort of dichotomy between the judging person and the judged objects and contents of the world. The rejection of this dichotomy is manifest in Heidegger's recognition of Dasein as the primordial locus and focus of truth. His view also requires him to reject the conception of truth as *adaequatio* (likening). In judgment there is no evident agreement between itself and what is judged; neither on the basis of picturesque correspondence nor on likening can it be established.

6. Truth Is Our Incomplete and Historical Self-Disclosure: How Incompleteness and Histority Instead of Destroying or Obscuring Truth Foster the Spirit of Quest for It

According to Heidegger, truth is uncoveredness, Dasein's disclosure. In truth Dasein is disclosed but not fully. Undoubtedly, truth appears before humankind, but it remains always partly hidden. The most clear forms of truth are to be found in our *thrownness,* being definitely situated in the world, *projection,* our self-presentation as potentiality, and *falling,* being shadowed, disguised, hidden, and lost in and by other things and beings. Heidegger finds in our *falling* the prevalence of untruth. This notion is also intended to show that humankind is required not only to uncover truth but also to keep its shine undimmed against the truth of a fall or hiding or disguise.

By our very nature we are partly in truth and partly remain lodged in or exiled into untruth. However, this lodge or exile is not permanent. "The borderline" between truth and untruth is available to us only from within; there is no external criteria to define or show it. For we can never totally uncover or cover up the things of the world, the possible objects of knowledge. Without any reference to Dasein the truth of the world does not assume any significance. At times Heidegger says that truth belongs to the basic constitution of Dasein, and again at times he affirms that Dasein is essentially in the truth. Perhaps what he means is this: Dasein is in truth *and* truth is in Dasein; this conjunction is made possible by the uncovering or disclosive character of both Dasein and truth. To highlight the Dasein-based character of truth he goes to the extent of asserting

> Newton's laws, the principle of contradiction, any truth whatever—these are true only as long as Dasein *is*.Before there was any Dasein, there was no truth; nor will there be any after Dasein is no more. For in such a case truth as disclosedness, uncovering, and uncoveredness, *cannot* be. Before Newton's laws were discovered, they were not 'true'; it does not follow that they were false. . . . To say that before Newton, his laws were neither true nor false, cannot signify that before him there were no such entities as have been uncovered and pointed out by those laws. Through Newton the laws became true; and with them, entities became accessible in themselves to Dasein. . . . Such uncovering is the kind of Being which belongs to 'truth'.[12]

To say that truth is Dasein based is to suggest, among other things, that it is *historical,* questionable, and corrigible. Ontologically speaking, Dasein is historical. It is so in two different (but inseparable) ways. The past is in it as present and the future in it as potentiality. Given this basic historical character of Dasein, the concepts of permanent science and eternal truths are ruled out by Heidegger. Scientific truths are essentially fallible and derivative, derived from the historical character of truth.

A related misconception is also sought to be dispelled by Heidegger. The Dasein-based nature of truth might be construed as *subjective* in a relatively superficial sense. Perhaps it is. For without making any reference to Dasein's state-of-mind one cannot make any sense of what truth is. In this limited, very limited, sense, truth may be taken as subjective. But it is *not* subjective in a very important sense. Dasein's encounter with the entities of the world is not independent of Dasein's state-of-mind. It is not within Dasein's discretion to face or not to face the entities of the world. Dasein itself has its being in the world. The fact that Dasein can uncover entities in themselves and in an intelligible way is not an act of choice of Dasein, rather it belongs to it. The alleged subjectivity of truth is often due to confusion between Dasein and subject, between Dasein and substance. Reification of Dasein as subject or substance contributes to this misconception. Without a correct understanding of the human-world relationship one cannot grasp the nature of truth.

Truth is not a creation of Dasein, humankind's being. It is in a way presupposed by it. But the term presupposition is in need of elucidation here. Truth is presupposed by *us,* our Dasein as being. Only in this sense can we not understand ourselves without disclosiveness, our disclosiveness to ourselves, which itself is truth. Truth is not a presupposition external to or above us. Truth is a part of self-disclosure and makes it possible for us to presuppose. Truth not only makes our presupposition possible but also enables us to understand the ground of other entities and their interconnections through their disclosure. Neither Dasein nor the entities of the world can hold back their truth. Because this truth, as already pointed out by Heidegger in his existential analytic, is not eternal, it has to be construed as essentially historical. Histority of Dasein-relative truth is consistent with the "fanciful idealization" of permanent science and "reification" of human reality as subject or substance. This idealization or reification also tends to foreclose the basic disclosive character of Dasein.

If Dasein is in the truth, so is certainty. Heidegger ascribes double significance to the notions of truth and certainty. Truth means Being-disclosive; the uncoveredness of entities is derived from it. Similarly, the basic significance of certainty consists in Being-certain. In its derivative significance any entity could be taken as certain provided it is present to Dasein as certain. All entities cannot be equally certain irrespective of their kind and Dasein. "The kind of truth, and along with it, the certainty, varies with the way entities differ, and accords with the guiding tendency and extent of the disclosure."[13]

Certainty gets reduced or lost to the extent the disclosive tendency of Dasein is obstructed, held back. What is historical, fated to be disclosed, to undergo change, cannot be held back, suppressed, without distorting its nature, without consigning it to the ashes of untruth. In this formulation of truth and uncertainty, Heidegger, following Nietzsche, introduces the concept of death as an element of tragedy. Truth, as understood by him, has a poignant and tragic sense in it. The disclosiveness of Dasein in which truth is available, shiningly perceived, does not come of always automatically or as a matter of course. One has to overcome the routinized factuality of the daily life. When one is covered by the routines, modes, and manners of the daily life, one fails to be disclosive of truth. Similarly, when one gets lost in the lives of others, one is not engaged in uncovering one's own self. When Dasein, humankind's being, is enveloped by the lives of others and their routinized ways, its own true nature and historical fate becomes obscure to itself. As a result, it is prevented not only from seeing the shining face of truth but also becomes uncertain of what is going in within itself and what lies ahead of itself. Curiously enough, in the poignancy of *lonely* death consciousness Dasein has the clearest view, rather preview, of truth.

7. Life, Death, Authenticity, and Truth: A Reconstruction

That death and truth are very intimately related has been undoubtedly emphasized by Heidegger. But exactly how *skepsis* comes into the picture has not been spelled out by him. It is not difficult, however, to reconstruct or interpret it. In facing death, in death consciousness, the human being is most authentic. Here the concept of authenticity is not to be taken only in its moral or axiological sense. The other connotation of authenticity, truthfulness, is also being clearly indicated here. In death consciousness one cannot remain hidden behind others or among things (of the world). Each person is fated to die. The circumstances of life and the facts of the world are all in a way paving the way to death. But in an important sense one is free to be *oblivious* of one's Being-toward-death. However, this is not to suggest that human obliviousness of these facts and circumstances can in any way bring their workings to a stop. Though one's fate is unstoppable and untransferable, one may not be always and adequately conscious of it. This lack of consciousness is a mark of relative inauthenticity of life, relative exile from truth. It is in and through one's Being-toward-death that one can be free from this authenticity and return from this exile. To live authentically means to own up to one's morality, to live even conscious of the death possibility. How lucidly Being discloses to us depends on how deeply can we own up to our mortality. Death is our most intimately known possibility. To use Heidegger's word: "Death is Dasein's *own most possibility.*" In this possibility one is lonely, nonrelational. In this possibility death ceases to be fearful to life. In this possibility death ceases to be a victory for life. By recognizing death as true one experiences the sort of certainty that has nothing to do with liberation or destruction. In death one's Being is proved neither victorious nor vanquished. In a deeper sense humankind can never be certain about death as an *actuality;* it is not available as a content of experience. We know of it as a part of our being, as a *possibility* of our being. Toward one's own *possibility,* own most *possibility,* one cannot be definite or certain. Death is one's mode of being *there*. But one can never be sure of the nature of that mode. This can be described both as a sort of indefiniteness and glimpses of freedom.

Neither in relation to life nor in relation to death can one be very sure and definite. In Heidegger's style of thought both life and death are available to humankind mainly as *possibilities,* possible *phenomena.* But neither life nor death as *phenomenon* is present to us as a given and fixed fact. Both are present to us as open-ended *phenomena.* Even in our most authentic moments of life we are "disturbed" by the force of freedom and cannot stand still or idle. Though it is said that our death consciousness is very authentic, this authenticity is not given to us in a fixed form. The sense of indefiniteness or uncertainty that seems to attend invariably our sense of life and our sense of death is rooted in our inalienable or unbarterable freedom. This ontological freedom is the epistemological counterpart of *skepsis.* Whereas *skepsis* environs knowledge, corrects it, and propels it forward, freedom unites our being with the past, surveys the present, and anticipates the future. But it must be pointed out that freedom retains something of the past in the present, fails to survey something that the present hides from us, and prevents us from being sure about our own fate. Somewhat analogously, *skepsis,* although it critically reviews the past, does not discard it completely; its insight into what is currently available is partly appropriative and partly dismissive. *Skepsis,* in spite of its inherent inconstancy and uncertainty, anticipates the future, explores it, and what is more, appropriates it. Because Heidegger's "epistemology" is embedded in his ontology, one finds the positive role of *skepsis* contained in Dasein's freedom. The Dasein that is doubtful is free from doubt. The Dasein that is free is free to doubt. By its very nature Dasein is skeptical and free, explorative and creative.

One of the basic points of Hegel's dialectic toward which I for one feel very drawn is its emphasis on the incompleteness of any view, even of the review of the concerned view. That logic is concerned not merely with the laws and truth values of thought and that it has within it an inherent incompleteness are brought to our notice by the dialectic of Hegel's logic. From the Greek conception of *logos* Hegel came out very boldly and creatively at least in one important respect. The tension of the pre-Socratic philosophy often framed in terms of conflict and cooperation, that is, dialogue, between the defenders of the Parmenidean *Being* and the Heraclitean *Becoming* is somewhat relieved in the Aristotelian *telos*. Kant's antiskeptical anxiety, mainly due to Hume's criticism of Newton, results in an antihistorical structure of knowledge, which is clearly antidialectical as well. Hegel wanted to retain the antiskeptical orientation of Kant but give a very positive and creative interpretation to dialectics. Though he uses dialectics as a method his view on the method is inseparably related to and expressive of the very nature of thought, thought as reality. His dialectic is marked by development and histority. By interpreting *logos* historically and dialectically he wants to show that the details of thought, both subjective and objective, in spite of their incomplete, uncertain, and even irrational look, are essentially systematic or organic.

Hegel's concession to "dialectical skepticism" displayed by developing and incomplete thought, reflections on thought, views and reviews, has been interpreted in two different, almost opposite, ways. On the one hand, the transcendentalists like Husserl, basically interested in establishing philosophy as rigorous science, tell us that Hegel made too many and unnecessary concessions to the skeptic's criticism of valid scientific knowledge. On the other hand, old positivists like Comte and new positivists of this century are severely critical of Hegel on the alleged grounds of his *speculative* retreat to the Absolute Reason not available in or testable by human experience. It seems that Hegel could anticipate these two possible criticisms against his position. For, he knew very well that he cannot easily get out of Kant's formal and close structure of the metaphysics of experience (backed up by the ultimate validating principle of the synthetic unity of apperception) without introducing the concepts of dialectical *history* and *human* experience. The implication of dialectical history is clear: reality as a total *whole* is not available to the experience of the finite human being, but by human understanding varying historical distances are covered and the past events and their rationalities are recovered. In spite of all its baffling details, history does not appear incurably skeptical to Hegel. True, historical events in their discrete particularity may appear irrational, contingent, and therefore uncertain, but Hegel tries to show us that in their *unity* all events exhibit definite rhythms and patterns. Neither all rhythms nor all patterns are available to our experience as a *totum simul,* as a whole. They are disclosed to us, discovered by us, gradually, dialectically-historically.

8. Hegel's Defense of (Historical) Experience of (Transcendental) Consciousness, Marked by *Skepsis* and the Cunning of Reason; Heidegger beyond Hegel; Transcendence as Human and without the Absolute

Hegel's emphasis on history may well be construed as a genuine concession to the claim of human experience and human performance. It is only human, not God's, *experience* that is in need of gradual supplementation and justification. It is only human, not God's *performance* that is in need of completion and justification. If God's own self-consciousness is taken as the paradigm of knowledge, the certainty of knowledge, the concept of *gradual* history becomes superfluous. Only in the human context do the concepts of gradual disclosure and historical discovery make good sense.

It may be recalled here that in Kant's "Idea for a Universal History" one finds that history is swallowed up by idea, details are lost in unity. Analysis shows that this notion of history is more in the nature of an ideal (to be realized in and through history) than in the nature of description of what is historical. In fact its objective is to avoid and preempt historical skepticism.

Undoubtedly, a similar perception underlines Hegel's own historiography. The main point at which he differs from Kant is that he wants to show that the Kantian ideal is not realizable without reference, explicit or implicit, to human agents, their experience and performance. Simultaneously, he is aware of the possible charge of skepticism that could be raised against the introduction of the irrational and fallible human factors in the "scheme" of history. One feels that his dialectical method is itself a partial answer to this criticism. In terms of this method he wants to show two related things: (1) the human details of history are rationally intelligible, not entirely a handiwork of the "Cunning of Reason"; and (2) even the inverted cunning works of Reason are, after all, works *of Reason.* By (2) Hegel seeks to take away the skeptical elements of historiography introduced or at least implied by (1). But the positivist critic of Hegel may affirm that, in fact, like Decartes and Kant, Hegel, strictly speaking, is committed to the paradigm of self-consciousness as knowledge; that is, *God's* self-knowledge as the paradigm of knowledge. Given this implicit commitment, Hegel's apparent concession indicated in (1) is bound to appear shadowy. In that case human history, the experience and performance of *human* beings, is fated to be incomplete, uncertain, and occasionally irrational. Hegel's accent on the Absolute Reason as the paradigm of transcendental knowledge, transcendental in relation to the limited reach and scope of human knowledge, is bound to berate the claim of human knowledge itself. The human cognitive capacity determined by our ontological finitude appears to be perpetually haunted by *skepsis*. *Skepsis,* then, turns out to be the basic "driving force" in humankind, driving us to know more and more of God's self-knowledge. Viewed from the other end it is God's *drawing force,* dialectically drawing humankind and human knowledge to His own plenitude and perfection, the identity of Knowing and Being.

Hegel's way out of the pre-Absolute skepticism may be construed in two different ways, historically and logically. Once we bear in mind the dialectical impetus of the *logos,* the surface duality between "the historical" and "the logical" seems to be two aspects of the same reality. By implication, this means that "the prehistorical" and "the prelogical," despite their pre-Absolute difference, are essentially pro-Absolute, dialectically disclosive and gradually approximative to the unity of the whole or reality. When with reference to Hegel's thought it is said that its history is "shadowy" and that its development is "logical," again two different aspects, ontic and epistemic, of the same real situation are sought to be simultaneously highlighted. Ordinarily speaking, "the epistemic" is inseparable from "the human"; but it need not be necessarily construed so. For knowledge, valid knowledge (*episteme*), may as well be interpreted as the *human* articulation of Reality that itself can hardly be equated with any human being or even all human beings taken together. Neither the human articulation of Reality nor our experience of it entitles us to identify "Reality" with "humanity." Reality is humanity and much more. The matter may be put in another way. By *Being* we do not necessarily mean this or that human being or even all human beings taken together. The very expression "taken together" is misleading in the context, because, this togetherness, the unity of the whole, is not available to our experience. The other aspect, the "epistemic," also need not be interpreted narrowly. Knowledge is not necessarily a possession or acquisition of this or that human being. It has a status of its own. It is a phenomenon on its own right. Its "ontic" status is not parasitic on the biographical locus of some person or other. By this what we mean is that "the ontic" and "the epistemic" aspects of Reality, though not incompatible under the aspect of the unity of the whole, God or Absolute, deserve independent analysis and understanding.

Hegel clearly sees that any part of Reality as a part and apart from other parts of it is bound to appear "stray" and "unconvincing," fostering a sense of skepticism in us, in our attitude toward those *parts* of reality. When parts of reality are particularly spoken of, we are invited, without announcement, to participate in an empiric-phenomenological discourse. In this discourse skepticism pertaining to parts is brought to our focus of attentive consciousness, requiring us to look into them, behind them, around them, and at what surrounds them. This invitation of *skepsis,* Hegel tries to show, is not to debunk our experientially available knowledge of particulars but as*sure*, to provide a transcendentally validating perception of this knowledge. Through his dialectic Hegel silently introduces a strong transcendental argument highlighting the view that truth lies in the unity of the whole and that the conclusive validity of knowledge is to be found only in that unity, not in the pre-Absolute parts of that whole or unity. In this ontological formulation of the principle of transcendental unity. Hegel fully exploits the Kantian insight to be found in the principle of transcendental unity of apperception.

The unity of understanding, though definitive, is not highest in terms of certainty and intersubjectivity. Intersubjectivity of what is understood, of knowledge, is ensured by this assumption about understanding under the transcendental principle of apperceptive unity. Because of Kant's commitment to the dualism between experientially available phenomenal objects and things-in-themselves he did not really know how to make one's knowledge sharable, in principle, with others. At the empirical level the problem of other minds, sharability of same knowledge by different minds, could not be plausibly solved. This obliged Kant to fall back upon the apperceptive capacity of the transcendental self. For, he thought, the transcendental unity of knowledge available to one is no different from what is so available to others. In spite of our empirical coexistence and somatic proximity, we, human beings, are undoubtedly different from each other, individuated by space and time, but, transcendentally speaking, our access to and sharability of the "same" scientific knowledge can hardly be questioned. In other words, according to Kant, our empirical situatedness and the resulting limitations are by no means a roadblock in our way to the highest possible scientific knowledge.

This formal-structural solution of Kant does not appear to be very convincing to Hegel. The Kantian way of solving the problem of other mind, other human knowledge, does not appeal to him. The latter takes sociohistorical situatedness of human beings and human knowledge much more seriously than the former. The "straight" transcendental solution of the problem suggested by Kant fails to take note of the rationality of the historically situatedness of diverse and seemingly unrelatable persons and human knowledge. So, Hegel proposes to solve the skeptical problem of knowledge of the minds, other people's knowledge, keeping fully in view the "limitations" due to historical situatedness of human agents, whose experience and performance, thought and action, are our main concerns. This problem has been viewed by Hegel both from the standpoint of individual human beings and that of the social collectivities.

Hegel's own solution to skepticism lies in the Reality as a whole, in the unity of whole. But the point to be noted here is that Hegel is offering his solution to skepticism not at the "place" of its origin. Whether we speak of epistemological skepticism rooted in the ontological finitude of human agents or of ontological skepticism itself, assimilating epistemology by implication within the finitude of human reality or Dasien, it is not to be found in Hegel's whole. The whole of which Hegel speaks knows no *skepsis* within it. Doubt as a mode of consciousness has no place in it because it is Absolute Reason or complete self-consciousness of God. None of these two concepts leaves any room for doubt, question, or correction. Therefore, Hegel's logic, the dialectic of logic, cannot stop anywhere other than in the unity of whole. In a sense Gadamer is right in observing that "the movement of logic is homeless: it can stay nowhere."[14] In the context of the unity as a whole the dialectic gets lost or makes no sense. For the unity as a whole, as Absolute Spirit, knows no dialectic movement within it. The dialectic characterizes an abstract Idea's development toward concrete Spirit or Reason. This movement encompasses both itself as a whole and the historically situatedness of human beings within it. Therefore, the dialectic is evident both in the development of the Idea as a whole and human beings as its different foci of articulation and execution.But the fact remains that the Hegelian solution to the problem of skepticism is made available to us where human knowledge is not haunted by *skepsis*. Do *we* seek a solution when there is no problem?

In response to this veiled criticism of Hegel's transcendental argument it may be pointed out that it is primarily based on the assumption of *human* epistemology and the *finitude* of humankind. By implication the Hegelian affirms that knowledge need not be narrowly construed only as a human phenomenon and that its basic identity is to be seen as a part, as a phase, and as an articulation of the Reality as a whole, self-conscious and self-developing Reason. In this metaphysical or ontological formulation of the issue the empirico-ontologic aspect of knowledge, though recognized, does not receive the primacy of focus. Hegel's "substantive" transcendental argument, somewhat like Kant's "formal" one, puts an end to the life and "dialectical" journey of the haunting *skepsis*. It seems that the end or death of *skepsis* marks the homecoming of knowledge.

10

The Antiskeptical Ontology of Kant and Hegel and the Proskeptical Thesis of Duhem and Quine

1. Different and Conflicting Notions of Knowledge: "Knowledge" and the Realm of Doubt Considered from the Anthropological and Theological Points of View

That knowledge, construed ontologically and as a synonym of *reality,* is free from doubt or *skepsis* need not be questioned. For example, when Hegel says that knowledge is God's self-realization, it is indeed the homecoming of Reality (in the form of knowledge). But our question is, can *human* knowledge, scientific or philosophical, be ever completely doubt free?

Evidently here we are plagued by a very serious ambiguity in the term *knowledge*. The proclaimed homecoming of the *skepsis*-free "knowledge" is not what we, human beings, are seeking, having, and questioning. It is not the knowledge that we have in and through experience and can improve by evidence or experiment. In other words, what we mean by knowledge has no place, leaves no trace, in the oceanic vastness of Hegel's Absolute Spirit or in Kant's transcendental unity. The unity that takes no note of the diversity of human knowledge is hardly human. Skepticism being essentially a feature of *human* knowledge must not be solved in a way that denies its very *human* root or locus. Notwithstanding their claims, neither Kant's nor Hegel's "epistemology" is anthropological in our sense.[1]

From Plato and Aristotle to Kant and Hegel God's self-knowledge has been taken as the paradigm of knowledge. Needless to say, this historical generalization does not take into account some variations in the conceptual articulations of the common issue of the concerned thinkers. The basic point of their agreement lies in rejecting or downgrading the significance of *human* knowledge as we find in different cultures and in different ages. It seems that they all felt more or less reluctant to recognize the changing historical character of human knowledge, its answerability to negative evidence and critical experience. They wanted to uphold an image of knowledge that is *skepsis* free, not open to question and correction. Science as an open-ended *skepsis*-informed form of knowledge so acceptable to the modern world becomes downgraded in their view, though they do not, could not, outrightly reject it. Human knowledge is taken as shadow of God's knowledge, providing only "stray" glimpses, a deformed silhouette, of the latter. So their attempts have been to redefine human knowledge as a dim analogy of, an approximation to, God's knowledge. Human knowledge is sought to be fashioned into God's to give it respectability, that is, certainty, freeing it from the shadowy influence of the skeptic.

In their bids to stave off skepticism the classical antiskeptics (as noted earlier) either identify reality with knowledge, being with knowing, or accept God's own reflective knowledge of reality, of itself, as the highest possible form of knowledge. A weaker form of the latter approach is to be found in Kant's transcendental realism, which conceives God not in his pure transcendent "existence" but in relation to the "systematic unity" of the objects of possible experience. To him this is the "highest formal unity." The most comprehensive "purposive unity of things." To meet possible criticism of his notion of God Kant presents God as "a principle . . . as applied in the field of experience."[2] He has no hesitation in *regarding* God as the "author of the world." But, like Newton, he refuses to see him as directly responsible for the mechanical or physical connection of the objects of the world. The connection that he tries to understand in terms of God as "a regulative principle (or idea) of reason" is teleological. Because he wants to avoid the fallacy of using this regulative idea as a constitutive one, the constitutive use of regulative idea leads us to what he calls, following the ancient dialecticians, "the error of *ignava ratio*." If God as supreme constitutive *idea* of reason can itself constitute and guarantee the rationality of the world and our knowledge of it, it is pointed out, our employment of it as an explanatory principle of what we know of the world makes hardly any sense. In other words, from the idea of God Kant finds no passage to its real existence in the world available to our experience. If the idea of God as the highest ideal unity of things is given up,

Kant fears, we may feel tempted to try to comprehend the incomprehensible, to search the unsearchable, "rising to dizzy heights where [reason] finds itself entirely cut off from all possible- . . . experiences."[3]

Kant's sober antiskeptical strategy, not affiliated to the unprovable existence of God *in* the world of experience, has a basic merit in it. It does not aim at achieving a facile victory in its battle against the critic who maintains that the infirmities of reason must not be lightly dismissed by resorting to speculative flight of abstract reason or by metaphysical invocation of panlogical God. The empiricist, who affirms the possibility of knowledge, sticks at the same time to the view that human knowledge by its very nature is questionable and corrigible, and invites Kant and other proempiricists to examine the basic issues of knowledge on the *natural* grounds of experience and without committing themselves to "incomprehensible" and "unsearchable" principles. In between God and the world Kant wants to carve out a definite place and role for human reason. By implication, he does not endow God with the responsibility of lending directly "its own perfect rationality" to the world, suffusing every part of the world with it.

The Kantian cannot, therefore, possibly endorse the audacious Hegelian approach to show or prove the *direct* march of God into the world, in human history. Though in principle Hegel recognizes the role of the human agent as the knower and actor in the world, in practice he leaves very little for him to know and act. Given the Hegelian scheme of things, the God-world relationship, the human agent can hardly know anything certain without following the ways of God's own knowing, the ways of his own self-realization in the world. Similarly if Hegel is right, humankind cannot do anything rightly that is not ordained by God; that is, does not conform to His own ways of doing things (in the world). In contrast, Kant wants to show that *human autonomy* is an important concept both in the sphere of knowledge and in the sphere of action. Admittedly, he, like Hegel, is deeply interested in refuting skepticism. But he wants to show that his refutation of skepticism does not entail denial of human autonomy, human freedom to know what is not the case, illusion for example, and to do what is not good, lying for example. Once this freedom is denied, human knowledge can hardly be accorded the place of its own in the unity of the whole (world). Kant is interested in recognizing the human root of skepticism, whereas Hegel portrays it as a deformity, an inversion, a sort of cunning concealment of God or Reason.

2. Skepticism's Different Facets and Construals

When we speak of the problem of skepticism it seems that most of us are not very clear about the nature of the problem. If one looks carefully into it one would find that "the problem" in fact is a cluster of problems. Conceptually it may be, and in fact historically it has been, viewed and reviewed from various points of view: epistemological, ontological, sociological, praxiological, and so forth. In contemporary Anglo-American philosophical quarters skepticism seems to be basically an epistemological problem. The available literature on the subject, in spite of its close association with philosophy of science in general and the nature of scientific mode of knowledge in particular, conveys a distinct impression that skepticism and what goes with it, for example, relativism, fallibilism, verificationism, consequentialism, and operationalism, are not respectable and should be avoided and, if possible, banished from the domain of knowledge.

It is interesting to note that the antiskeptic critics of different shades, though unfavorably disposed to classical metaphysicians, are not averse to using their ideas and arguments in the attempt to contain skepticism. In this respect the contemporary continental trend as represented by Husserl, Heidegger, Sartre, and their followers, to my mind, is at a comparative advantage. On the one hand, they are remarkably modern and show deep concern with the practical problems of life, individual as well as collective; and, on the other, they are steeped in the tradition and recall influential ideas of the classical metaphysicians. The history of wars and revolutions, science and technology, evidently cast its shadow and light on these thinkers. Unlike the analytic philosopher's primary, if not exclusive, concern with the narrow epistemology or scientific methodology or a blend of the two, the phenomenologist or existentialist philosopher shows a much larger professional perception of the life and the world. The latter's attitude, notwithstanding its classical affiliation, has a definite appeal to the common persons who have not decided to spend their best creative time in the library or laboratory.

The point I am raising here is that issues of skepticism are not confined to epistemology as it is narrowly conceived by most of our analytic thinkers but is concerned with other aspects of human life, its problems and prospects. It was the merit of the classical metaphysicians to bring to our notice the intimate relationship between the individual and the world, of our life with reality, despite our normal forgetfulness of it. The skepticism that we generally associate with knowledge or scientific knowledge and try to analyze within its narrowed confines, rightly understood, is a part of our larger life in which, besides knowledge, practical problems, theoretical issues, and various other values and disappointments are effectively present. To situate the problem of skepticism in the larger perspective of life, relating it to our ontology and praxiology, is one of the main achievements of the phenomenological-existential thinkers and artists of our times. That knowledge or its theory, epistemology, is not an isolated phenomenon and therefore cannot be understood as such has to be adequately realized by us. Otherwise our investigation into skepticism would be one-sided and our conclusions incomprehensible.

Those metaphysicians of experience who, like Kant and Hegel, propose to be fair to the insistent demands of *human experience* and at the same time try to rise above its skeptical pull are obliged to show a tension in their systems of thought. When they speak of experience they cannot take the term exactly in the sense we understand and use it. Our experience is open to be affected by our human finitude, fallibility, and limitations. If we are committed implicitly or explicitly (implicitly like Kant, and explicitly like Leibniz) to the paradigm of *God's experience* or, more specifically speaking, to that of God's *self*-experience, the whole perspective of skepticism assumes a new meaning of form. If God is identified with consciousness, the expressions like "God's experience" are most likely to be interpreted as mediated, humanly mediated. God's experience, then, in effect turns out to be the experience of God's people, of human agents. If the mediated character of this form of experience is denied, if human experience itself is deemed to be *sovereign,* more than merely *autonomous,* that is, its metaphysical affiliation to God is denied, the term *experience,* in that case, assumes a degenerate positivist sense, open to the charge of skepticism. For this reason, Leibniz, for example, was critical of "new philosophers [like Newton] who pretend to banish final causes from physics."[4] According to him, it would be wrong on our part to think that this best possible world, marked by the most perfect laws of nature, has been designed by God only for our, human, understanding. The world has not been fashioned only to confirm the hypothesis framed by the scientists. If we think so, we are liable to fall into error of those statesmen who ascribe immense foresight to the plans of kings or of those commentators who ascribe too much erudition to their authors. Our (scientific) hypotheses regarding the structure of the world should not be likened to God's reflections on it. The former are liable to error, the latter are not. Besides, our knowledge is more or less confined to "simple ideas and modes," nominal essences (of Locke). But we are very much uncertain about the *real essences* of the world.[5] Although Leibniz speaks of the uncertainty of our empirical knowledge, he need not be taken to talk only in terms of epistemology. Because he thinks that, in spite of its limitations, our experiences somehow are parts of a whole and, even if that whole is not experientially available to us, we have at least partial and "unconscious" access to it. In other words, there is a continuity and unity between our *partial* experience of the whole and the existence of the *whole* itself.[6]

We have already noted a similar line of argument of Hegel that has been successfully highlighted and exploited by Heidegger. Heidegger's interpretation of the Hegelian expression *experience of consciousness* may or may not correctly reflect the intention of the author but its relevance is undeniable. One can easily quote supporting metaphysical thesis on the issue from Hegel's *Phenomenology, Logic,* and *Philosophy of History.* In this connection one is particularly reminded of his view of the organic relation between parts and wholes of Reality. Skepticism is rooted in the otherless independence of ego; that is, when the ego, engrossed in its abstract freedom, is unable to realize the negative nature of this freedom and fails to attain its *other*ed and concrete character. Skepticism, to Hegel, is not merely an epistemic "*partiality*" *but* also an ontological "instability," a sort of "aimless fickleness . . . of going to and fro . . . from one extreme to self-same self-consciousness, to the other contingent, confused and confusing consciousness."[7] The mind is thrown into uncertainty not only because of the partiality of this knowledge but also because of the uncertainty of its aim, its *telos,* its own place in the world as a whole.

A comparable trend of antiskeptical strategy is discernible in Kant's formulation and use of the principle of transcendental unity of apperception. According to him, if the content of human perception or intuition is found to be not concretizable by schema, objectifiable by categories, and certifiable by the transcendental unity of apperception, it is "confused and confusing," blind and uncertain. Objects of the world that are *sensible* are also *intelligible.* The "things" that are sensible are, in a way, simultaneously presented to our understanding. For example, to an ordinary observer the surface of the earth *appears* flat; but it is known to be spherical to the scientist. From an observational point of view what appears flat turns out to be, is known to be, spherical from a theoretical point of view. When Kant speaks of the *conjoined* function or simultaneous employment of understanding and sensibility, he, in fact, is highlighting the importance of the *transcendental* and *unitary* employment of reason.[8] The different powers of reason are functionally systematic or, at any rate, convergent. Occasionally, Kant goes to the extent of characterizing their interrelationship, the interrelationship of the different powers of reason, as *harmonious*. It is true that Kant concedes that the actual availability of the harmonious and unitary form of reason is "hypothetical." But, at the same time, he affirms that "we must seek it in the interest of reason . . . we must endeavour . . . to bring systematic unity into our knowledge." Going a step ahead, it is further claimed by him that the unity that we seek is already presupposed by us. Besides, this unity is not a merely heuristic postulation. The unity found in the different powers of reason, the "parsimony in principles," is not to be taken only as "an economical requirement of reason." It is, says Kant, "nature's own laws." Unless this unity of reason is there, we are left with no criterion for deciding empirically what is true and what is false, what is understanding and what is misunderstanding. Unless we accept this principle of unitary reason, we cannot, Kant maintains, travel beyond the plateau of skepticism, the resting-place for reason.

> The law of reason which requires us to seek for this unity, is a necessary law, since without it we should have no reason at all, and without reason no coherent employment of the understanding, and in the absence of this no sufficient criterion of empirical truth. In order, therefore, to secure an empirical criterion we have no option save to presuppose the systematic unity of nature as objectively valid and necessary.[9]

It is to be noted that the unity of reason to which Kant is drawing our attention has two aspects; epistemic or subjective and natural or objective.

The supreme unity of the powers of reason is to be found in the transcendental unity of apperception. In it lies the hidden key to open up the systematic unity of nature available to experience. To put the matter from the other end, the systematic unity of nature remains unknown without the presupposition and employment of the unitary principle of transcendental apperception. Nothing "contingent," nothing "fickle" or "wandering," is left within the realm covered by the said pincer movement of the antiskeptical plan of Kant. This logic of antiskepticism has been construed both epistemically and ontologically. In the case of Kant this logic is primarily formal. In the case of Hegel it is both formal and historical.

Skepticism, for Hegel, is an essential part of consciousness's journey toward self-conscious certainty. In the process of this journey *human* consciousness does not, rather cannot, get permanently stuck anywhere. Whether it wanders or rests somewhere, it is a journey to a definite goal, it is being-toward. When Hegel sounds very definitive in his reference to what he calls *natural consciousness,* he gives the impression that even at the level of nature, of the understanding of nature, human reason is in a way definite of its *telos,* despite the incomplete or negative character of Reason's presence in natural consciousness. The one-sidedness of consciousness in its encounter with nature may appear entirely negative, a sort of nothingness. But Hegel is not at all prepared to admit this view. For, his logic obliges him to maintain that the one-sidedness of nothingness of a particular mode of natural consciousness is not empty or devoid of content. Not only does this sort of consciousness have a content of its own, but, in addition, it is to be borne in mind that this content is the outcome of a definite dialectical development of consciousness in its relation, relation of understanding, to nature. In other words, what goes before a negative mode of consciousness and what lies ahead of it are all equally definite. The negation in dialectial transit, to use Hegel's expression, is *determinate* negation. The negation or one-sidedness that is symptomatic of skepticism is given a very positive place in Hegel's theory of knowledge as well as in his theory of reality. If the one-sidedness of natural consciousness seems to conceal the nature of true consciousness, this concealment, in a way, is the content of concealing consciousness itself. By implication this takes away the purely negative effect or one-sidedness of concealment, of skepticism. If Hegel is right, skepticism is the partially unconcealed self-concealment of consciousness.[10]

In other words, the skeptical mode of consciousness has, paradoxically enough, an antiskeptical element or moment in it. The experience of consciousness that is haunted by a sense of uncertainty has simultaneously a sense of certainty in it. The consciousness that is experienced, experienced skeptically, transcends the bounds of its skeptical moments. Human consciousness, set out in its journey toward truth, cannot be held up permanently by the doubting modes and moods of our natural experience, the experience of natural objects. The *freedom* of a truth-seeking consciousness cannot get lost or exhausted in its skeptical-natural wanderings. If the skeptical modes of consciousness, their one-sidedness, are taken to be final, the cognitive journey of consciousness is bound to be halted. This rather pessimistic view of the nature of knowing or experiencing consciousness is not acceptable to Hegel. Stripped of its details, Hegel's view of consciousness, epistemological as well as ontological, has no room for intrinsic doubt, still less of ignorance. All doubt, all ignorance, on scrutiny, turns out to be veiled promise of definite knowledge. The immediate experience of this or that object may turn out to be dubitable or illusory but reflection shows it as a part of a plenum of consciousness, explorable by experiencing consciousness itself. It is positive and has a contribution to make toward our search for knowledge. In brief, in the name of skepticism Hegel is not prepared to accord any ultimacy to the skeptical mode of consciousness.[11]

3. Are the Unitarian Approaches to Knowledge Necessarily Antiskeptical in Their Inspiration?

It is of deep interest to enquire into the reasons for which both Hegelians and Kantians attach so much importance to the unity of perception (or intuition), understanding, and apperception, on the one hand, and that of reality, nature, or both, on the other. To establish the antiskeptical case very firmly one notices that the defenders of the idea (or ideal) of the absolutely certain knowledge ascribe a priori certain characteristics to our mind as well as to the world. When we speak of the mind we mean different capacities and different levels of mental activities, from simple perception to comprehensive apperception. The epistemic unity claimed to be evident in different human knowing powers does not appear to be sufficient to the antiskeptic to prove the case of absolutely certain knowledge. To make the case plausible and strong, the antiskeptic claims, in addition, that nature or reality itself is also unitary. Whereas Hegel draws no hard and fast line of division between nature and supernature or reality, Kant, even after recognizing it firmly, shows his anxiety to minimize its impact on his ambitious program of the *ideal* unity of the said two realms, phenomena and noumena. Hegel's God does not know within it any real dichotomy between nature and supernature, between the empirical and the transcendental: their *unity,* though empirically undemonstrable, is speculatively not only unavoidable but also available. Kant's formulation of the issue, in spite of its seeming modesty, is as ambitious and optimistic as Hegel's. Whereas Hegel speaks of *unity* between the natural and the supernatural without recognizing any division between them, Kant speaks of the *affinity* between them notwithstanding their division. To make his case thorough on the issue, Kant tells us how our understanding is facilitated by reason in terms of the principles of *homogeneity, variety* or *specificity,* and *affinity.*[12] So that our understanding is not lost in the variety of nature's particulars reason advises us to look into their homogeneous character. In fairness to Kant, at this point, one has to admit that he, by recognizing the principle of variety or specificity, puts a sort of check on the uncritical tendency toward unity, seeking unity where perhaps it is not. The law of affinity, marked by general continuity, tries to combine the wayward or bewildering tendency of the variety of what we experience with the homogeneity of what we understand. True, Kant admits that the systematic unity that reason seeks cannot be *constituted* by it; it is used as a necessary regulative principle.

A careful and close look into Kant's formulation of how the ideal and systematic unity is exploited by our reason shows his indebtedness to Leibniz's law of continuity and anticipates Hegel's concept of metaphysical unity amidst historical diversity. In different ways each of these classical rationalists is strongly opposed to the very idea of according any place to what is contingent or accidental in nature or reality. In different ways they are led to the more or less identical conclusion that human reason or understanding cannot have anything skeptical in it because that entails implicit recognition of the presence of the "irrational" or the "contingent" in nature or reality. To them nature and reality are so thoroughly rational and rationally intelligible that they cannot think of how skepticism can be recognized except as an aberration of our understanding, a wayward wandering of human reason, or a purely temporary and inhospitable resting place of it.

It seems that when we say that we are certain about a piece of knowledge, our certainty is partly based on positive evidence and partly based on our self-consciousness, the consciousness of our being, the state of being. One who is unsure about one's own self is likely to be skeptical even if provided with positive evidence relevant to one's knowledge. The converse of this situation also seems to be true. Sometimes we feel certain (in respect to a piece of knowledge) not because we have enough positive evidence for it at our disposal but because we are confident. Confidence and diffidence, strength and weakness, and similar other attributes of our being have a lot to do with our knowledge, our knowing. Knowing cannot be isolated from being, nor being from knowing.

When some moral philosophers say that we cannot *be* good unless we know, really *know,* what good is, what is implied is this. Being cannot be separated from knowing. Knowing is a particular mode of being. But here *particular* means a partial expression of being. Being has its other particular expressions—hoping, feeling, willing, and so on. Disregarding or oblivious of other modes of being, knowing itself cannot attain that sense of certainty which may be described as paradigmatic self-assurance. From the other end it can be said that self-ignorance cannot be complete; that is, self-obliviousness can never be complete. Obliviousness or ignorance is a sort of concealment of being. But in this sort of concealment being is at least partially unconcealed or disclosed and, therefore, the so-called ignorance or obliviousness cannot be complete.

4. Is Human Knowledge Bound to Be Skeptical? Why Is God's Knowledge Claimed to be Doubt Free? Knowledge without God: Epistemology without Theology

The problem of skepticism or the problematic character of skepticism is essentially human. When it is said that skepticism makes sense or becomes intelligible only in the epistemic context, what, in effect, is highlighted is the human context of validation or vindication of knowledge. The question of validity of knowledge, with which our traditional epistemology is concerned, is basically rooted in the limitations, the growing character, of our knowledge. In the case of God's knowledge this question of epistemology does not arise at all. Because most of the received notions of God claim that God is all-knowing, all-powerful, all-good, and suffers from no limitation or negation. For example, in the context of an all-good God the Kantian idea of goodwill does not arise at all. Kant tells us that God's will, unlike others, does not give rise to any sense of duty. God is necessarily all-good and, therefore, cannot possibly *will* to be good. "Holy will" cannot be deemed to be under any obligation; it is duty-free. "To be good" presupposes a *lack* of goodness. The very notion of God rules out this element of lack or negation in God. Similarly, it cannot be said of God's knowledge that there is an element of lack or negation, that is, something skeptical, in it. To God one cannot rationally ascribe either lack of goodness or lack of knowledge. Whenever in speculative metaphysics dealing with the notion of God the issue of *skepsis* is raised, implicitly *human* knowledge is referred to. For example, when the Hegelian speaks of the human will to know, this willing person is deemed to be partially differentiated from and not completely affiliated to the unity of God. Although the being of humankind is said to be organically related to the Being of God, yet the difference between the two is taken to be real, at least for the time being. Skepticism is therefore characterized as temporary, initial, or historical and not final. In the finality of God's unity, in the wholeness of God's Being the question of *skepsis* loses its meaning. The human agent is situated in reality, within God's scheme of things. This is a limited situation. It is within the life of human knowledge that the question of skepticism arises. Unless its answer is sought and obtained within the limits of the human situation, metaphysical or transcendental, antiskeptic strategies appear to be more or less irrelevant. The metaphysical theories of Descartes, Kant, and Hegel exhibit a distinct tension between the empirical and the transcendental, between the epistemic and the ontic, between the human context and the divine context of knowledge.

Even if the metaphysician's transcendental God hypothesis is dispensed with, it is difficult to view *skepsis* as entirely free from the said epistemic-ontic tension. Human knowledge, even in its most widely acceptable scientific form, is not free from some transcendental implications. True, the classical thinkers, the followers of Kant and Hegel, for instance, will not readily recognize scientific knowledge as the paradigm of human knowledge. Rather, they try to show how in the larger *metaphysical* context scientific knowledge assumes added significance and concreteness. However, for the sake of argument if one insists that science embodies the best possible form of human knowledge, the problem of skepticism does not disappear; the traditional problem only takes a new form; or, one might say, the old problem is formulated in a new way. Even in this new formulation skepticism shows the "old" epistemic-ontic tension between the unity of the whole (reality of God) and the human or limited empirical availability of that unity in and through human experience.

Those metaphysically minded philosophers who are invariably suspicious of the naturalistic mode of consciousness or the scientific mode of knowledge have their own way of tackling the problem of skepticism. The immediacy of natural consciousness is regarded by them as a part of a not-so-immediate or of even large and distant part of reality. Bringing skepticism close to the immediate character of human experience, they try to show how the immediate parts of reality available in sense experiences and expressible in observational statements are organically and logically related to and embedded in the whole of reality. This whole may be construed both ontologically and methodologically. In the Hegelian world-view scientific statements, especially the lower-order observational statements, are interpreted as excerpts of the transcendental unity of reality itself. However, in the Hegelian system the distinction between the empirical and the transcendental is only a matter of degree. Every empirical statement has in it a transcendental element. Every particular empirical statement is said to be a *particularization* of a relatively universal segment of reality. Not only Hegel but also Kant highlights the transcendental affiliation of the lower-order empirical (or intuitional) statements. The spatio-temporally determined parts of reality, which are immediately and experientially available to us, are not cut off from their transcendental moorings. On the contrary, their determinate character is explainable only on the assumption that they are situated in and parts of a large whole of reality, nature, or God, whatever we call it.

This ontological mode of doing away with the deficiency of skepticism is sought to be supplemented epistemically by postulating some transcendental and cognitive capacities of the human mind. Some innate capacities are ascribed to the human nature and by virtue of which, it is claimed, we can rise above our sense immediacy and know what is distant, very small, or very large. Some extreme forms of innatism construe the human mind as a miniature universe. The converse is also claimed to be true. Leibniz's monadology provides the classic example of this view. Every individual mind has everything of the universe in it. This presence of the universal in the individual, though not *clearly* evident, is said to be ontologically, that is, in principle, undeniable.

5. Doing away with the Analytic-Synthetic and Physics-Metaphysics Distinctions; Methodological and Ontological Holism

The antiskeptic epistemological position has its methodological analogue. Some empiricists are of the view that the distinction often drawn between analytic statements and synthetic statements, between parts and wholes, on close scrutiny, are found to be untenable. Due to some semantic confusion or lack of thorough investigation into the nature of the empirical, it is argued, we are led to the conclusion that the analytic-synthetic distinction is absolute. According to these empiricists no statement can hold its truth ground *against* all possible experiences. The statement that can hold its ground *against* every possible experience has nothing to do with any experience whatsoever. To say this is to delimit the scope of empiricism. The "meanings" in terms of which some statements are decided and declared to be analytic must not be deemed to be Platonic or transcendental in the bad sense; that is, not in anyway rooted in or influenced by experience. By following this line of argument Duhem[13] and Quine[14] try to do away with the traditional distinction between metaphysics and physics, between physics and mathematics, and even that between psychology and mathematics. On the one hand, Duhem affirms, "we cannot . . . derive from a metaphysical system all the elements necessary for the construction of a physical theory," and on the other, he readily concedes that the progressive precisification of physical theory has been made possible because of the metaphysical "belief in an order transcending physics."[15] In brief, according to him, physics is a "better defined and more precise reflection of metaphysics." In a somewhat similar vein Quine, as noted earlier, speaks of the hidden empirical substructure of everything in the world—from tables, chairs, and mountains to numbers, classes and atoms. One of the methodological upshots of thoroughgoing radical empiricism is to conclude that there is no crucial evidence or experiment and that, therefore, no theory can be decisively refuted or falsified. Evidential statements, verifying or falsifying, can only bring about some change, central or peripheral, in the system of interlocked statements of experience of varying scope. It is to be noted that in both ontological holism and methodological holism there is no place for an islandish or isolated truth claim of any statement, particular or universal.

This sort of methodological holism entails skepticism and a rejection of realism. At least this is how the Duhem-Quine thesis has been implicitly criticized by Popper and others who show relatively stronger realistic inclinations. Realists like Putnam are also opposed to this weaker type of realism, "dangerously" close to skepticism. Once it is maintained that there is nothing in reality that can turn out as decisive negative evidence against a theory, it is suggested, by implication, that nothing is mind independent in reality that can establish its claim as a clear disproof of a scientific theory. In other words, what is being denied is the existence of *some* structural features of reality that, expressed in linguistic statements, can be said to be conclusive disproofs or decisive falsifiers of the theories purported to describe reality itself.

A stronger formulation of this view would be this: reality, to the extent it is knowable, is so *thoroughly* constituted or constructed by *our* mind's conceptual apparatus or categorial framework that the question of the existence of falsifying evidence or external disproof does not arise at all. This formulation of methodological holism, partially ascribable to Kant, is bound to remind us of ontological holism of Hegel. The point of striking similarity may be put in this way: reason cannot possibly give rise to what is irrational, or the *ground* of valid knowledge cannot have in it anything that can possibly invalidate it. The understanding that makes nature possible cannot implant in it anything that could possibly show up itself as negation of knowledge of nature, negative evidence in relation to the theory of nature. The very ways in which nature is constructed and structured rule out the possibility of the unreason's presence in it. The *cunning* of Reason, as Hegel observes, is not really "of reason." By its very nature reason is *cunning*less. Yet the fact that we at times *experience* it is because of the lack of inadequacy of reflection on the concerned experience. Somewhat in a similar, primarily epistemic, vein Kant maintains that "error is brought about solely by the unobservable influence of sensibility on the understanding, through which it happens that the subjective grounds of the judgment enter into union with the objective grounds and make these latter deviate from their true function."[16] Because in Kant's scheme of things error is judgmental and judgment partly depends on what it receives from the sensibility, the error is due to error in judgment, in mistaken interpretation of the sense-experience.

If a methodological holist like Quine is right, we can never be free from the possibility of doubt in knowledge or judgmental error. Kant assures us that the situation is not that desperate. Our native apparatus of knowing can well save us from the possibility of falling into the pit of error under the influence of sense-experience. Hegel goes a step beyond Kant and affirms that, ontologically speaking, methodological misuse or misappropriation of reason may give rise to only a temporary disfiguration or suppression of reality; but reality itself provides the criterion of unity enabling us thereby to identify it as the *appearance* of reality as distinguished from reality itself.

In a way Quine, like Kant, is a constructionist. Both agree that what we call the world of science is human construction. The Quinean process of construction, as noted earlier, is somatological. The nature that is studied by human beings and finds embodiment in science is not quite different from human nature itself. Quine draws no hard and fast distinction between human nature and the nature studied by humans. In a way it is nature's self-reflection. But, unlike Kant, Quine finds nothing in human nature, nothing necessary and a priori, that might enable it to know nature by constructing or constituting it in an infallible manner; that is, ruling out the possibility of doubt in the form of knowledge.

Naturalism may be, in fact historically has been, defended mainly in two different ways. Unlike Kant, some naturalists draw no sharp distinction between the natural and the supernatural, between the empirical and the transcendental. Their view is that what we regard as supernatural or transcendental, on analysis, is found to be grounded on what is natural or empirical. This ground may be shown either reductively or nonreductively. Naturalists like Quine favor reductionism. They do not find any fundamental distinction between the nature of the knowing subject or mind and that of the known object or matter. In his bid to establish the *natural* character of epistemology Quine takes pains to show how the human organism, itself a natural object, placed among other natural objects, and interacting with them, acquires certain properties, linguistic and behavioral, that qualify it to be designated as a subject, even as a *knowing* subject.

However, this is not the way all naturalists do away with the commonsense distinction between the natural and the supernatural, between the empirical and the transcendental. Admittedly, naturalists, unlike dualists or gradualists, if seriously and logically pressed, will deny any *fundamental* distinction between the two. But their analysis of the supernatural or the transcendental assumes a nonskeptical, at least seemingly nonskeptical, character. Hume and his followers, for example, are *theoretically* opposed to the validity of induction, the result of extrapolation or generalization from what is given in particular sense-experiences. But in *practice*—in history, morals and politics, for example—they do not exhibit any strong skeptical inclination. Remember that many followers and even critics of Hume have pointed out two quite different trends in his thought; viz., skeptical naturalism and nonskeptical naturalism.[17] Reid, for example, argues that Hume, the epistemological skeptic, espouses a sort of moral theory that is more or less free from skepticism. Several moral sense theories of the eighteenth century in England and Scotland defended the view that humankind has in it a natural or native moral sense that, unlike the physical senses, never yields to skeptical conclusion. Also remember in this context that the Humean fork, the defence of the analytic-synthetic distinction, is inconsistent with the radical empiricism of Quine. Following this train of reasoning one is justified, at least partially, in holding that Kant, acknowledgedly roused from his dogmatic slumber by Hume, did not entirely part with the latter.

6. Skepticism Pertaining to Natural Law Statements: Nomic and Accidental Necessity and the Inductive or Deductive Method to Substantiate the Certainty Claim

An analogous controversy is found regarding the nature of the universality of laws. Philosophers fall apart on the question of whether the universals of law are (1) accidental, human-constructed, and contingent or (2) embedded in the very nature of reality, transcendentally universal, and necessary. It is not easy to decide if law statements are universal in the weaker sense (1) or in the strong sense (2). The problem has been formulated by philosophers in very many ways down the centuries. When we speak of a scientific law of the form "Every A is B," what is it that we wish to say? The traditional empiricist answer to the question, intimately associated with the name of Hume, is that A and B are constantly conjoined and this constant conjunction is not restricted by space and time. The scientific law, though a generalization from the limited experience of fact, is not limited in its scope. In a way it is both factual and expressive of the structure of reality. But this claim raises some ontological issues and unless we are in a position to form definite views on the claim it is difficult to substantiate whether the proclaimed general relation between A and B could be said to be *necessary*. For, the notion of factual necessity has itself been construed in two different ways causally (the weaker sense) and nomically (the strong sense). If it is said that "every A nomically must be B," should we think that the structure of reality is such that it unconditionally rules out the possibility of "a situation of A and not-B?" Or, do we merely posit that such a negative possibility is ruled out by the very meaning of the law? Or, in other words, are we entitled to hold the view that "an A and not-B situation" is ruled out only by the *formal meaning* of the law statements and not otherwise; that is, factually. The Humean account of nomic necessity, expressed in scientific laws, has found favor with such thinkers as Ernst Mach, Karl Pearson, Harold Jeffreys, and R. B. Braithwaite. They all endorse the Humean view "that universals of law are objectively just universals of fact, and that in nature there is no extra element of necessary connection."[18]

By claiming that law statements are not spatio-temporally limited in scope it is implicitly affirmed that induction, which involves going beyond the given experience, is justified. For, the past states of affairs on which the concerned law statements rest may be cited as evidence in favor of it. Additionally, the evidential value of such states of affairs is not merely historical but also predictive or prospective. The open-ended law statements, therefore, can be expressed also in the form "If A, then B." Whether this strategy for vindicating law statements goes far enough is disputed. For, many logicians and methodologists are of the view that the natural or the structural necessity underlying laws cannot be vindicated merely in terms of unrestricted universality. Laws may be unrestricted yet conditional. Mill affirms that the generality of causal laws is not sustainable merely on the basis of invariability of succession, but what in addition is called for is (satisfaction of the requirement of) unconditionality.[19]

In addition to the "invariable succession" requirement when the "unconditionality" requirement is insisted upon, the idea is to firm up the cognitive status of laws or theories purported to explain laws. The assumption seems to be that the Humean view is vulnerable to the typical anti-inductive criticism, whereas the latter is not. It is not correct to think (1) that when A and B are found to be *constantly conjoined* or *invariably* succeeding one another the sort of generalization, "If A, then B," we get is open to refutation by contrary evidence and, (2) that, in contrast, when we have *unconditionally* related A and B, the law statement of the form "A is B" we get is something empirically very reliable and not open to predictive or prospective contradiction. The attempt to draw this distinction between the Humean view of law and the Millian one, though inspired by the idea of meeting the anti-inductive criticism raised against the former, seems to have proved unsuccessful. It is difficult to deny that the latter view of the law, much like the former, is based on induction. Neither induction nor probabilistic reformulation of it can possibly give us laws that are not open to falsification. If falsifiability of laws is ontologically ruled out, induction is replaced by deduction as *the* method of discovering or establishing laws.

Can we epistemologically vindicate laws merely by methodologically preferring deduction to induction? The question seems to be deeper and ontological in character. The theories that seek to explain laws are required to describe or presuppose, directly or indirectly, the structure of reality sought to be mapped or indicated by laws. Unless reality itself is believed to be *logically necessary* or rational *without exception,* it is difficult to have a true theory of law statement subsumable under it that can be justifiably proclaimed to be true always and under all circumstances. Besides, this *panlogical* statement regarding the structure of reality has to be accepted as an a priori and necessary truth.

The success of the attempt to vindicate laws by preferring deduction to induction is dubious. For the major premise of the deductive argument, which is taken to be true, is either a metaphysical assumption, or a policy statement, or itself an inductive truth. Each of these alternatives of the so-called deductive way of refuting skepticism regarding law statements is open to the well-known objection. First, if it is metaphysically assumed, as it has been by Hegel, for instance, that the reality as a whole is rationally structured and becomes gradually and historically available to human experience, this assumption appears to be metaphysical in an unwarrantable sense. Because parts of this assumption are (1) the substructures of the assumed structure of reality are isomorphic and (2) all possible experiences of the structure and substructures available in the form of evidence are destined to be confirmatory, and *never* otherwise, of the concerned laws and their parent theories. This strong metaphysical assumption need not be necessarily of the Hegelian form.

Empiricists of different hues speak instead of the *uniformity of nature* (Mill) or the *limited independent variety of nature* (Keynes). If we think that the uniformity of nature is an overarching principle and that it suggests the presence of *uniformities* in nature that are empirically disclosable, in a way (without ceasing to be empiricst) we are exploiting a metaphysical insight defended in different ways by Leibniz in his monadology and Hegel in his panlogism. A similar interpretation may be offered of the Keynesean principle of the limited variety of nature; for it too assumes that the variety of structures and substructures of nature being limited, the evidence pertaining to them are bound to be convergent in the long run. Additionally, the Keynesian principle also assumes that no configuration of a part of nature or state of affairs can be isolated or identifiable and explainable in terms of its local laws. Positively speaking, both reason and experience suggest that local states of affairs are related to and have to be explained by certain global laws. In a manner of speaking, these global laws are theoretical surrogates for the metaphysical assumptions espoused by Leibniz and Hegel.

To avoid inductive skepticism—that is, skepticism associated with the inductively established truth claims of the laws of nature—the sort of assumption that is often made smacks of metaphysicalism, idealistic or realistic, Leibnizian, Hegelian, or Keynesean. It is difficult for a consistent empiricist to espouse the metaphysical assumptions that, strictly speaking, are not based on experience. If our empirical evidence exhibits a convergent trend or propensity and it is found to be *analogous,* we are willy-nilly led to affirm that the said trend and the analogous characteristics of the available evidence are in fact grounded in some such principles as the uniformity of nature or the limited variety of natural kinds. Mill has logically brought to our notice how difficult it is for the thoroughgoing empiricist to accept any of these principles or, more correctly speaking, postulates. The structure of inductive inference seems to be vitiated either by the fallacy of *petitio principi* or that of *regressus ad infinitum.* There is an important difference between the principle that is secular or metaphysically neutral and a postulate that is metaphysically or ontologically committed.

Unless the structures and substructures of nature intended to be related by laws are analogous in some way or other, they can hardly be brought together by the inductive method. In brief, induction without analogy is impossible. Hume affirms that all sorts of inductive reasoning from causes to effects logically require, besides constant conjunction, resemblance between the evident objects and the available or projected ones. To quote him, "Without some degree of resemblance, as well as union, 'its impossible there can be any reasoning." This view has been endorsed, though from a different standpoint, by Keynes. "Some element of analogy must . . . lie at the base of every inductive argument."[20]

Analogies are of two kinds, positive analogy and negative analogy. Induction proceeds neither by positive analogies nor by their numbers as such. Superficial analogies are not of much consequence in giving us right knowledge about the objects covered by laws. Negative analogies and negative evidence are relatively more important in strengthening the relations among the law-governed objects. The reliability of law statements increases by (1) reducing the resemblances (between the objects) in respect of unessential properties and (2) increasing the differences known to exist between the same. Keynes does not attach more importance to pure induction; for it does not help us to get to the bottom of the general skeptical problem, to increase certainty of our knowledge. He finds negative analogy very important; for it gives us information regarding the fundamental characteristics of the objects. When he speaks of "the *variety* of circumstances" in which analogy between the concerned objects has to be studied, he recalls Newton and reminds us of Popper who, later on, highlighted the importance of negative evidence and disfavors the idea of "the crude unregulated induction of ordinary experience."[21]

Both J. M. Keynes and C. D. Broad are of the view that induction proceeds on the structural assumption that nature is a finite system and that it has some basic *generating* properties from which other properties are *generated*. The finiteness of the system and the relation between the generating properties and the generated ones are essential ingredients for inductive progress of knowledge, reducing the element of uncertainty likely to accompany it. The point may be clarified *via negativa*. If we are pushed to the assumptions (1) that nature is an infinite system and, (2) that the relations between the generating properties and the generated ones are quite indefinite and uncertain, we are left with no lane to *proceed* along with the inductive path. If the variety of nature knows no limit, we cannot *rationally* decide how to raise probability value or content of our law statements to be yielded by induction. If every part of nature is quite unlike every other part, it is not clear how to discover any law or induce any generalization regarding them. One might even raise the question of whether there can be a Nature at all the parts of which are quite unlike each other. Remember here that to the thoroughgoing empiricist the Leibnizian way out of the problem framed in terms of (monadological) perspectival (or representational) unity is not available. Those who are persuaded that skepticism regarding inductive generalization cannot be removed by resorting to some "metaphysical" assumptions about the structure of reality search for other escape routes. For example, it has been argued by both Mill and Ryle that the major premise of the inference is to be construed as a *rule* or *license* of inference, not as statement about the structure of nature. According to this view, we inductively infer one particular statement from another using the *relevant* rule or license. Rules logically behave like laws. These rules or laws are not true or false, provable or disprovable, in the way the statements to which they apply are. The legitimacy of these laws or the rationality of these rules is established in terms of their ability to ensure our passage, inductive passage, from some particular matter of fact to other particular matter of fact. Law statements are explanatory, teach us how to do certain things and perform certain operations. "Law-statements are true or false but they do not state truth or falsehoods of the same type as those asserted by the statements of facts to which they apply or are supposed to apply."[22]

It is to be noted that the Millian-Rylean view construes laws *instrumentally*, removing the possibility of their falsifiability by taking away their factual or ontological commitment. As per this account laws are not addressed to, and are not descriptive of, particular structures or substructures of reality, (experience of) reality can never render them false or even partially damage their *success* claim. To avoid "metaphysicalism" the empiricist becomes instrumentalist in the context of justification. Laws are sought to be justified not by testing them against appropriate experiences but by watching their performance as rules of inference. If rules of inference *fail* to enable us to infer from particular statements of fact other statements of fact, their reliability, not first-order truth claim, goes down.

A somewhat similar view has been defended by Carnap also. Raising the question, Are laws needed for making predictions? he answers it in the negative. According to him, we infer particular from particular. For example, having seen many white swans and no nonwhite swans we infer that the next swan will be white and we will be willing to bet on it. Although we do so certainly we make use of our previous experience of white swans and perhaps without even considering the question whether all swans in the universe without exception are white. According to Carnap, "we see that the use of laws is not indispensable for making predictions. Nevertheless it is expedient, of course, to study universal laws in books on physics, biology, psychology etc."[23] Here Carnap, like Mill and Ryle, has taken an instrumentalist view of law. Against the appropriate background of a law statement embodying or reflecting past experiences regarding white swans one feels (inductively) confident to predict that the next swan will be white.

7. Law Statements in the Light of Quantum Mechanics: Realism and Probabilism; Reichenbach's Interpretation of Alternative (Not Dualistic) Descriptions; Evidential Indeterminacy and Holism from Reichenbach to Quine

Besides Carnap, among those who have worked seriously on the problem of induction and probability in the recent past one must count Reichenbach. A defender of the frequency theory of probability, he, like Carnap, maintains that there is nothing wrong in probability statements about a single event "if daily experience supplies us with a number of similar cases." The events of our daily life constitute a series that admits of the frequency interpretation of probability. "To speak of a meaning of probability for a single event is a harmless or even useful habit, because it leads to a correct evaluation of the future as soon as this language is translated into a statement about a series of events."[24] The logicians who defend the frequency interpretation of probability interpret probability statements in a way that is different from scientists' statements about the facts of the world. When someone casting a die predicts that the probability of turning up of face "six" next time is 1/6 or that the probability of turning up of face "non-six" next time 5/6, one does not claim that one's prediction must come true. A statement of this sort is called a *posit* by Reichenbach. A posit cannot be said to be true or false in any ordinary sense. Only its probability can be *rated*.

Reichenbach claims that his interpretation of a predictive statement as a posit solves the problem of induction and vindicates the empiricist conception of knowledge that has been badly damaged as a result of Hume's criticism of induction. In the process, however, it is conceded that the form of knowledge vindicated by the frequency interpretation of probability is not demonstrative. This does not prove that the concerned predictive statement is true, only that it is a good or the best available posit. Reichenbach's way of justifying induction is claimed to be nonprobabilistic, for the theory of probability presupposes induction. The proof that Reichenbach constructs is purely mathematical. The calculus of probability is constructed in axiomatic form. All the axioms of probability are purely analytic statements. The only nonanalytic principle found in the application of the calculus (of probability) pertains to the determination of a degree of probability by means of an inductive inference. Having found a certain relative frequency for a series of observed events and having assumed that the same frequency will be repeated approximately in further extension of the series, the rating of a posit is inductively determined.

Because the use value of a posit depends on the limit of the frequency, naturally the question arises of how to know it. Reichenbach admits that the assumption of the frequency limit is not provable. But it is argued that the use of inductive inference helps to find it out. The inductivist's approach has been likened to the fisherman's interest in catching fish. The latter, without knowing where (or even whether) he would be able to catch fish, knows this much that if he wants to catch fish he must go on casting his net.

> Every inductive prediction is like casting a net into an ocean of the happenings of nature; we do not know whether we shall have a good catch, but we try, at least, and try by the help of the best means available. We try because we want to act. . . . Posits are the instruments of action where truth is not available; the justification of induction is that it is the best instrument of action known to us. . . . All knowledge is probable knowledge and can be asserted only in the sense of posits; an induction is the instrument of finding the best posits.[25]

Reichenbach contrasted his theory of probability with the rationalist's one. The merit of his proposed solution of the problem of induction, he claims, depends on his *empiricist* principle of indifference. The rationalist, as we know, offers a *synthetic self-evident* interpretation of the principle of indifference and consequently arrives at a synthetic a priori inductive logic. This is rejected by Reichenbach who claims himself to be a consistent scientific empiricst. Though he recalls his own synthetic principle, he hastens to add that on the basis of it the truth claim he makes of his posits is purely analytic. The consistent empiricist can make only "analytic contributions to knowledge." The "quandaries of empiricism" found in Hume's skepticism are said to be due to a "misinterpretation of knowledge" and claimed to be avoidable by a "correct interpretation" of knowledge as obtained in modern science. Referring to mathematical physics Reichenbach says that "the search for certainty had to die down" and "gone is the ideal of the scientist who knows the absolute truth."

Evidently, Reichenbach was writing under the strong influence of quantum physics and the principle of uncertainty. In a way he has tried to make the simple point that certainty is not in science and the area where it is claimed (by the "rationalist") to be available is not scientific but metaphysical. The scientist and the scientific-minded philosopher are advised to be content with probability and not search in vain for certainty. Skepticism in this sense is not only acceptable but commended as the most respectable form of human knowledge. Reichenbach's antiskeptical probabilism based on some "empiricist," *not* self-evident, principle of indifference, that is, unbiased Nature, can hardly refute skepticism in a convincing manner. Nor does he claim it as would the transcendentalist and the self-evidentialist. But about the certainty of the analytic statements he endorses the Humean line and, therefore, his instrumentalism cum skepticism is confined to the domain of synthetic scientific statements. Rejecting the alleged analytic-synthetic distinction, Quine, the critic might say, enlarges the domain of cognitive uncertainty. In effect his methodological holism leaves no area of knowledge unsuspect. If for Leibniz's God every statement is analytic and therefore certain, for Quine's person every piece of knowledge—mathematical, historical, and metaphysical—is synthetic, that is, informative, and therefore suspect (at least in principle). The fall out of God-human asymmetry in epistemology is nowhere so manifest as one perceives it here.

However, it is clear from Reichenbach's version of logical empiricism that it is not at all incompatible with realism. On the contrary, unlike Mach and other empiricists of stronger persuasion, he points out that denial of *more or less* definitely identifiable objects of the spatio-temporal world takes us back to the fold of speculative idealism, if not solipsism. Realism must not be compromised in the name of economy of thought. For all scientific thoughts and theories, elaborate or economic, need to be verified. What verifies a theory is not experience as such but experience rooted in and traceable to objective reality. Observational and experimental verification does not entail psychological or logical constructionism. Observational data and logical techniques, though necessary, are not sufficient for the purpose of scientific theory construction. In addition, objective reality has to be posited.

The preceding explication of the relation between verificationism and realism, Reichenbach rightly points out, needs further refinement. Observation statements cannot conclusively substantiate the truth claim of our theoretical statements about the objective world. Observation statements can only *probabilify* the latter but cannot prove their certainty. An element of evidential indeterminacy or underdeterminacy is always there in all empirical statements. Reichenbach's view on the subject is bound to remind one of Duhem and Quine.

> The statement that there is an objective physical world can only be maintained as highly probable, and not as absolutely certain. We have good inductive evidence for the existence of a physical world. . . . And it is meaningful to speak about an objective physical world because statements about such a world are inductively derivable from observations. . . . The physical world is not uniquely determined by observations. . . . There is a plurality of equivalent descriptions, and the usual realistic language in which we describe the physical world is more or less one among these descriptions.[26]

The availability of alternative descriptions of "one and the same" objective world suggests, among other things, that no description is uniquely or conclusively true. To admit that other possible descriptions may also be true amounts to dilution of realism to some extent—but only to *some* extent--because realism becomes meaningless unless it is at least assumed that the world has a probabilistic structure of its own that is inductively supportable. If this condition is rejected, it is not clear how we can possibly make even *probable* statements about the objects of the external world. Even the frequency distribution of probabilities, on scrutiny, suggest its ontological rootedness. Strictly speaking, we cannot make any probability estimate that is absolutely reliable for the purpose of action or making a decision. The difference between *analytic* estimation of probability and cognitive content, if any, of the evidential basis of estimation must not be forgotten. Because I have discussed this point earlier, I will not go into it here in detail.

I would like to mention another point referred to by Reichenbach regarding the evidential indeterminacy of every statement about the things and beings of the world. In their eagerness to deny the independent existence of the external world some idealists go to the extent of asserting that the ego constitutes the so-called external world. Some of them try to buttress their view in terms of the observer-observed relativity, a corollary of Heisenberg's principle of *uncertainty* and Bohr's principle of *complementarity.* This idealistic interpretation of quantum physics is totally unacceptable to Reichenbach. The idealist, when consistent, is a solipsist. As a solipsist one cannot deny one's own personal existence. Curiously enough, this existence, as the Buddhist and Hume have tried to show, proves ever elusive. But in a way, inferentially in this case, the radical idealist has to be *more or less* sure about his or her own existence before denying that of external reality. In this situation one has to either invoke the Vedantic argument of self-knowledge (*ātma-sākṣātkār*) or the Cartesian *cogito* argument or some such argument. Obviously to the empiricists like Reichenbach none of these arguments is available. So they take a different route, inductive evidential route, to establish the existence of the self.

The sort of argument that enables the physicist to posit the existence of the external world is inseparable from positing our own existence. The supposed cut between the self and the world defended differently by the Sāṁkhya, Kant, and Wittgenstein is rejected by Reichenbach. The counsels of despair, that our self-existence is elusive and that the world is unknowable, are equally unwarranted. The observation statements available with us are good enough inductive evidence to posit the existence of the world. To posit is not to prove. It is a relatively weaker cognitive achievement. With reference to self-existence comparable achievement is also inductively available. From observation statements about my own body and other bodies, their actions, and memories, we do have inductive evidence to believe in the existence of my self and other selves. None of these evidential statements in isolation can be regarded as good or probable evidence. Only their totality is to be taken as inductive grounds justifying our belief in the selves. This justification is probabilistic and not certain. In this way, by incorporating the mental existence into the physical existence, the dualistic cut between the two is sought to be healed. In the process the traditional problem of epistemological passage from physical reality to the human mind and that of inferability of the physical world from mental data are claimed to have been solved.

There is no doubt that Reichenbach and philosophers of his type stand very close to Quine's physicalism. It is true that by trying to heal the cut between the world and the mind and that between the waking life and the dreaming life the former has significantly pushed forward the cause of physicalism. Two points, however, need to be clarified. If one wants to make the empiricist proposal to probabilify inductively our belief in self-existence and in the existence of world, one leaves some very important questions unanswered or only partially answered. Must we remain satisfied only with the probability structure of the said two forms of existence? Or can we get into their (more definite) causal structure? Should we not try to answer the natural question of whether our knowledge, probabilistic or causal, of self and of the world are at a par, that is, equally reliable, or not? Reichenbach states that the "myth" of two worlds must be given up. But he does not tell us why self-existence should be incorporated into the physical existence and not the other way around. Of course he does not conceal his intellectual anxiety over two possible consequences of accepting the second alternative; viz., solipsism and transcendentalism. Solipsism is neither verifiable nor refutable. Transcendentalism is enemy of skepticism, corrigibilism, and, by implication, opposed to the spirit of scientific progress. Those who like Reichenbach are opposed to the search for certainty strongly favor "the principles of knowledge [which] change together with its content."

Second, it seems that, conscious of Reichenbach's problems with the numerical determination of the weight of evidential statements, Quine abandons the former's idea about how relevant experiences affect or modify a theory. By induction the sort of weighing that Reichenbach proposes to assign to evidence appears to be suspect mainly due to two considerations: (1) the highly idealized character of weight-determination and (2) the discrete character of the concerned evidences. Because of (1) calculation of weight becomes practically very difficult and because of (2), left without numerically weighted individual evidence, one cannot determine the weight or cognitively meaningful content of the concerned theory itself. Both the theory and its evidence are embedded in other theories and other evidence, other than the ones given. For example, a theory of language learning is intimately interconnected with various theories of biology, psychology, and has for its background many other conditions and considerations. In this very complex situation it proves extremely difficult to determine which individual and new theory or evidence adds or subtracts what weight or content from the individual components of the current corpus of knowledge. The issue is clearly put by Quine in this way:

> in a scientific theory . . . a . . . sentence is ordinarily too short a text to serve as an independent vehicle of empirical meaning. It will not have its separable bundle of observable or testable consequences. A reasonably inclusive body of scientific theory, taken *as a whole,* will indeed have such consequences. The

> theory will imply a lot of observation conditionals . . . each of which says that if some observable conditions are met then a certain observable event will occur. But, as Duhem has emphasized, these observational conditionals are implied only by the theory *as a whole*. If any of them proves false, then the theory is false, but on the fact of it there is no saying which of the component sentences of the theory is to blame. . . . The scientist does indeed test a *single sentence* of his theory by observation conditionals, but only through having chosen to treat that as vulnerable and *the rest, for the time being, as firm.*[27]

This formulation of Quine's position seems to me very perceptive. First, it stands firmly committed to methodological *holism* and yet it recognizes the vulnerability of *individual* sentences of a theory. To admit clearly that the working scientists are often obliged to test a single sentence of their theories is in a way a sort of concession to the main point insisted upon by methodological localists like Popper. Though minor in nature, it is significant. Second, I like this formulation because it does recognize the commonsense fact that adverse evidential reaction on a single sentence does not have any *immediate* impact on the rest of the theory. By putting the matter in this way Quine takes care of the two main aspects of theory-construction: to recognize the need of confining attention to what happens to an *individual* component of a theory because of its relevant evidence; and also to admit that the theory as a whole remains vulnerable in principle. One does not hesitate to accept Quine's point that says that a sentence is too short a text to disclose its empirical meaning. But the more problematic issue is this: which totality of sentences, of what size, consisting of how many sentences, and in what context may text be considered a reasonable and independent vehicle of empirical meaning? This is like the dilemma of deciding how many hairs must there be *for* or *against* describing someone's head as bald. The skepticism that surrounds ostensive definition is also relevant to the point. It is easy to say in logic what a fallacy of composition or division is like, but there are always some "marginal" cases that lend themselves to different interpretations. Given these difficulties, it is not at all surprising that some philosophers like Quine prefer to follow a middle path. Their preferred starting point of philosophizing physics consists of middle-sized objects and not atoms or galaxies.

I will say more on this point after we briefly peruse Duhem's way of tackling this and related issues. Historically speaking, Quine's methodological holism or monism may be more due to Carnap than to Duhem. But, conceptually speaking, one can hardly deny the striking affinity between the positions of the two. Duhem certainly anticipated Quine in more than one way.

8. Why Crucial Experiments Fail to Be Crucial? The Points of Agreement and Difference between Duhem and Quine

What is often referred to as the Duhem-Quine thesis is resented by some writers. They think that some contemporary philosophers of science and epistemologists in their eagerness to highlight the points of agreement between Duhem and Quine do not take adequate note of the points of difference between them. The main intention behind bracketing Quine with Duhem seems to attribute to both of them a thesis that clearly admits that all scientific theories and laws are more or less uncertain, open to revision, and therefore never conclusive. In a sense this is true. However, one may argue that this truth is trivial and hardly disputed by the working scientists themselves. Only when science is philosophized to immunize it against all possible attempts to falsify it, does the issue of corrigibility of science views become a matter of dispute. The issue may be broached conservatively or radically. When serious Duhemian scholars like Stanley Jaki oppose the very label *Duhem-Quine thesis* I guess they have primarily two things in mind: in its received form it is inconsistent with the published works of the French philosopher and historian of science; and its partial plausibility must not blind one to many basic points of difference between Duhem and Quine.[28]

First, Quine himself writes in his "Two Dogmas of Empiricism" that the origin of his methodological holism is essentially due to "Carnap's doctrine of the physical world in the *Aufbau* [which asserts] that our statements about the external world face the tribunal of sense experience not individually but only as a corporate body." This version of methodological holism pertains primarily to the empiricist dogma of reductionism or methodological localism and is basically phenomenalist in character. To Duhem, science without metaphysics is a muddleheaded approach and therefore to be rejected.

Second, Duhem's religious and metaphysical commitments are clearly at variance with the spirit underlying Carnap's *Aufbau* and Quine's *From a Logical Point of View*. However, this does not negate the limited but important point that both Duhem and Quine clearly affirm that scientific theories cannot be tested in isolation.

Let us look into the point a little more closely. Duhem maintains that a physical theory is not experimentally tested by the precise observation of natural phenomena as such. The *interpretation* of phenomena and their symbolic representations are invariable attending conditions. The scientist is not directly concerned with the perception of concrete facts. Rather, the main concern "is the formulation of a judgment interrelating certain abstract and symbolic ideas which theories alone correlate with the facts really observed."[29] The scientist's role is not confined to creating an artificially clear and precise language to formulate the data of his or her experience. On the contrary, the language of science itself is contingent on the creation of a physical theory. Between abstract symbols and concrete facts o science there may be a sort of correspondence but to expect "complete parity" between them is ruled out. More positively speaking, an element of disparity between the practical fact of observation and the symbolic *theoretical* fact seems to be unavoidable. The same theoretical fact may correspond to an infinite number of distinct practical facts. To put the matter from the other end, the same practical fact may correspond to an infinite number of logically incompatible theoretical facts. This shows, among other things, that practical facts have no compelling or clinching effect on the scientific theorist. In a way the theorist always retains the freedom to formulate the perceptually available facts in more than one way.

These considerations lead Duhem to his well-known thesis that, strictly speaking, there is no crucial experiment in physics. One or two points need to be cleared up before this point is explicated. Extensive use of mathematical language and deduction ordinarily gives one the impression that physical theories are extremely precise and, therefore, correct. Correctness and precision are deemed inseparable. But this view does not take note of the problem of translation between the theoretical fact and the practical fact, between theories and perceptual data. Theories are never conclusively or fully determined by the evidential or explainable facts, retrospectively or prospectively, inductively or predictably; that is, deductively. This truism holds good irrespective of the nature of the language we use, mathematical or commonsensical. When a practical fact is mathematically symbolized, the symbol stands not for a single fact but for a bundle of several facts. The contents of the bundle are not exhaustively enumerable. The bundle works as theory. To ascertain the truth of its contents we have to resort to mathematical deductions or anticipation. Thus the inverted (or bundled) facts as theory in the process of experimentation are reorganized into a mathematical formulation and then tested deductively. This sort of mathematical deduction remains perpetually open ended. Neither the mathematical language used for the purpose of testing nor the facts obtained through deduction can conclusively ensure the decisive truth of the concerned theory.

> A mathematical deduction, stemming from the hypotheses on which a theory rests, may . . . be useful or otiose, according to whether or not it permits us to derive a *practically definite* prediction of the result of an experiment whose conditions are *practically given*. This evaluation of the utility of a mathematical deduction is not always absolute; it depends on the degree of the sensitivity of the apparatus used in observing the result of the experiment. . . . [And it] also depend[s] on the sensitivity of the means of measurement used to translate into numbers the practically given conditions of experiment.[30]

The basic point against which Duhem warns us is this. In our anxiety to get a practically definite prediction or deduction we must not assert a rigorously formulated true proposition. The price of definiteness or accuracy is very high. We are almost unconsciously required to ignore the complexity of attending conditions. So the consequence that by its seeming definiteness tends to deceive us is not of much value. Our anxiety for rigorous truth must not make us insensitive to the complexity of the practically given conditions, including the inexactitude of the means of measurement. It is enough if the theoretical premises used by us for the purpose of deduction are found to be "approximately true" and the consequences thereof are "approximately exact."

This Duhemian formulation, to start with, may appear loose. But perhaps he is right in pointing out and emphasizing the fact that the whole exercise of testing has two complementary aspects. First, it is assumed that the basis of deductive testing is not absolutely true. Had it been so the question of testing it would not have arisen at all. Second, the deductive consequences or predictions by which the theoretical premises are tested are not themselves irrevisibly given to our sense-experience. The consequences are partly due to theories and party due to methods of measurement. In addition, one may point out the unavoidable inaccuracy involved in representing symbolically what we get in sense-experience. Being fully aware as he is of the seeming "looseness of this explication," Duhem insistently asks us to believe that this is the best possible way of ensuring scientific and realistic rigors of mathematical deduction.

One must not be deceived by the apparent rigors of mathematical formulation and symbolism. Giving this explication, he asks us to bear in mind that mathematics is not made "simpler and cruder." On the contrary, it is refined, made realistic, and brought closer to both the theoretical situation and the practical conditions of scientific research.[31]

If *symbolization,* mathematical or otherwise, is a step necessary for theoretical simplification and correlation of the complex phenomena of nature, *interpretation* is necessary to make use of the concrete empirical data. The second requirement indicates that the data as such, that is, without interpretation, cannot be used either in theory construction or testing. The experimental physicist is not directly concerned with empirical data but with the formulation of a theory that systematizes and makes use of those data and, simultaneously, keeps them in their systematized form open to possible enrichment, correction, and modification by further data. In brief, data enter into a theory through interpretation, and again through interpretation they are of experimental value for scientific theory. But from this "interpolative" or double role of data one must not conclude that they are of no *critical* significance and can be easily incorporated into the concerned theory.

Notwithstanding his explication and qualifications, one may feel that the Duhemian picture of the physicist is more artistic than scientific. In a sense perhaps it is true, but that does not entail the rejection of this working procedure of the scientist. The latter can represent a theory as an explanation sketch, an impression used by Hempel in a comparable context. The sketch may be filled in by additional data. But the data available for filling in the sketch may go well or ill with it. The theoretical picture may be clarified or obscured by the added data.

However, if the added data are so interpreted as to make them compulsorily consilient with the theoretical sketch, then their use value turns out to be more or less additive or illustrative and not critical. The burden of Duhem's argument is to show that even interpreted facts can play a role that is critical and modificatory. Unless Duhem succeeds in showing it in terms of his theory, it proves difficult for him to establish his basic point that history of science is not only evolutionary but also progressive. Interpretation as such cannot be an input for progress, unless the input has its critical thrust and modificatory edges. Because the critic may always say with some plausibility that by *very liberal* interpretation any critical or infirming evidence may be transformed into a supportive or confirming evidence.[32]

Duhem's emphasis on interpretation of empirical facts leads his critic to accuse him of failure to get into the *depth* of reality and weakness toward phenomenalism and conventionalism. His repeated and approving reference to Mach's methodological maxim of economy of thought has lent support to the critic's accusation. Abstraction and generalization involved in theory construction are described by him as a "double-economy." Apparently, he is overimpressed by the principle of economy as recommended by Mach: "all science seems to replace experience with the shortest possible intellectual operation." The lesson that Duhem somewhat hastily draws from this methodological rule is rather radical in nature. This appears to him as good enough justification of condensing an infinity of facts, actual and possible, into a single law and unifying a multitude of laws into a single theory. Even after making a liberal concession to the Machian position Duhem turns back and critically realizes that consistent application of this rule would enable the physicist to assimilate and manipulate empirical knowledge in an unrestricted way. Obviously he does not feel happy about some consequences of this approach to science. He is afraid that espousal of this heuristic method would reduce scientific theory to nothing more than an economic tool.

I recall in this context Lakatos's stricture against what he calls the *monster-barring method.* If all inconvenient facts or negative evidence are interpreted in a very liberal and convenient way, Lakatos thinks, every theory can be "saved" and its refutation avoided. Additionally, if theoreticians are allowed to supplement their theories by an endless number of axilliary hypotheses, negative evidence can hardly falsify them. The "monster" of criticism or refutation may easily be "barred," if we resort to the earlier twofold strategy, liberal interpretation of facts and ad hoc supplementation of theory by axilliary hypotheses. This is a Popperian line of argument against methodological holism intended to vindicate the cruciality of crucial experiments.

In fairness to Duhem one must note that though he appreciates the Machian principle of the economy of thought, he does not accept it in toto. On the contrary, he has his own ontological commitments that, as reflected in empirical facts, make it difficult for theoreticians to deny their importance and manipulate them in any way they like. Somehow or other facts have an element of recalcitrance that does not allow theoreticians to assimilate them easily within their theories. Some favorable commentators of Duhem rightly point out that though he is theologically committed to a sort of "revealed" metaphysics, that does not deny him the autonomy he needs in scientific inquiry. At least that is his own conviction. Instead of *metaphysics* he ought to have used the more neutral word *ontology.* Ontological obduracy of empirical facts is responsible for his opposition to the Machian type of thoroughgoing instrumentalism.

In analysis, his position is found to be like that of the conventionalist or the pragmatist. In fact Louis de Broglie finds in him a sort of conventionalism that stands very close to the positivist and pragmatist conception of nature. In this respect he has been apparently influenced by his contemporary physicist and mathematician Henry Poincaré. Both are opposed to idealism, yet their realism is not robust or metaphysical. The sort of conventionalism they defend is not at all inconsistent with a modest version of realism. On the one hand, they do not distrust the verdict of sense-experience or practical facts; on the other hand, the statements about practical facts are not recognized as foundational in any strict sense. Because even the reports of direct experience are open to multiple interpretations. From this one must not rush to the conclusion that at the level of theory what survives our tests today would be able to do so in all times to come. On the contrary, it is prudent to maintain that an accepted hypothesis of today may have to be abandoned tomorrow in the light of an adverse verdict of experience. The realistic import of this point is transparent. At the same time, the historical hindsight of Duhem prompts him to admit that whatever theory is accepted, however well-tested, it may be found today, it has to be abandoned at some time or other.

But the main question he puts into focus is, What is the nature of abandonment or refutation? A theory is never accepted or rejected in isolation. When scientists speak of acceptance of a theory what they leave unstated is that it has to be consistent with other well-established theories. Similarly, prospectively speaking, when they assert possible modifications or even refutation of a theory what they take for granted is its embeddedness in other unrefuted theories.

> Whatever the nature of the hypothesis . . . it is never in isolation contradicted by experiment; experimental contradiction always bears as a whole on the entire group constituting a theory without any possibility of designating which proposition in this group should be rejected. . . . When certain consequences of a theory are struck by experimental contradiction, we learn that this theory should be modified but we are not told by the experiment what must be changed. It leaves to the physicist the task of finding out the weak spot that impairs the whole system.[33]

The holistic construal of theories has two main implications: the fallibilism and provisional character of individual theory, and the realistic constraint that underlies this fallibilism and provisional nature of theory. The "weak spot" of a theory in an obvious sense is within the corpus of theory itself. But in an important sense its root, its evidential cause, lies outside it. The internal structure of a theory and its external constraints or supports are complexly related, through a number of intermediaries: theoretical facts, practical facts, and principles of measurements and calculations.

Why is truth enigmatic? Why do determining truth and avoiding falsity prove so problematic? These and similar questions are bound to engage philosophers' attention. The reason why time and again Quine is to return to the question of the "exact" relation between theories and things is not quite different from the reason underlying Dummett's concern with bivalence and dissatisfaction over the alleged inadequacy of Putnam's revision of bivalent classical logic.[34] Which one of the two uneasily related values, simplicity of theory and fidelity to evidence, should be given preference in theory construction?

On the issue we may opt for one or more of the four main positions.

1. The simplicity of the theory has to be accorded the highest importance, but the structure of the theory as a whole must be more or less responsive to the impact of evidence.
2. Evidence is partly internal and partly external to the theory and therefore the theory builder's attempts to preserve or improve the simplicity of the theory does not depend much on such efforts.
3. Theory is our construction or conjecture but what confirms or infirms it is external to us and our theory, no matter whether it is knowable or not.
4. Unless evidence is available in a more or less *determinate* way its impact on the theory cannot be correctly or clearly indicated.

Position 1 may be attributed to Duhem. Position 2 is very close to Popper's theory of truthlikeness. It would not be perhaps unfair to say that position 3 substantially reflects Putnam's views on reference, truth, and internal realism. Position 4 seems to be close to that of Dummett. I must qualify my statements linking the preceding views with some philosophers' names by saying that this is merely typological and not very exact. The positions distinguished are not themselves clearly demarcated from one another. Quine, for example, perhaps would readily agree to both 1 and 3. In different ways Quine, Dummett, and Putnam are all in favor of revising bivalent classical logic. Putnam is of the view that quantum logic and a commitment to realism easily go together. Replaceability of classical logic by quantum logic is not a matter of factual or mathematical discovery, it depends mainly on how one interprets logical constants and formulates the theory of meaning. According to Dummett, for instance, "meaning cannot be conceived as given in terms of conditions for the possession of truth-values which attach determinately to statements independently of our knowledge."[35] Dummett's formulation of truth in terms of justification or availability of truth conditions gives one the impression that his realism is rather thin. The actual truth conditions are not easily available within our knowledge. And as Dummett insists on these conditions, Putnam fears, his realism is bound to get diluted. Therefore Putnam's suggested way out consists in *idealizing* the epistemic truth conditions. By highlighting the requirement of idealization, Dummett comes close to Quine's position.

Quine is not primarily interested either in defending or rejecting the bivalence thesis of the classical logic. It attracts him, but the attractiveness of "simplicity" must not blind one to the realistic difficulties it ignores. The paradoxical cases like "bald head and hairy head" suggest that to provide precise criteria for determining the correctness of using words to mean objects or situations is not very easy. This is due to the open-ended texture of empirical terms. And that also partly explains the inadequacy of ostensive definition. In an indirect way this underlies the partial indeterminacy of translation of one language into another. For example, the language of physics cannot be exactly translated into that of geometry. In the process, some loss of meaning seems to be unavoidable. However, this does not lead Quine to deny the factual content of the theory. The diversity of the contents of the theory and the inexactitude of the terms used in it do not necessarily take away the theory's simplicity or the austerity. Inscrutability of reference does not disturb him deeply. By translating the free-floating referring expressions of alien language in our own equally "unfounded" language, Quine assures us, we commit no avoidable sin. Ontologically speaking, no language is specially privileged; that is, firmer or clearer than the other. What minimize the indefiniteness of referring expressions are the language users' beliefs. No language exhibits its anatomy; not even its physiology. If the sartorial look of some appear tidier than the rest, the credit goes to the tailor; that is, the logician or the litterateur, depending on the context.

Quine's realism, unlike Strawson's for example, does not take "referring expressions" as its paradigm. Rather it rests on the reality of classes and universals. However, by doing away with the asymmetry between the subject terms and the predicate terms, strictly speaking, he stands equally distant from or close to their ontological commitments. Positively speaking, the symmetry thesis implies colivable coherence between what subject stands for and what predicate stands for. On the one hand, to indicate his realistic leanings Quine traces the sensory roots of the language use; and on the other hand, to preserve the simplicity of the theory or total science he highlights the realism of universals and classes. What directly or indirectly irritate our bodies are out there in the world, not easily internalizable without disrupting our web of beliefs. The transition from (fleeting) occasion sentences to (durable) eternal sentences is necessary for theory construction. The initial conditions of the theory are provided by the real world. But the theory itself is often found to be remote from the originary conditions and mediated by a host of beliefs and a lot of lower-level theories. From sense objects one may (rather is obliged to) move to abstract objects like universals and sets, facilitating the regimentation of language. If extreme sense proximity makes our perceptual grasp blurry or misty, far distance gives rise to fancy and doubt. Quine is convinced, the doubt cannot be removed by counterfancying the transcendental argument. Science must not be exempted from its answerability to what it owes its origin, the external world. If science fails to predict its behaviors, it must return and critically look back on itself. Its critical self-search is likely to be rewarding when its house is in reasonably good shape; that is, when the structure of science is simple. The more familiar we get with the space-time framework *of objects,* the smoother becomes our learning process. The process of induction is increasingly facilitated, elaborate, and conscious, "and in the fullness of time we even rise above induction, to the hypothetico-deductive method."

Whatever may be one's method (inductive, deductive, or narrative), of structuring the theory, the factors underlying its possible changes are twofold, internal and external. On the one hand, besides being internally coherent, its relation with its parent conceptual scheme has to be minimally consistent. On the other hand, it has to be responsive to the evidence, pro and con, relevant to it. In determining both the *relevance* and *weight* of evidence the *initial* say goes to the scheme within which the theory in question figures. This does not mean that *final* weighting of evidence is purely an internal affair of the parent scheme and the affiliated theory. If one *decides* what is true purely in terms of internal coherence of theories or beliefs and ignores the views, beliefs, and arguments from the outside world, including other things, beings and cultures, one is sure to face a lot of serious problems.

First, the distinction, not dualism between the theory and what it is all about can be denied or belittled only at a high and avoidable cost. The role of realism of all sorts, robust and modest, is substantially given up. In addition, it is counterintuitive. Second, coherentism, whether it is of the Bradleyan or Davidsonian type, fails to yield correspondence and is unfair to the external world. Bradley's idealism marked by the self-contradictory nature of the world of appearance, at least refrains from making any prorealistic claim. In contrast, Davidson and his followers, notwithstanding their uncertainty over the epistemically mediated relation between the conceptual scheme and worldly contents claim to have achieved success in containing skepticism. Reference to Quine, Rorty or Dummett, to my mind, does not improve the situation. Skepticism can hardly be contained by coherentism.[36] For, after all, coherentism itself cannot be coherently formulated in a surveyable form. Although I am favorably disposed toward Davidson's antifoundationalist thesis, I cannot endorse his formulation of the third dogma in terms of so-called anticonfrontation. The Davidsonian form of anticonfrontationalism reduces the world into the position of a non-playing player in the game of knowledge seeking.

Finally, as I have already argued, the world that provides all sorts of evidence and information cannot be totally internalized. Merely because of the effective existence of multiple epistemic intermediaries between the theoretician and the world we cannot deny either the distinction between the two or the inexhaustibility of the latter as the source of further information and "troublesome" evidence. What can possibly "disturb" the simplicity and coherence of our theory and is found to be not easily internalizable or really recalcitrant must not be suspected, discarded, or conveniently interpreted, emasculating its nature and destroying its identity. This is not the way of "saving" the theory. This is the way of first weakening it and then deadening it. This is not the realist's way of combating skepticism.

Skepsis means critical scrutiny, revision, and search. If the world is virtually denied entry into the domain of theory or knowledge, creative skepticism is sure to be replaced by dead dogmatism.

11

Between Dogmatism and Skepticism

1. Dogmatism and Skepticism as Two Different Moods or Stances of (Otherwise) Normal and Rational Human Beings

The title of this chapter may justifiably appear quite misleading. One might get the impression that, if we reject skepticism, we are sure to land in dogmatism and that, if we refuse to be dogmatic, we are obliged to fall back on skepticism. This formulation of the dilemma appears to be oversimple and we are led to believe that between skepticism and dogmatism there is no third or fourth (or several other) "resting" positions. As we ordinarily see, between the extremes, black and white, of the philosophical spectrum is a spacious gray area. Besides, one needs to be clear about the areas to be marked black and white. Just by giving a label *black* to a particular position, say, skepticism or even dogmatism, we hardly characterize it correctly. By doing so we primarily express our own preference or, one might say, prejudice.

Most philosophers, like most of us, people of common sense, have their moments of "dogmatism" and also those of "skepticism." We have already noted how basically antiskeptical philosophers are required to give an account, often a very long one, explaining why they are rationally obliged to accord a place, however limited that might be, to skepticism in their theories of knowledge. We have moments of doubt in our lives, both theoretical and practical. Because our pursuit of knowledge turns out to be *interested* for various reasons, we cannot own or disown a view without someone or another, strong or weak, sense of *satisfaction* or *dissatisfaction*. An element of will is always there to goad and to be goaded by our knowledge. Will as articulated in action gives rise to *expectation*, ending in satisfaction or frustration, depending upon the nature, positive or negative, of the actual outcome. Even when our will to know is not articulated in action, that is, remains "suspended" within our body-mind complex, it seeks its own goal, some sort of *relaxation*. If the goal-seeking tension does not find any satisfactory outlet, it generates *frustration*.[1]

A philosopher who starts with some such expression as "I am sure" and affirms something thereafter, has no special cognitive right to rule out all possible ways of being questioned about the content of this affirmation. Innumerable references could be given showing that very careful philosophers, just like the "prereflective" person in the street, are claiming that they are "sure" about the existence of external world. From the long history of science, philosophy, and philosophy of science we find an equally large number of very careful thinkers questioning and criticizing this realistic affirmation and its different formulations. Some assert that the world of science is a matter of inference; others claim that it is a matter of construction; Still others firmly maintain that the existence of external world, though extremely attractive and useful, is nothing but a hypothesis and, strictly speaking, not provable. Thus we find that one's firm assertion is seriously challenged and doubted by the other. Similarly, we find, one's denial or doubt appears to be unexamined and unfounded by the other. This game of "Yes *or* No" can be played indefinitely.

But what, in practice, do we *do*? Do we play the game of "Yes *or* No," that of "Yes *and* No," or that of "Yes *but* No"? Reflections on both theory and practice clearly suggest that we do not play the game of "Yes *or* No." In our theoretical moments we may affirm something, affirm it very strongly, because of some *practical* reason, but act otherwise. As mentioned earlier, even in our theories, taken as a whole, we have to confess our moments of "wonder," "speculation," and even "doubt." To take the point from the other end; when we act with a sense of certitude, if challenged, we often confess that the said sense of certitude was more willful than warranted. The critic may characterize this attitude as dogmatic. Taken in isolation, perhaps it is. Wittgenstein has rightly pointed out that to speak of "doubting doubt" makes no sense. Either we doubt or we do not. Either we agree to the proposition expressing a doubt or we reject it or we may hold back judgment about it. In any way, by "doubting doubt," if there is anything like that at all, we do not add or subtract anything from the original doubt, the content of doubt, to be more precise. Similarly, cannot we say that by being dogmatic about a dogma we do not add anything to its content, that is, what the concerned dogma is about, or subtract anything from it? "Dogmatic dogma" is hardly anything more than "dogma" itself or its reiterated form. It is like "doubted doubt" a mere verbiage. Then, are we called on to produce grounds in favor of or against doubt so that it does not prove vulnerable to some such charges as "a mere verbiage?"

To answer these questions the first thing to be carefully considered and defensibly decided is whether the mode of consciousness, named doubt, that is, skeptical in character, is totally different from or constitutes an integrated continuum with the mode of consciousness known as knowledge. Whatever might be the ontological status of consciousness—physical, biological, or independent of both, or interactive with them, for the time being or for the purpose of expounding our argument—we are assuming that knowledge and doubt are different modes of same consciousness. When we speak of the "degrees of doubt" we give the impression, maybe unintended, that something in doubt makes it nonhomogenous. In the context of knowledge when we speak of "degrees of knowledge"—opinion, imagination, knowledge proper, and so on—we somehow stand committed to the view that knowledge is graded, that truth claims of different grades of knowledge are differently satisfiable, that there is difference between truth and truthlikeness, for instance. If we closely look into truthlikeness or verisimilitude, we realize that our cognitive approximations to truth (as ideal) are distinct and clearly different. If doubt is construed as integrated with knowledge and forming a continuum, it is difficult to avoid the conclusion that in (what we call) doubt is a cognitive component.

There are philosophers who maintain that doubt and knowledge, though seemingly antithetical or antipodal, are in fact two extremities of the same cognitive consciousness. What is implied by them is that between doubt or skepticism, on the one hand, and certain knowledge, on the other, are other kindred (but not quite unalloyed) modes of consciousness. Some philosophers like to define this doubt-knowledge continuum in terms of varying clarity and distinctness. "Certain knowledge" is claimed to be free from the traces of doubt and "uncertain doubt" marked by glimpses of knowledge. Some of the writings of Leibniz, for example, lend credence to this view.[2] Other philosophers like Descartes are inclined to define this continuum in terms of varying will. Error or mistake is attributed to will.[3] But although discounting the possibility of universal doubt, what in effect the Cartesian concedes is that our will cannot be *completely* free from knowledge and capable of *autonomously* generating what is false. In other words, falsity cannot be a content of false consciousness. Some *cogito*-like principle must be there to sustain false consciousness and disclose its falsity. An all-enveloping false consciousness or universal doubt is said to be an empty speculation or a "mere verbiage."

Whether one traces "the origin" of doubt to obscurity, lack of clarity, or will, it seems undeniable that knowledge as available to us, embodied and imperfect human beings, is more or less ambiguous and uncertain. Even when something is *logically* shown to be immaculately correct or clearly proved, we are *psychologically* haunted by a sense of uncertainty. This "sense" may be condemned as rootless or merely psychological, neurotic; even then we ourselves cannot completely get rid of it. Knowledge does not appear to be a simple case of having or not having, of owning or disowning. There is something else in it. We may call it an element of decision. Even presented with a "valid piece of knowledge," a well-established theory, we may, often in fact do, feel undecided about its acceptability. What is logically or provably accepted by others is not ipso facto accepted by us. We may have in us some extralogical considerations preventing us from deciding its acceptance. In other words, logical proof by itself is not epistemologically compulsive. An element of decision or, one might even say, an ethic of decision is always at work in our mind that has a lot to do with our acceptance, hesitant acceptance, rejection, hesitant rejection, reviewed rejection, and the like. We, fallible creatures, doubt or waver not only in our commonsense moments of life but also in our scientific and mathematical moments. Nothing is absolutely doubt free in any of our cognitive enterprises.

2. Different Meanings of Doubt: Epistemological and Ontological Issues

The very possibility of a doubting situation may be, in fact has been, understood in different ways. First, the doubt of a doubter may be rooted in his or her consciousness so that the intended object of knowledge is not fully or completely available. It is not easy to spell out what "complete availability" means or amounts to. One might plausibly maintain that for finite and fallible person "complete availability" is hardly anything more than a useful *ideal* to aim at and approximate. Second, one's doubt may be due to one's reflective attitude. Every time one reflects on the content of one's knowledge, one finds something new, missed, or added. This changing character of the content revealed in reflection often leads us to doubt whether our knowledge is really reliable. Third, the doubt may be rooted in the elusive character of our knowing mind itself. If the mind that is supposed to be the locus or owner of knowledge itself, on scrutiny, turns out to be indefinite or fluxist in character, our skeptical disposition tends to assume a more radical character, irrespective of the nature of the object or the content of knowledge claim. A theoretical way out of this form of skepticism is to be found in the view that mind itself or its sustaining principle, self, or soul, or conscience, is self-shining; that is, not vulnerable to error or mistake. Fourth, another form of skepticism affirms that the question of doubt arises only in respect of the objective world, external things, and has nothing to do with the self-evident self. What is suggested is this: when we are obliged, instead of the *self,* to fall back on the *other* in search of evidence *for* our knowledge claim, we are landed in doubt. According to this view, often ascribed to Descartes, doubt is traceable to self-other dualism or dichotomy. Unless some such bridging principle as Cartesian "veracious god" or Leibnizian "pre-established harmony" or Husserlian "intentionality" is invoked and used, the dualism and the resulting skepticism appear to be unavoidable. However, this is not to commit ourselves to the correctness or otherwise of any of these principles. Fifth, skepticism is often said to be due to the mixed-up character of our "cognitive" phenomena. Affection, volition, conation, expectation, and so forth are found to be inextricably mixed up with and influential on our cognitive enterprises and achievements. Pure knowledge is a myth. "Impurity" of knowledge is very natural and human. It is only when we pin our *faith* on the paradigm of pure and perfect knowledge of the infinite God do we start feeling unhappy with our so-called skeptical mode of knowledge. It is therefore not surprising that many epistemologists invoke some or other form of transcendental argument or certificatory principle to show that human knowledge under the aspect of the universality of God's knowledge can be freed from the blemishes of "natural" doubt or uncertainty.[4]

Once the difference in the construals of skepticism is duly recognized, the principles summoned and deployed to tackle the problems of skepticism are bound to prove diverse. There is no unique way of solving the problem of doubt. Because the term *doubt* itself is not used in an unequivocal manner, different perceptions of the nature of doubt invite different methods or approaches to solve it. The very fact that "the nature" of doubt does not appear identical to epistemologists, scientists, or even people of common sense may easily be interpreted as an additional argument in favor of the skeptic. The resulting outcomes being diverse and, at times, even conflicting lend additional force to the skeptical view(s). The typologies of skepticism and antiskeptical theories or practices offered to refute skepticism has no clinching effect and cannot close the controversy forever. None of the suggested refutations can possibly yield a categorically imperative ethic of decision in respect to the acceptance of a particular theory or a prescribed course of action. Whatever might be our available arguments and evidence for and against a (factual) proposition or a (valuational) proposal, our choices often turn out to be hard. This view may well appear as vague, general, and abstract. But it seems to me that even if we take a down-to-earth concrete case, the picture is not likely to be different. In this context we are reminded of the classical problems of ostensive definition or naming properly. The elusiveness of an ostensive definition confirms the skeptic's sober stand. The seeming steadiness or rigidity of the (logically proper) name-nominatum relationship is a postulational way out at best and a useful fiction at worst.

It is true that in some commonsense cases like the existence of our family members, household furniture, commonly visible trees and houses, we do not ordinarily entertain doubt. In this sort of case we speak of shared experience, successful communication, and the like. But a little reflection is enough to convince us that even from this (supposedly) doubt-free area of commonsense experience the haunting ghost of skepticism cannot be completely exorcised. Arguments from illusory perceptions, the problems of (noncomplementary) perspectival descriptions and the like, when deeply probed, lend strength to the skeptic's modest sense of practical realism.

Whatever might be the philosophical persuasion of the skeptic, it is clear from the typology of skepticism that the skeptic's view is somehow related to the finitude and fallibility of the knowing subject and the incomplete availability of the object to be known.[5] True, the skeptic may be asked about the reason for such uneasiness with a self-proclaimed view of human fallibility and the finitude. The critic may squarely pose this question: Where do you get this notion of finite and the fallible person? Is one condemned to espouse this notion? Does it mean that the skeptic is tacitly committed to the unspecified existence or essence of some infinite and infallible human nature? If this nature is really infinite and infallible, must we characterize it as human? Or, in that case, should we characterize it as divine?

The skeptic's response to these questions may be somewhat like this. When I say, "my knowledge of myself is incomplete," it does not bind me to the view that there is a self-view of mine available to me that is transparent and complete. Unnecessary importance need not be attached to the "relativistic" undertone of the terms like *in*complete. When one speaks of the incompleteness of one's self-view, one is expressing only the open-textured character of one's own *self*-knowledge. From our own life experience we find that not only our views about *other* things and beings undergo change but also our self-image. This all-comprehensive change, both objective and subjective, in spite of its comprehensiveness, continues to influence us. Furthermore, this changing perception of self-image and other-image also proves communicable. This changing character of the knowing subject and the object known (or knowable) does not demolish the case of the skeptic. There is no incoherence between the assertion that I know myself and the assertion that the knowledge that I have of myself undergoes change. Logically speaking, the only intuitive requirement to be satisfied is this: the "I" that changes and is an "object" of knowledge is less durable or stable than the "I" that knows it and to which the former is presented. This logical requirement thus in no way ontologically obliges the skeptic to fall back on the view that the "real knowing subject" is immutable or changeless. Our knowing is like our motion on board of a moving boat or flying airplane. It is a well-known theory of physics that a body may appear stationary or motionless while in fact it is moving. But all depends on the relative motion of the viewing body (or the body-mind complex) and the viewed body.

The critic of the skeptic, in a bid to hold his or her ground, points out that the skeptic's arguments are piecemeal or fragmentary in nature, and that they cannot be brought together and viewed as a systematic whole. Pressing this point further it has also been argued against the skeptic that such arguments are ad hoc and, what is worse, inconsistent. The skeptic's logical responsibility is said to be minimal. For, it is alleged, the skeptic is content with only pointing out the inadequacy and the nonconclusive features of the opponent's arguments and gives the impression of having no particular view to defend because that might compromise the basic skeptical stand.

From this line of argument one might think that the skeptic's position is primarily, if not exclusively, epistemological and that it has no ontological commitment at all. One's access to reality or reals cannot be understood except in terms of experience. The reach of experience, being limited, is open to contradiction. At this stage the argument may be exploited in either way, proskeptical or antiskeptical, depending on what interpretation is put on "uncontradicted experience." In the Indian tradition we find that Buddhists like Nāgārjuna and the defenders of Nyāya-Vaiśeṣika like Vātsyāyana and Jayanta Bhatta are engaged in a debate on the point. The latter strongly maintain that all forms of valid knowledge, perceptual or nonperceptual, have their unmistakable objective references that attest to the existence of a universe of reals in and around us. Our experiences of the reals, unless contradicted, have to be taken as valid. Somewhat like the phenomenologist, the Nyāya-Vaisēṣika philosopher thinks that experience has always in it an objective reference that imparts a definite character to experience. In other words, experience is necessarily experience *of* something and that something determines the shape of experience itself. If the skeptic wants to maintain that the open-textured character of experience is a ground for doubting it, naturally, it has been argued, the question arises: How can an experience without a definite shape of its own be possibly contradicted by another experience (without a more or less definite shape of its own)? In other words, contentless or shapeless experience, that is, experiences without objective reference, however provisional that might be, cannot clash or contradict another experience of the same kind.

This line of argument and counterargument, if carefully scrutinized, shows that the rival schools are committed to different notions of skepticism. In spite of their seemingly destructive skeptical dialectic, *catuṣkoti nyāya* for example, the Buddhists still maintain that they believe in some reality, *tattva,* which should not be taken as mere void or nihil. The commonsense association of the term *śūnya* (as empty) has given rise to the unwarranted conclusion that the Buddhists' skeptical epistemology has made it impossible for them to own any ontology, theory of reality, whatsoever. All that the Buddhist dialectic tries to show is the "futility" of our propositional affirmation, negation, affirmation and negation, affirmation or negation, and so on, of what there is. The unpredicability of reality does not amount to its total denial. The Buddhists seek to show that the predicates we use to characterize reality, on logical scrutiny, are found to be pointless or untenable. Predication means conceptualization or categorization. According to the Buddhists, *śūnya* (as reality) is neither conceptualizable nor categorizable. It is sui generis. It has to be realized.

As we know, this view is rejected by the Nyāya-Vaiśeṣika philosophers, who understand reality as consisting of innumerable reals. But the reals, according to them, are of different types and categories. They make it plain that these are *not* to be taken as ideal constructions or somehow dependent on the human mind, nor are the reals the articulations of God's mind. On the contrary, they affirm, these categories, or *padārthas,* viz. substance *(dravya),* quality *(guṇa),* action *(karma),* universal *(sāmānya),* particularity *(viśeṣa),* and inherence *(samavāya),* are absolutely objective facts.

Whether one becomes conscious of them or comes to know of them has nothing to do with their being. It is purely accidental that one knows these reals. Though Kaṇāda and Praśastapāda speak of only the six positive categories, a seventh category, negation or nonbeing *(abhāva),* has been recognized by the later exponents of the system like Udayana, Śridhara, and Śivādityạ.

The basic metaphysical view of the Vaiśeṣika is shared by the Naiyāyikas. But the latters' scheme of categories is quite different from that of the former. They recognize as many as sixteen categories: means of valid cognition *(pramāṇa),* objects of valid cognition *(prameya),* doubt *(saṁśaya),* purpose *(prayojana),* probative examples *(dṛṣṭānta),* established conclusion *(siddhānta),* principles of a syllogism *(avayava),* hypothetical reasoning *(tarka),* conclusive knowledge *(nirṇaya),* arguing for arriving at truth *(vāda),* arguing for victory *(jalpa),* merely destructive argument *(vitaṇḍā),* fallacious reasons *(hetvābhāsa),* quibbling *(chala),* pointless objections *(jāti),* and vulnerable points in an argument *(nigrahasthāna).* These categories, on analysis, are found to be primarily logico-epistemic. I say "primarily" because the objective reference of knowledge is strongly insisted on by the Nyāyas. Though the main importance is attached to the means of valid cognition, yet by their very nature they are objectward. The other fourteen categories are viewed under the aspect of and as subsidiaries to the means of valid cognition. Intimate knowledge of the subsidiary categories is very necessary to decide the appropriate means of cognizing them. In this connection, how doubt is to be known is also to be discussed and decided. Note here that, unlike the Vaiśeṣikas, the Naiyāyikaṣ do not claim their enumeration of the categories to be exhaustive. Whereas the Vaiśeṣikas' main interest is ontological, the Naiyāyikas seem to be basically concerned with the issues of knowledge and how knowledge is related to reality. It is therefore often known as a *pramāṇśāstra.*

In the Nyāya-Sūtras of Gautama doubt has been defined as a conflicting judgment about the precise character of an object, arising from the recognition of properties common to many objects, or properties not common to any of the objects, from conflicting testimonies and irregularities of perception and nonperception.[6] Here one observes five kinds of doubt. First is the recognition of common properties; for example, seeing a tall object in the twilight it is difficult for us to decide whether it is a person or a post; we are in doubt because the property of tallness belongs to both. Second is the recognition of properties not common; for example, having heard a sound we keep on wondering whether it is eternal, we do so because the property of sound abides neither in human, beast, and so on, which are not eternal nor in atoms, which, according to the Nyāya, are eternal. Third are conflicting testimonies; for example, merely by study we cannot decide whether the soul exists, because some systems of philosophy affirm that it does, whereas some other systems deny that it does. Fourth is the irregularity of perception; for example, we perceive water in a tank where it really exists, but water appears also to exist in the mirage where, really speaking, it does not exist. Fifth is the irregularity of nonperception; for example, we do not perceive water in the radish where it really exists or on dry land where it does not exist.

From the Nyāya view of doubt it appears that the origin of doubt is primarily epistemic; that is, it is to be found in cognitive judgment. Though every piece of knowledge has its object, the latter cannot be held responsible for our doubt. In other words, if the object of knowledge could be grasped in its purity, doubt would not arise in the knowing mind. In the analysis of the skeptical situation, the Naiyāyika presupposes an "unencumbered" human cognitive capacity that can clearly, unquestioningly, reveal the nature of object to us. The reals do not give rise to doubt. Ontology is not to be blamed for our skepticism. For skepticism is peculiarly human; its roots are to be found in some blemishes, imperfections, or confusions of our mind. The uncompromising realist absolves reality from the responsibility of generating skepticism. The knowing mind is itself responsible for its skeptical states. Doubt, though a knowable object, is not per se a real object. It is due to a sort of (intra- or inter-) category mistake committed by the knowing mind.

The Naiyāyikas' effort to draw a sharp line of distinction between the pure or unencumbered knowing mind and the pure reals to be known is the central point of debate between the realists like them and the skeptics and the relativists like Nāgārjuna and Hume. The latter are firm in their view that the reals, whatever they might be, have no intrinsic and objective character of their own and that in the process of grasping them our mind can never operate purely; that is, in an absolutely unencumbered way. In other words, the knowing mind is believed to be always loaded with theories, biases and prejudices, beliefs and expectations, *saṁskāra* and *kalpanā*, which cannot simply be wished away. It would be naive to maintain that conception-free perception can successfully grasp the reals as they are.

Careful scrutiny of the views and arguments ascribed to different skeptics and dogmatists suggests that the words *skepticism* and *dogmatism* are more in the nature of label or ideal type than truly reflective of the concerned specific views. Hardly any philosopher worth the name will admit to being a dogmatist. It is interesting to note in this connection that the Sanskrit word for *dogma* is *sarvatantrasanmata,* the view endorsed and shared by all schools. In this particular sense many philosophers, almost all, will agree to admit, for example, that there is external world and that there are other human beings. The dispute arises and is limited to the ontological or the epistemological status of what is believed by all philosophers. The same belief or dogma may be understood and explained differently. And that it is actually done so is a well-known fact even outside the domain of philosophy, which is often accused of useless, if not insincere, logomachy. It is, therefore, not surprising that in response to the question, What is wrong with skepticism? we often hear the answer, it is not only paradoxical and untenable but also insincere. The alleged paradoxicality is ascribed to the skeptic's *assertion* of his or her own *thesis,* whatever that might be. If to the skeptic all views are untenable, it is then difficult to say how one's own view or thesis can escape this stricture or fall outside the domain of untenable views. In that case the consistent critic is advised to remain silent, not to *assert* or *deny* any view. Note here that if some skeptic or other becomes really silent for the sake of consistency, the minimum that is conceded by implication is a subscription to the principle of consistency. But we need not pursue this quibbling because the consistent (and, therefore, silent) skeptic's view will *not* be there before us to discuss its tenability or otherwise.

A somewhat comparable line of argument may be offered to show that the alleged dogmatist is not dogmatic in the foolish sense. If someone maintains a view strongly, affirming it repeatedly, and refuses to give any argument in support of this view, we may, in fact we do, characterize the person as a dogmatist and the view as dogmatic. However, this does not prove that the dogmatist has no argument or, at any rate, consideration in support of that view. The view may appear so clear and transparent, so self-presenting, that when challenged to defend it the dogmatist does not take the challenge seriously at all and, what is worse, thinks that the challenger is insincere. Neither dogmatism nor skepticism should be taken literally.

In this connection, it is instructive to recall, as alluded earlier, that Nāgārjuna's nihilism has been interpreted in quite different ways. Some authors, following the Nyāya tradition, maintain that Nāgārjuna is a thoroughgoing skeptic and that he has no ultimate ontological commitment whatsoever. He has often been criticized as a mere logomachist *(vaitaṇḍika).* It is undeniable that occasionally some Buddhists have characterized themselves as *vaitaṇḍika;* that is, they have no view of their own but refute the proponent's view using the latter's own logic. Whatever they might have said, it is clear to our rational understanding that those who have nothing to say of their own are not to be expected to take the pains of thinking and writing out their own views or even their reactions to others' views. The positive implication of our interpretation of the position of the nihilist-Buddhists like Nāgārjuna and Candrakīrti, that they had definite views of their own, has received support of some authoritative interpreters of Buddhism. Nāgārjuna's metaphysics has been likened by T. R. V. Murty to Kant's metaphysics.[7] The nihilist's apparent denial of metaphysics is to be understood in a constructive way. They seem to deny it because it is inaccessible to reason that ordinarily operates judgmentally; that is, aided by predicables. They believe in a higher cognitive capacity, intuition *(prajñā),* which enables them to realize what is real *(tattva).*

It is also clear that Stcherbatsky's *Conception of Buddhist Nirvāṇa* that it would not be correct to think that *nirvāṇa,* the highest intuitive realization, means denial of the reality of the phenomenal world, *saṁsāra.* If the latter is totally denied, the question of the highest realization or liberation does not make any sense. On the contrary, the transcendental world assumes its meaning in and through the phenomenal one. Unless the phenomenal world is somehow posited, the question of its gradual cancellation and that of attaining the higher reaches of the transcendental life do not become intelligible. In this discourse on the reality or otherwise of the phenomenal and the transcendental worlds, strictly speaking, both the expressions *phenomenal* and *transcendental* should be taken as bracketed. In the process of unbracketing them the Mahāyāna Buddhist tries to show the practical necessity as well as the logical inadequacy of affirming the reality of the phenomenal world. The intuitive grasp of the transcendental reality, embodied in the state of *nirvāṇa,* does not mean rejection of the phenomenal world. In the subsequent concept of the cosmic body *(dharmakāya)* of Buddha, one notices an ingenious attempt to reconcile and unify the reality of the phenomenal and that of the transcendental.[8]

This constructive interpretation of negativism and relativism is bound to remind one of Hegel's view of the progressive dialectical method and Bradley's destructive one. We have already noticed earlier how Hegel's dialectical progress of knowledge may be, and actually has been, construed all along to have a skeptical strain in it. It is well known that Bradley's destructive logic of *appearance* does not totally deny a positive concept of reality—reality as pure sentience.

The basic point to be borne in the context of the debate between the skeptic and the antiskeptic is the precise form of skepticism (or antiskepticism) to be considered and, if possible, refuted. In addition, we are advised to bear in mind our own (as well as others') criteria of acceptance (or establishment) and rejection (or refutation) of a particular view and its supporting arguments. We accept (or reject) many views and arguments on nondemonstrative or persuasive grounds. This only shows that all our *theoretical* decisions are not based on pure logic. Many of our views and beliefs rest on *practical* considerations, cultural affiliation and the like. In other words, before we make a realistic effort to tackle the problem of skepticism we must recognize at least two facts: the knowing subject is not an isolated enterprise free from its all attending conditions; and the object to be known is not an isolated entity, unrelated to other entities and singly available to the knowing mind. It is here I would like also to differentiate the two aspects of the skeptical situation, ontological and epistemological; that is, we should explore *ontological* localism and holism, on the one hand, and *epistemological* localism and holism, on the other.

According to some early schools of Buddhism (the Sarvāstivādin and the Vaibhāṣika) both *saṁsāra* (the phenomenal world) and *nirvāṇa* are real. The elements of the former—past, present, and future—are conceived as equally real. These elements are of two types, the eternal ones and the transient ones (in their manifest form). In their nonmanifest stage the transient elements become lifeless but still contain the potentiality of living manifestations. This view of the Vaibhāṣika stands close to the Sāṁkhya doctrine of the undifferentiated matter *(prakṛti)* in its eternal form containing the possibility of its future modifications within it. Both the views are in a way anticipative of the modern cosmological view to the effect that, although different forms of energy may be destroyed due to heat or cold, energy itself remains eternal. The state of enlightenment (Buddhatva) is not to be construed materialistically, as a sort of *ucchedavāda,* for that would be contrary to the very spirit of the teachings of Buddha. The death or extinction that the Buddhist speaks of is in accordance with some moral law *(karma)* and not physical law. Among the elements of the phenomenal world the Vaibhāṣika recognizes matter, mind, and force. The knowing mind, like other elements, also becomes extinct in *nirvāṇa*. But, as pointed out earlier, this is *not* an extinction that is total, absolute, or without the possibility of return or reformation. Knowledge of this mind or that mind may get lost in *nirvāṇa,* but *nirvāṇa* itself is not an eternally lost case. In this way an antiskeptical formulation of the Vaibhāṣika view is quite possible.

Even the Sautrāntikas, the realist Buddhists, who take *nirvāṇa* as the absolute end of the manifestation, the end of passion and life, are unwilling to deny *nirvāṇa* in general. For that amount ot accepting *nirvāṇa* without a Buddha. Their denial is purported to refute the materialistic concept of *nirvāṇa* and, in this respect, their position is close to that of the Mahāyānist school. But the difference from the Mahāyānist in the context of knowledge is not to be minimized. The recognized reals of the Sautrāntikas are not confined to sense data, consciousness, and volition, but also include objects of sense perception.

In the idealistic (Yogācāra) school of Dignāga we find a new epistemology in which existence and cognition, object and subject, coalesce. That this is not an easy concept of epistemology to grasp is conceded by Dignāga and his followers. For example, in his analysis of solipsism Dharmakīrti admits that though we are obliged to accept the presence of the Buddha as *dharmakāya,* or the cosmic body, in all subjects and objects, it is not easy to explain this compresence.

It is well known that the Mādhyamikas like Nāgārjuna and Candrakīrti have expounded very radical views of ontology as well as epistemology. According to them, the defenders of the basic doctrine of momentariness, both *saṁsāra* and *nirvāṇa* are unreal, separately unreal. Kumārila has criticized them not only for denying the existence of external objects but also for denying the reality of human ideas. Śaṁkara feels exasperated with the radical nihilism of the Mādhyamikas and does not know how to argue with them. For they reject the very possibility of cognizing the absolute reality by logical means or methods *(pramāṇa).* However, Vacaśpatimiśra sympathetically interprets the Mādhyamika position as an affirmation of the inadequacy of logic in solving the question about the very elusive nature of existence and nonexistence. The unspeakability *(anirvacanīyatā)* or the inadequacy of every predicable in relation to the nature of the absolute reality has been pointed out time and again by Śaṁkara himself. It is not at all surprising that the modern scholars like Keith and Stcherbatsky are of the view that Nāgārjuna's real aim was to show that human intellect in its bid to grasp the nature of highest reality lands itself into hopeless antinomies. In other words, if these interpreters are to be believed, we have to remain content with the knowledge of what is less than truly real. By its very nature human intellect seems to be condemned to skepticism, a sort of "imperfect" knowledge of the "imperfect" reality.

The question is, Is this skeptical position the best possible one we are capable of attaining? Or can there be a knowing mind of an altogether different kind, which is capable of having knowledge that is not plagued with irremediable antinomies? In search of answers to these questions we are reminded of two kinds of knowing mind: the human mind and the (notionally all-perfect) God's mind. Epistemic localism of human mind is said to be absent in God's mind. Consequently, it is claimed, the latter is free from the skepticism and relativism of the former.

Epistemic localism and ontological localism are often found to be going together both in India and the West. As we have already noted, some (idealist-realist) Buddhists like Dharmakīrti maintain that the real is what is available in pure sensation. It is to be regarded as particular. It is reality in itself. It is pure affirmation without any trace of negation in it. This notion of reality *(nirvikalpakaṁ)* is distinguished from what is ideal *(savikalpakaṁ)*. If the real is particular, the ideal is universal. If the former is available in pure sensation, the latter is grasped conceptually (in understanding). If the real is "in itself" *(nirapekṣa)*, nonrelational, the ideal is "in the other" *(sāpekṣa)*, relational. Whereas reality is affirmative, ideality is negative.

This view of reality and ideality is likely to remind one familiar with Western philosophy of Berkeley and his famous dictum, *esse est percipi*, essence of thing consists in its being perceived. In this case what is extraperceptual, lies beyond the scope of pure sensation, that is, external, is ideal. The distinction between the ideal and the real need not be overemphasized. The internal is causally dependent upon the external. Their identity is insisted on only by the extreme idealist. The external is real and efficient, particular and moving, instantaneous and positive. The causally produced internal image is universal, immutable, nameable, and negative. The postulation of the existence of the external object corresponding to and causing sensation is basically psychological and not logical or absolute.

The works of Dignāga and Dharmakīrti indicate a dialectical, critical, and antidogmatic trend. Their logic marks a departure from the realistic ontology of the early pluralists (Sarvāstivādins). In a way it is also critical of the realist (Vaiśeṣika) school strongly affirming the reality of the universals. Dignāga tries to show that there are three different possible ways of understanding the world. Or we may speak of three different worlds, planes of reality: the world of things, the world of ideas, and the world as a whole. At the logical plane we have a pluralistic world consisting of matter and ideas available in sensations and conceptions. The intermediate world knows no matter in it and consists only of ideas. Even matter itself is an idea. This is again a clear anticipation of Berkeley's immaterialism. Epistemologically speaking, it also indicates the limits of human cognition. The reality of the world as a whole, ontological holism, is not logically available to us. It is metalogical and does not recognize the law of contradiction. This law is said to be applicable only to the world of plurality, necessary for understanding the pluralistic reality of matter and ideas, sensations and conceptions. In and through introspection we get acquainted with this world of multiplicity. The monistic world is metalogical and to be grasped through intuition that is not bound to the law of contradiction *(prajñā)*.

Dignaga's *Prajñāparāmita-piṇdārtha* has been written from the monistic point of view and his *Ālambana-parīkṣa* contains a defense of idealism. His accent on *prajñā* or intuition is purported to vindicate the veritable knowledge of the world as a whole, monistic reality. In this connection, once again, we are reminded of Berkeley's resort to *notion*, a self-evident and intuitive epistemic capacity, for establishing the existence of the all-comprehensive God. Interestingly enough, Berkeley also tries to reconcile the empirical world of plurality with the intuitive world of unity. Neither Dignāga nor Berkeley would be prepared to accept that the world of unity is *transcendental* in any ordinary sense. What they would perhaps emphasize, instead, is that the world of unity is metalogical, has to be grasped through notion or intuition. This parallelism may be viewed as further extended. Both Dignāga and Berkeley would be prepared to recognize the limits of knowledge at the level of the reals given in sense perception. But they are opposed to the idea of extending skepticism to the level of the monistic world that, strictly speaking, is neither nameable nor *logically* provable.

> Dignāga and Dharmakīrti call themselves idealists, but they are realists in logic and idealists and even monists in metaphysics. In logic reality and ideality are divorced, but the "Climax of Wisdom," says Dignāga, "is Monism." In the very final Absolute subject and object coalesce. "We identify," says Dignāga, "this spiritual Non-duality, i.e., the monistic substance of the Universe with the Buddha, i.e., with his so-called Cosmic Body." Philosophy here passes into religion.[9]

Under whatever head, philosophy or religion, we critically examine the relation between *many* objects of the (seemingly?) external world and *one* (transcendental?) reality, the affinity between the Buddhists like Dignāga, on the one hand, and the Advaitins like Śaṁkara, on the other, is unmistakable. It is no surprise that

Śaṁkara, otherwise a severe critic of Buddhism, has himself been criticized as a camouflaged *(pracchanna)* Buddhist. Both the Advaitin and the Mahāyanist have only accorded a dream status *(svapnavat)* to the empirical or external world. This is not the place to go into the subtle difference between Dharmakīrti and Śaṁkara on the ontological status of dreams and the objects available in dreams. In the context of skepticism the point of interest to us is the cancellability or corrigibility of our dream world by the more durable, often claimed to be eternal, real world. In a sense even the philosophers ordinarily known as idealists-monists, from Dignāga and Śaṁkara to Hegel and Bradley, are prepared to offer a detailed account of the empirical world, although at the end they try to show why ultimate ontological and epistemological status cannot be accorded to it. It would be rather naive to think that any serious philosopher, whether an idealist of this sort or that, can possibly fail to recognize the reality of the sense world (studied by natural sciences).

If unreality is defined in terms of corrigibility or cancellability, it seems clear that our scientific world somehow gets dissolved into the world grasped by Reason or *prajñā* (intuition). The parallelism between Buddhism and Advaitism is instructive in their common bid to defend the "lower" or the empirical half of the skeptic's case. That is, the empirical world, in spite of its seeming concreteness, turns out, on scrutiny, unreal or dreamlike.

It is also instructive to note that at the "lower" level both the skeptic and the antiskeptic are comparably puzzled by the "appearance" of the "unreal" world. Whether it is criticized as dreamlike or transient or self-contradictory, that it is *somehow* there is hardly disputed. The dispute centers around the questions like, In what sense is it there? or in what sense does it disappear or is it found to be self-contradictory or does it get sublated?

Answers to such questions are marked, broadly speaking, by two sorts of skepticism, constructive and destructive. The constructive skeptic highlights the inadequacy or the *local* character of skepticism. By implication, what this skeptic means is this: our (finite and fallible humans') knowledge is bound to be dubitable for two reasons, our own limitedness or situatedness, and the limitations under which objects are presented to us. Having conceded this, to start with, the constructive sceptic takes a somewhat firm epistemic stand at the "higher" or transcendental level. At that stage the skeptic affirms the availability of valid knowledge in which the limitations and constraints of the lower level are stated to be absent, just absent. When pressed to elaborate how the imperfections and blemishes of knowledge of the "lower" level disappear at the "higher level," the constructive sceptic has nothing very *logically* clear to tell us. At this stage we hear some such expressions as "ultimate reality is unspeakable *(anirvacanīya),*" "it is like a positive-negative illusion, suppressive of what is there (truth) and expressive of what is not there (falsity)." The overused expression *dialectical illusion* is also heard in this context. Needless to add, it has different and divergent connotations.

The radical skeptics like Nāgārjuna introduce "destructive dialectic" right at the "lower" or empirical level. Even in response to such a simple commonsense question as, Is the table on which I am writing real? they are prepared to entertain doubt, at least in principle. Here the point to be noted is "in principle." Naturally at this stage Hume comes up in the mind of one who is familiar with the history of European skepticism. The "destructive" skeptic appears destructive only at the level of *principle,* but in *practice* is found to be just like us; that is, a believer in the existence of commonsense things like tables and chairs, pens and paper. If *in effect* my table supports my paper enabling me to write on it with my pen, the skeptics, like me, are prepared to say, "it is there" or "there is no sense in doubting its existence." This is a simple but convincing test of reality or existence. The Buddhist analogue of it is *arthakriyākāritva,* the principle of effectivity.

If this interpretation of the destructive skeptics' view is accepted what they say does not appear to be very puzzling or bizarre. On the contrary, their criticism of the transcendentalists' defense of certain knowledge in terms of self-evidencing reason or intuition *(prajñā)* merits careful attention. First, that there is such a level of reality to be known is questionable. Second, that there is some self-evidencing method of knowing reality is equally open to question, certainly *in principle.* Finally, equally questionable is the assumption that there is some *knowing subject* who can perfectly or infallibly grasp knowable objects. In other words, the identity or mutual adequacy of the knowing subject and the knowable object can never be established or convincingly proved. Consequently, the skeptic disarmingly submits that certainty of transcendental knowledge rests on a dogmatic and untenable assumption.

Interestingly enough, a variant of this "destructively dialectical" argument is often used also by transcendental defenders of certain knowledge. In different ways they put across the position. Some say: "the empirical world *(saṁsāra)* and the transcendental world *(nirvāṇa)* are identical at bottom; but this identity is not demonstrative or logically provable." Others say: "the highest reality and the highest knowledge are identical at bottom; their difference is merely an appearance or superficial." Still others say: "perfect knowledge is necessarily of God, though He, by His unlimited power and grace, can occasion it in and through human beings."

In different forms this problem, apparently favoring the case of the skeptic, appears in the writings of the later Buddhists and Berkeley. Those *(kṣaṇabhangurvādi)* who believe in the doctrine of momentariness, momentariness of the sensible-real, on challenge, are obliged to fall back on the concept of Cosmic Body *(dhārmakaya)* of Buddha. The real, available in sense perception, is destined to perish. It makes its existence felt by its effect. Its efficacy or effectivity gets exhausted in a moment and, consequently, *dis*appears. If everything disappears, then to account for even the *appearance* of the empirical world *(saṁsāra)* and the regularities and continuities found in it prove very problematic. The initial explanation that this appearance of a permanent world is due to the constructive power *(kalpanā)* of the mind, the prorealist critic might say, does not go deep into the heart of the issue. For even the appearance of existence and permanence of the sensible-reals and their continuity beg a deeper explanation. In terms of the Cosmic Body of Buddha, the most comprehensive form of reality, a surrogate of God, the deeper explanation is sought to be framed.[10]

3. How Practical Dogma and Theoretical Doubt Go Together: Hume, Kant, and Buddha

The problem of how dogma and doubt go together, understandably, is encountered also by Berkeley. Being committed as he is to the principle of *esse est percipi,* he endorses a phenomenalist account of continuity. That is, the *perceived* existents may be endowed only with a phenomenalist continuity. When they cease to be objects of some mind's perception, their existence and continuity turn are dubitable. In other words, phenomenal continuity needs to be backed up by some "higher" and more permanent ontological or, at any rate, phenomenological continuity. Whether Berkeley's account of God can take adequate care of the threat of possible skepticism arising out of unsupported phenomenology is a moot question that has been discussed by many commentators like Luce, Collins, and Bennett. If the empirical world, marked by continuity, is left unsupported by a continuously perceiving God, it may fall into pieces or get reduced to a "false imaginary glare." When, for example, Philonous raises the question, "How should those principles be entertained, that lead us to think of the visible beauty of the creation of false imaginary glare?" his opponent Hylas concludes in haste: "My comfort is, you are as much *sceptic* as I am."[11]

Hylas may be wrong but he has a point. If the existence of external objects is made conditional on its being perceived by God's mind, the truth claim of scientific theory is bound to come under a shadow of doubt. The person who does not believe in the existence of God finds it very difficult to accord any serious recognition to the world of sensible reals, the world of science. Keeping this possible criticism in mind, Berkeley affirms that from his principles it does not follow that "things are every moment annihilated and created anew." Far from that, he claims that his position does *not* entail skepticism and disbelief in senses. On the contrary, he attaches great importance to the deliverance of the senses. Further, he claims that his principles are totally opposed to skepticism. The mainstay of his antiskepticism is God, "the [eternal] home of the perceivable when it is unperceived by man."

The Buddhist concept of the Cosmic Body, the Body of all bodies, and Berkeley's concept of God as the eternal home of the humanly unperceived bodies are clearly analogous in their intent. They are indicative of a rearguard action to save phenomenalism and empiricism from their natural drift toward skepticism, theoretical skepticism.

In this context we are again advised to look at the Humean strategy to contain the said drift without invoking highly speculative concepts like the Cosmic Body or God as the eternal home of the unperceived. Hume's insistence on the desirability of seeking the origin of our scientific or credible knowledge, as distinguished from irrational beliefs and opinions, in sense impressions is in a way mere reiteration of what in fact we, people of common sense, are doing in the course of our natural lives. Like other empiricists, he is clear-sighted enough to recognize, besides perception, inference, inductive inference, as a source *(pramāṇa)* of knowledge. That our knowledge is not confined only to the sensible reals and that it comprehends as well nonsensible objects within its scope are clearly acceptable to Hume. The more the knowing subject is distant from the object of sense perception, the less certain is he or she about the truth claim of the resulting knowledge. Though the Pyrrhonian in Hume gladly concedes the superior epistemic claim of "strong and lively impressions" over that of "weak and abstract ideas," his view, as noted earlier, cannot be characterized as radically skeptical. Neither the Buddhist defender of the doctrine of momentariness nor Hume is so unwise as to deny the reality of the world of common sense and that of the natural scientist. If as philosopher (of common sense or science) he critically looks into and raises questions about the limits and tenability of what is ordinarily taken for granted, his position does not deserve to be berated, still less condemned, as insincere and skeptical. On the contrary, one may well recall, in this context, the quite opposite (and perhaps equally radical) view as contained in a poem by Thomas Blacklock and referred to by Hume himself:

> The wise in every age conclude,
> What Pyrrho taught and Hume renewed,
> That dogmatists are fools.[12]

At the level of common sense, as we all ourselves know, we accept many things at face value. For example, when I close my eyelids and cease to see the book before me, I do not doubt about its unperceived existence. Though unperceived, I, true to my common sense, believe that it is there, despite my theoretical familiarity with the Berkeleyan dictum: the essence of a thing consists in its being perceived. What is true of me is more or less true of all other biologically normal and rational human beings. It has been rightly perceived and highlighted by Hume that we have in us two seemingly opposite attitudes, one "skeptical" and another "dogmatic." It is not at all surprising that David Hume has been considered both as "one of the greatest skeptics in the history of philosophy" and as a consistent critic of that "fantastic sect," the Pyrrhonian skeptics. Skepticism and naturalism go together. Hume's philosophy illustrates it very persuasively.

Little reflection is necessary to show that there is no real paradox in our *trust* in the *commonsense view* of things and the *philosophical distrust* or doubt about those very things. As a matter of principle, at the *theoretical* level, Hume draws our attention to the fact that when we go beyond our experience our knowledge claim becomes vulnerable to question and doubt. His criticism of the general truths, laws, and theories of science, the results of inductive inference, has indeed a strong point in it. Though we accept these general truths, we cannot deny the possibility of their refutation. So, in effect, *refutability* and *acceptability* may well go together in our lives. For example, although many of us, philosophical realists, entertain *doubt* about the external world, the past events and the future ones, that does not prevent anyone from *believing* in their reality. Called upon to explain the grounds of their beliefs, they refer, among other things, to commonsense notions, traditional ideas, customary beliefs, and the like. Each one of the grounds, if closely looked into, is found to be questionable. But that awareness does not prevent us from using these grounds to entertain certain beliefs as if the latter are quite rationally acceptable. This dilemmatic nature of the relation of "doubt" and "dogma," refutability and acceptability, is brought out clearly by Hume himself.

> The sceptical doubt, both with respect to reason and the senses, is a malady, which can never be radically cur'd, but must return upon us every moment, however we may chase it away, and sometimes may seem entirely free from it. 'Tis impossible upon any system to defend either our understanding or senses; and we but expose them further when we endeavour to justify them in that manner.[13]

In brief, Hume is a Pyrrhonian as a theoretician or philosopher, but as a practical man he does believe in all things we all do. In fact, Hume's trust in custom, tradition, habit, and so on, is allowed to work as a sort of transcendental argument. That custom, tradition, and so forth, are products of long experience of many generations of people of the same cultural milieu is well known. They are like the silt left behind by the stream of passing experience, hardened or softened by the passage of time and always open to accretion and erosion. Their durability and wide acceptability commend themselves strongly to our understanding. But that does not exempt them from questioning and correction. Noncrucial counterexamples do not render them totally invalid. Traditions, like theories, undergo change but manage to maintain their continuity.[14] This is true of the different forms of cultural traditions. Even revolutions, political or scientific, do not destroy the continuity of scientific or political tradition. The talk of a "gestalt switch" in the context of "paradigm shift" is more an instructive metaphor than a description of the actually changing or historical trend.[15]

The parallelism between the Humean use of the "transcendental" argument to Kant's invocation of full-fledged transcendental argument is instructive. In both cases the argument is exploited to contain the threat of skepticism. According to Hume, external objects are known by the knowing subject through the sense representations of the former. But the relations between the objects and their sense representations cannot be shown to be causally *necessary*. The same sort of argument is applicable to indicate the *contingent* character of the relation between the knowing subject and the introspectable representations of the same. In the absence of the availability of demonstrative proof between external objects and their representations, on the one hand, and the knowing subject and its representation, on the other, Hume feels obliged to confess his *theoretical* inability to claim that his knowledge of the self, the knowing subject, and also the external world is beyond doubt. His atomistic psychology and the laws of association do not entitle him to have access to any *cogito*-type certitude in respect of self-knowledge and intentional certitude in respect to external objects. Strictly speaking, all representations are mere *re*presentations; that is, lack in certifying the *immediate* availability of the knowing subject or the known objects. By implication, Hume concedes that not only the subject but also the object are, by their very nature, elusive. Our *theoretical* skepticism regarding them is thus sought to be justified. Yet, Hume, like all others, is *practically* prepared to accord respectable cognitive status to both psychology and physics, the disciplines concerned with mind and the world respectively.

In spite of his inclination toward theoretical skepticism, the reason for which Hume defends the respectability of the cognitive systems is due to his practical faith in the wisdom embodied in our *social system*. Using modern idioms one might say that Hume was anticipating what we now call *sociology of knowledge*. We are threatened by skepticism only at the level of a *psychology of knowledge* when we are led to believe that "floating" sense impressions of the mind and the world are all that we can know. Uneasy with the "floating" character of sense representations, the Humean is often pushed to the skeptical conclusion that our knowledge, both psychological and physical, is "unsupported" and hangs awkwardly, as it were, in the air. Hume's consistent phenomenalism ostensively refuses to have the benefit of any sort of foundationalism, at bottom or from above. Consequently, he resorts to what he thinks is the next best position, sociological foundationalism. It is from this point of view one has to understand his frequent reference to tradition, custom, and so on, as the *rational* basis for accepting those conclusions of science that, theoretically speaking, are questionable or dubitable.

Hume supplements this argument by highlighting the role of feeling, instinct, and habit in human nature. The human is not only a reasoning animal, but has other concrete and persistent propensities, instincts, feelings, and so forth that must not be lightly dismissed or totally divorced from our reasoning ability. When this *total* picture of the human being is viewed as the background of what we call sociology, we get a clearer and more correct view of our cognitive systems. In brief, it is largely to Hume that the Western philosophy owes its clear understanding of the relation between a given social system and the prevailing cognitive systems.

I call this reconstructed Humean argument *transcendental* because it is placed above the phenomenalistic or the skeptical accounts of mind and the world. A social system is not "floating" or "unsupported." It is self-supporting and more durable than the cognitive systems to which it lends legitimacy. This whole approach of containing skepticism may be designated as the primacy of practical reason.

Before I take up Kant's use of the *formal* transcendental argument it is interesting to recall that in different systems of Indian philosophy the importance of social practices, *lokācāra, lokavyavahāra, lokarīti,* have been given a very important place as the test of the truth claim of knowledge. In more than one way it anticipates Hume's skepticism and naturalism. In India not only the phenomenalists but also the realists show this orientation. In Nyāya espistemology realism and pragmatism, a sort of consequentialism, are found to be very ingeniously reconciled. Both the Buddhists and the Naiyāyikas like Vacaspati are of the view that the grounds of knowledge are extrinsic, antecedent or subsequent, to it. One piece of knowledge, or judgment, for its validity depends on another or several other pieces of knowledge. Though in fact because of overwhelmingly confirmed or extremely familiar character of some judgments we accept them as validating grounds, *in principle* nothing prevents us from questioning them. Potter has a point in characterizing the Naiyāyikas as "fallibilists": "they hold that no empirical judgments are necessarily or indubitably true."[16]

It goes without saying that Kant's invocation of transcendental argument is not confined only to the aim of containing skepticism of sense, of imagination and of understanding. Strictly speaking, the question of skepticism arises at the level of judgment. To speak of subjudgmental skepticism makes sense only in the context of Leibniz's epistemology, which draws no sharp distinction between, say, perception and apperception. In Kant's case, an antiskeptical strategy is validated step by step, grade by grade. Anything like the Leibnizian law of continuity is not available to him, which could possibly ensure continuity between sense perception, imagination, understanding, and apperception. Though sensibility as sensibility does not work judgmentally, what it achieves by spatio-temporally punctuating or determining the material stuff given to it from without is obviously judged or reflected on by the faculty of understanding. True, that in sensibility some sort of synthesis is achieved is a deliverance of judgment (of understanding); but what is achieved is due to sensibility marked by spatial coexistence and temporal succession. From this one must not rush to the conclusion that sensibility as a faculty is altogether different from imagination or understanding. As affirmed earlier, they are graded abilities. One might say, all these graded abilities and what they achieve are subsumable under and regulated by the supreme synthesis, apperceptive synthesis of the transcendental self. What is achieved in sensibility is recaptured and reinforced in imagination in the form of *schema*. From schema (of imagination) we move to object (of understanding). The achievement of imagination, especially of productive imagination, is recaptured more clearly and in a further reinforced manner, in understanding. The higher up we go in the scale of epistemic unity and achievement, the clearer becomes the evidence of reinforcement. The active character of the self asserts itself more and more as we move up gradually in the upper epistemic reaches, in the unity of understanding, and in the superunity of apperception.

The Kantian strategy is intended to stave off skepticism right from the beginning. According to his account, neither successive nor coexistent matters of sense perception are discrete and in need of being connected by some "external" (and generalized) laws of association. Besides, in Kant's case the "I think" principle is present at every stage. For example, when I say "I sense," in effect, what I maintain is "I think I sense." Similarly, when I say "I imagine," I might say as well "I think I imagine," and so on. My understanding is also subject to, that is, regulated by, the "I think" principle. Without this supreme principle the unity sought to be achieved with varying degrees of clarity and success cannot be achieved at all. The transcendental unity of apperception used by Kant as the most important principle to combat skepticism is not itself a self-validating principle. At least that is not Kant's claim. Neither does he try to get his Cartesian *cogito*-like "I think" principle affiliated to God nor does he affirm that the principle is an articulation of God in human beings as his unerring knowing power. He speaks of the possibility of the "misemployment" of the ideas of pure reason. This is said to be the natural dialectic of the human reason itself. That this concession might sound proskeptical is anticipated by Kant. He is not prepared to say that human reason, the highest tribunal competent to settle all the rights and claims of *speculation*, is itself the source of dialectical deceptions and illusions and the resulting absurdities and contradictions. Consequently, he feels (one might say, perhaps unnecessarily) obliged to offer its transcendental deduction.

The ideas that the "I think" principle puts to use are not

claimed to be objects absolutely but objects in the idea. In the formation of absolute objects concepts are necessary, that is, employed; but, for the formation of an object in the idea no concept is necessary, only a schema corresponding to which there is no object available is directly given. And the latter enables the knowing mind to represent to itself other objects indirectly, in their systematic unity. This unity appears to be there only in relation to the said idea. In this way Kant's (seemingly individual) "I think" principle is sought to be related to the concept of the highest intelligence. Here the concept is a mere idea having no object-reality as its referent. This schema of the concept of the highest intelligence provides the knowing mind the greatest possible systematic unity for the empirical employment of human reason. The supposed object of this idea gives in a way the object of experience as an item of the most comprehensive unity. Thus Kant felt entitled to claim "that the things of the world must be viewed as if they receive their existence from the highest intelligence."[17]

Because the ontological grounds of this highest principle of unity is not available to Kant, he readily concedes that it is only a heuristic concept. In other words, it is called on to serve only a regulative, not constitutive, purpose. The purely speculative reason, according to Kant, can give us the concept of God that, though deistic, provides us the conditions necessary for viewing all objects of the empirical world as a unity. This complex and tortuous way of offering a transcendental deduction of the ideas of human reason is obviously very convincing and foolproof to Kant.

But the question remains whether this deduction proves the *necessary* character of the ground from which the said ideas are claimed to be derived. On this crucial question Kant's responses are systematically ambiguous. On the one hand, time and again, he reminds us of the regulative and *heuristic* character of the *idea* of the highest unity of all knowledge of nature and, on the other, in the context of both cosmology and theology he keeps on telling us that the grounds of this systematic unity of the world must be postulated and that it is *necessary* in character. But to dispel the possible misunderstanding, "misunderstanding" from his point of view, it is said that this transcendental ground or the deistic supreme principle must not be credited with existence. It is only an idea. Though it is postulated, human cognitive access to it is denied. It is presuppositional and not propositional. Having said all this Kant goes to the extent of suggesting that we have a feeling of its presence, not existence, in the universe as a whole. The legitimacy of this idea is sought to be established in terms of increasingly discovered purposiveness in the world.[18]

When I speak of Kant's ambivalence on the question I do not accuse him of lack of clarity. The empirical realist in him is opposed to the idea of claiming the existence of an object corresponding to the highest *idea* of unity. The transcendental idealist in him is responsible for postulating this idea to ensure the highest possible unity of what one can possibly know and thus to chase away the last traces of skepticism from his account of human reason. In the process he gives rise to two types of philosophy, the instrumentalism of the Neo-Kantian philosophy of "as if," and the philosophy of objective idealism, ascribing existence to the idea of highest intelligence.

By drawing, maintaining, and using the careful distinction between thought and knowledge, between the object of thought and the object of knowledge, Kant ingeniously tries to have the benefit of both the worlds—the empirical and the transcendental, the things as available in and through intuition and the things as they are in themselves. Kant's noetic dualism leaves behind a tension, feeding subsequently two streams of philosophy, open-ended consequentialism cum instrumentalism and objective idealism of the closed variety.

Kant's elaborate and imaginative architectonics of knowledge and thought, in spite of its subtlety and comprehensiveness, proves inadequate in the face of practical experience of life. In his bid to get away from the Humean skepticism he tries to save reason from what is called its possible misemployment and the resulting illusions, contradictions, and absurdities. He wants to achieve this end from within the bounds of reason, constructing an elaborate and essentially *justificatory* critique of it. The basic structure of his critique, both in its theoretical and practical applications, is marked by the same spirit. It seeks to extend the scope of reason, in the light of sensibility (in the case of pure reason) or will (in the case of practical reason), but always trying simultaneously to ensure that it does not bump into anything alien to its nature or what may possibly contradict its claim. This extracautious and extrasuspicious attitude to all "unknown" objects (of knowledge as well as will) is symptomatic of his antiskepticism verging on dogmatism. From within reason he constantly polices its *positive* uses and keeps a vigil against all possible *negative* uses. Given his scheme of thought or critique of reason, neither theoretical reason can encounter and be contradicted by any counterexample nor may practical reason legislate nonuniversalizable principles or conflicting moral rules. Thus both constitutive and regulative uses of reason are prevented a priori from encountering any case contrary to their claims. Neither regulative ideas nor constitutive concepts nor their combination can yield anything, constituted or regulated, that would violate their generating conditions.

This unilateral and a priori use of reason is patently *justificationist* and anticritical. The empirical here is consistently subordinate to and supportive of the transcendental. Kant's transcendental justificationism, originally intended to be a revolution against Humean skepticism, ultimately turned out to be counterrevolutionary and an ally of dogmatism. Elsewhere I have tried to show how the Copernican revolution proclaimed by him has been betrayed by him.[19]

It is instructive at this stage to recall two immediate predecessors of Kant, Vico and Hume, who anticipated the dangers of developing a theory of human reason that is not truly reflective of the limitations of human nature itself. Vico, the author of *New Science* (1725), warned his contemporaries, who were working under the spell of Newtonian science, by pointing out that one cannot author anything, scientific or artistic, that is more certain that one could be of oneself. In effect, he was questioning the very *certainty* principle that Descartes proclaimed and Newton tried to defend in *Principia* and trying to establish his own new science of interpretation. Notwithstanding the adulatory attitude of his age toward the Newtonian paradigm of science, it was indeed very insightful on the part of Vico to have anticipated that the newly discovered laws of mechanics, simply because of their essential human origin, could not be as infallible as some others were uncritically trying to make out. That was the time when Locke's "underlaborer" concept of philosophy, that is, philosophy must *justify* science, had its heyday. Vico's own attitude to human knowledge of all forms was clearly antijustificationist and prohermeneutic. In a way, Vico, like Hume, his junior contemporary, was deeply interested in understanding cognitive systems as part of the environing cultural system. In their own different ways Vico and Hume dispensed with the ideal of *certain* knowledge. They refused to be unduly afraid of skepticism and to land in the lap of dogmatism.

It has been pertinently pointed out by, among others, Hume that both the skeptic and the dogmatist may follow, at least at the theoretical level, a sort of dogmatism. The dogmatist may refuse to be skeptical or critical about his or her convictions and conclusions even in the face of contrary evidence and examples, including the crucial ones. If we carefully scan this view, we will find that it is not being practiced uniformly even by its strongest defender. In a way one might say that systematic or consistent dogmatism, strictly speaking, is never to be found, either theoretically or practically. Perhaps it goes against the very nature of human beings, unless we assume that human nature itself is like all-knowing God's. Even this assumption is dogmatic or purely speculative; that is, not answerable to any empirical evidence, ordinary or crucial. Earlier I have said in a somewhat different context that "dogmatic dogmatism" is a mere verbiage, a vacuous expression. A dogmatist who refuses to learn anything from experience, personal or secondhand, could summarily dismiss its role in the context of knowledge. This dismissal would be patently dogmatic and again inconsistent with the actual practice of life. Besides, this sort of dismissal, coupled with the proclaimed pursuit of unquestionable truth, indirectly implies that the knowing subject is so constituted, endowed with such a priori faculty, that he or she can speculatively spin out infallible knowledge purely from within. This sort of extreme conclusion offends both our common sense and scientific sense. If we accept it, even philosophizing on common sense and science turns out to be otiose. For we are pushed back to the conclusion that philosophical reflection on or discovery of the methods underlying common sense and science have nothing to offer us, enabling us to doubt or criticize their findings.

Taken to its dogmatic extreme, skepticism also proves unrewarding and very unnatural. If everything is doubted, then the very possibility of the existence of something to be known is virtually questioned. To maintain that there is nothing to be known or that something believed to be knowledge is not really so leads us nowhere. What is worse, given this view, not only the validity but also the very possibility of knowledge are ruled out straight away. Unless it is somehow assumed, at least to start with, that knowledge is possible, epistemology becomes a futile undertaking, and the question of search for knowledge makes no sense. Frankly speaking, consistent skepticism, if there is a position like that at all, is not a part of inquiry concerning knowledge; that is, it falls outside the very scope of knowledge. It has been rightly observed that a strict skeptic neither defends a view nor refutes one but is a *muni,* keeps silent on all issues. Apart from the absurdity of this view, it is manifestly unnatural. Not that one in practice cannot be silent and remain so indefinitely. The point is, such a person, however normal or even admirable in other respects, would not be of consequence to us as a knower or knowing subject. The skeptic with whom we are concerned is a person who believes in the possibility of knowledge, subscribes to some sort of epistemology or other, and yet claims the right to question the views entertained by others and even oneself. According to Hume: "A true sceptic will be diffident of his philosophical doubts, as well as of his philosophical conviction; and will never refuse any innocent satisfaction, which offers itself, upon account of either of them."[20]

What in effect Hume says is that a true skeptic cannot be confident about the utility and outcome of that skepticism, nor be sure about the correctness of such views. Skepticism does not require the skeptic to be skeptical always and in all matters. Nor does it deny one the legitimate right to have one's conviction and defend it as rationally as one can. Practically speaking, the difference between the skeptic and the dogmatic is a matter of degree. They represent two different *ideal types*. In real life situations they are partly dogmatic and partly skeptic. This point seems to have been endorsed by Hume when he says:

> It seems evident, that the dispute between the sceptics and dogmatists is entirely verbal, or at least regards only the degree of doubts and assurance. . . . No philosophical dogmatist denies, that there are difficulties both with regard to the senses and to all sciences: and that these difficulties are in a regular, logical method, absolutely insolveable. No sceptic denies, that we lie under an absolute necessity, notwithstanding these difficulties, of thinking, and believing and reasoning with regard to all kind of subjects, and even of frequently assenting with confidence and security. The only difference . . . between these sects is, that the sceptic, from habit, caprice, or inclination, insists most on the difficulties; the dogmatist, for like reasons, on the necessity.

Though the "difficulties" referred to by Hume are admitted in parenthesis by the classical rationalist, their importance is systematically underestimated by them. However, Kant's case is slightly different. Despite his recognition, he sounds pretty confident that these difficulties can be overcome and tries to show how. Kant's antinomies are reminiscent of Hume's "difficulties," which is said to have roused the "critical" thinker from his dogmatic slumber. Kant tells us: "It is not the investigation of the existence of God, the morality and so on but rather the antinomy of pure reason . . . that . . . first aroused me from my dogmatic slumber and drove me to the critic of reason itself, in order to resolve the scandal of ostensible contradiction of reason with itself."[21]

Antinomies, illusions, contradictions, or *difficulties,* whatever term we use, the problem indicated remains more or less the same: there is a persistent and internal tension within reason itself. It is due to the "irremediable" fact that human reason is simultaneously exposed to two different influences, internal or internalized and external or externalized. We have some temporary or even persistent experiences that reason is not prepared to accept as the real state of affairs. We have in us some "native" inclinations or speculative tendencies that, even in the face of contrary experience, we are not ordinarily prepared to give up or disown. Being mainly concerned as he is with objects of experience (in the chapter on antinomy) and not with things-in-themselves, Kant seems to be caught in a dilemma between the competing claims of empirical realism and of transcendental idealism, between those of transcendental idealism and of transcendental realism.

The empirical realist is required to establish that objects are *of experience* (actual or possible). If at the empirical level we are called on to establish the existence of the infinite world, we are put into an insurmountable difficulty. For, in that case, we must have some sensible material out of which a concept at least of possible experience could be formed. But obviously that sort of (space-time determined) synthesis in sensibility or corresponding schema is just not available for understanding.

When Kant asserts that cosmical concepts are basically concerned with synthesis of appearances and, for that reason, are basically empirical in character, and when, further, he reiterates that these concepts deal only with things as objects of possible experience and not as things-in-themselves, he tries to show the incoherence of transcendental realism. His point is that every effort to prove that the world as a whole is a whole of all appearances and existing in itself would prove abortive. This indirect disproof of the claim of the world as existing in itself is obviously intended to establish Kant's favorite thesis of the transcendental ideality of *worldly* appearances. In other words, ontologically speaking, things cannot be both appearances and things-in-themselves at the same time. This sort of proof through antinomy is claimed by Kant to be of "critical and doctrinal advantage" and "not dogmatic."[22]

It is not difficult to guess Kant's intention to highlight the paradoxical character of reason's claim to be equally and simultaneously fair to the claims of the empirical and the transcendental. The inevitable frustration of the claim, to his mind, strengthens his theses of empirical realism and transcendental idealism. Although he appears to be indulgent, at least to start with, to reason's first concern with experience, gradually it becomes clear that reason is obliged to make use of principles that extend beyond the arena of all possible empirical employment. This native tension tends to dim the light of reason and invites contradictions. As we all know, Kant's own preferred way out of this impasse is to espouse realism at the empirical level without giving up idealism at the transcendental level. What in fact he strongly advises us not to do is to try to extend empirical realism to the transcendental level and defend transcendental realism. At the same time he appreciates this persistent tendency of reason in us. For example, being aware of our relative position in the space-time framework, we keep on conceding that our local space is a part of a global space and then, when we are asked about the position of the global space itself, we concede its relative position in the solar system, and so on. But when we are asked of the position of the universe as a whole in which solar systems and similar other innumerable systems are said to be "floating," we do not know how to answer it (in terms of space-time framework).

This conceptual inability of ours does not push us *logically* to the conclusion that this is an illegitimate question. In other words, the sense of legitimacy we feel in us while entertaining these sort of questions without being able to answer them is somehow *natural* but not logical. Thus in reason Kant finds contrary pulls, objectward and "subjective" consciousness centric. The first finds its expression in Kant's elaborate concern with the possible *objects of* experience and the second in the transcendental *consciousness of* objects. By vindicating the transcendental character of consciousness he tries to lift *objects of* experience from the privacy of individual life and paving the path of intersubjective or objective validation of science.

It is not difficult to see the tension between "transcendental *consciousness of* objects" and "*objects of* experience." The latter cannot be constituted by understanding alone without aligning and affiliating itself to the *regulative* ideas of reason. Behind both understanding and reason, besides other faculties, stands steady the principle of transcendental consciousness marked by its regulative and unifying or synthetic characters.

Here one notices the second tension operative within Kant's epistemology. The rules of constitution and the principles of regulation, though unable to work independently without inviting "difficulties" for themselves, are not perfect partners either; that is, they are not made for each other. No preestablished harmony of the Leibnizian sort is between them. The idea of God, called upon to do this work indirectly, proves somewhat unequal to the task. Consequently, the tension between *regulative function of the principles* and *constitutive role of rules* remains.

This appears to be a fallout of the "gap" between empirical realism and transcendental idealism. If the ideas of reason are given an empirical status, leveled down and treated at a par with the objects of experience, Kant faces two problems. First, as pointed out earlier, there is nothing sensible in our body-mind complex enabling us to transform transcendental ideas into objects of experience. The regulative ideas that make constitution of objects possible cannot themselves be constituted by understanding. To put the matter in a different way, the understanding that makes nature possible under the guidance of transcendental consciousness cannot claim to be the author of the latter. Even if this ambitious claim is conceded in principle, one does not know how in practice its consequences could be consistently worked out, because it implies not only the abolition of the proclaimed distinction between (empirical) realism and (transcendental) idealism but also, what is worse from Kant's point of view, a sort of naturalistic reductionism. In view of these difficulties it is not at all surprising that Kant attaches so much importance to his basic argument in the chapter on antinomy.

His critique of reason as a whole aims at giving a solution to the problem, a possible cure of the tension. Right from the beginning he tells us that the tension or perplexity found in reason is not its own fault, rather it is mainly because understanding in its bid to expand its area of conquest, cognition, summons some *principles* to appear before it as *objects of experience*. But by their very nature those principles are a priori and enabling conditions of what understanding itself tries to achieve.[23]

Thereafter at the end of the analytic he reminds us that empirical appearances must be construed as parts of *one* unified nature and that this unity itself has to be taken a priori and not sought and found as an item of the world of appearances. The unity we seek is in a way outcome of the unity we have already assumed.[24]

Again in the dialectic he emphasizes that the idea of the *whole* world is empty.[25] It naturally sounds odd that the empty idea can, and in Kant's opinion does, contribute to our (understanding's) constitution of the natural world of appearances. How can an empty idea possibly help us in having something cognitively informative?

The basic question that emerges out of Kant's ambivalent account of the relation between (the principles of) reason and (the rules of) understanding is this. Cannot understanding work in an *unruly* way; that is, departing from its own rules? Cannot reason be *unreasonable* in its ways of "guiding" understanding in relation to the latter's constitution of the world of appearances? Cannot reason be unreasonable directly or in a straightaway manner; that is, arrogating the nature constituting role to itself? To all these questions, as we know, Kant's response is "No." According to him, rules work somewhat blindly or at least in a shortsighted manner. Rules of understanding are more or less stuck in their immediacy. Principles of reason, though said to be empty in one context, are credited with the capacity of transparent vision in another. The only qualification attached to the transparency claim is that this capacity, though said to be native to us, is not necessarily available to us in practice. But this qualification, if deeply probed into, reopens all the basic issues in a problematic manner.

The transcendental (or metaphysical) realist maintains that reality—small bits, big chunks, or as a whole, in whichever way we put it—is there irrespective of our ways of experiencing it. According to this realist, if any fault or imperfection is found in knowledge, reality is not to be blamed for it. This position is obviously stronger than that of the empirical realist. The latter at least is prepared to admit that knowledge of nature being a product of the partnership between the empirical given and the structure of human understanding, if something goes wrong, the objective side must be held, at least partially, responsible for it. A defect or unreliability of knowledge, in that case, is in principle attributable to *how* our understanding receives or interprets the given.

Now this formulation of imperfect knowledge takes note of both sources of error. It may be due to the nature of the given and its mode of being given. Under different conditions, spatio-temporal ones for example, "the same given" may appear differently to the knowing subject. Even the illusory and fictitious *objects* have to be accorded an ontological status, however transient or cancellable that might be. Strange objects like *unperceived regularity* (regular behavior of very small and very distant bodies) and *continuity of irregularly perceived bodies* bring out the dual responsibility of the said cognitive partnership enterprise. The vulnerability of cognition's truth claim from both ends, *unsteady* nature of the object and *questionable* ways of receiving and interpreting it, is bound to invite the skeptical threat. Appearance may pretend as reality and experience may wrongly conceptualize what appears before it.

Being aware of this threat, the empirical realist, in this case Kant, defines "empirical object" in such a manner that it fails to pose any threat to what understanding does. The object is nothing but objectivized experience. Being a product of or constituted by experience, it cannot prove incongruent with experience itself. In a way it is reduced to the position of a passive or slumbering partner. Is not this slumber dogmatic?

This Kantian strategy to fend off skepticism is buttressed by another collateral thesis: the faculty of sensibility is passive and that of understanding is spontaneous. By implication it is being affirmed that constitution of object is essentially rooted in the constitutive power of understanding and its relatively spontaneous character. However, here the thin claim of realism is sought to be defended by pointing out that understanding cannot constitute object entirely on its own and that it needs some sensible contribution from without to act on. Otherwise this position gets dangerously close to that of subjective idealism, if not solipsism.

The weakness of the realist component of empirical realism can be brought out in another way, by highlighting the idealist component of transcendental idealism. What understanding or experience does (in the process of forming objects) it does under the guidance of some transcendental ideas. These regulative ideas have their seat in the transcendental nature of the self, which stands highest in the scale of spontaneity. Although understanding is at least partially influenced by the passive persistence of the given, transcendental ideas are said to be *freely* active, active in their regulative employment. This point receives Kant's special attention in his second *Critique,* where he is primarily concerned with the role of reason in relation to will. The spontaneous or free character of the transcendental self is postulated by Kant both in the context of pure reason (principles of knowing) and in that of practical reason (principles of acting).

Though the service of this principle of freedom (or spontaneity) is summoned by Kant to counter the threat of skepticism, it brings in its wake some unintended problematic consequences. If in the context of practical reason human freedom could be really absolute, that is, Good Will can really assume its best possible form, the latter could not be distinguished at all from God's (or Holy) Will. In Kant's own admission, Holy Will is to be distinguished from Good Will, which is essentially human and moral whereas the former is devoid of any moral import whatsoever. Without exposing his concept of Good Will to the charge of vacuousness leveled against it by Hegel, Marx, and others, Kant cannot stop short of admitting that it does have unrealized contents or objectives. Neither our moral lives nor our cognitive lives could have been lives of endeavor and adventure if we did not have in us a consciousness of lack or imperfection. Our freedom to know does not mean we can know *all* that we need or will. Nor can our freedom to will get us to the *peak* of the highest perfection from which we could legislate laws *for all* and freely acceptable *by all.* In the sphere of knowledge also we encounter various limits and constraints, notwithstanding the affiliation of our understanding to the transcendental unity of apperception, the supreme epistemic synthesis marked by the highest possible spontaneity and unity.

I am not claiming that the points I raise here are not somehow anticipated by Kant. In fact I draw upon him, especially his self-confessed difficulties. But in the process I am trying to bring out a point, though noted by him, but, strangely enough, repeatedly skirted by him. To speak, as Kant does, of "human nature," "experience," "understanding," and so on, in general terms does not help us much in grasping the problematic features of knowledge, especially the skeptical ones. Again to speak of dubitability or verity of human knowledge only in general terms makes little sense. The truth claim of this or that piece of knowledge disturbs us and therefore concerns us most. Cognitive enterprises, irrespective of their scope, small or large, are *specific* or, one might say, context specific and person specific. Unless we decide to define *universal knowledge* in the wrong way by dehistorizing or decontextualizing it, it is not at all clear how the basic problems of knowledge could be satisfactorily tackled. If *human* knowledge by its very nature would have been free from all problems, Hume's skepticism could not have even roused Kant from his dogmatic slumber, inducing him to undertake the stupendous task of working out the *Critique* of reason. But if Kant or, for that reason, anybody else is really determined to give us only the universal picture of knowledge, the only possible epistemic architectonic, it is bound to prove not only uninteresting and unsolicited but also untenable.

The untenability may be traced to the specific character of the knowing subject and of the object to be known. Unfortunately, an exaggerated fear of skepticism drives Kant to a position where he proposes to operate with a concept of human nature that is claimed to be universal and a concept of the object that is also stated to be universal. But there are many places where he gives us a contrary impression. For example, in the section on amphiboly he tells us that understanding is impossible unless corresponding to our language we have something particular given in our sense-intuition.[26] Unless a rule of synthesis is synthesis *of* something specific, what is synthesized remains elusive, if not unavailable. If rules cannot be illustrated by concrete examples, it is difficult to decide their very areas of applicability and unless that is somehow conjecturable by us we have them only in the abstract and not concretely.[27] When we do not know concretely, it is difficult for us to communicate it, either theoretically or practically.

Communication in practice involves effective identification of the meaningfulness of action, linguistic or behavioral Theoretical communication in and by language also demands of us to indicate definite relationship between words and objects or items of experience. Without satisfying the *specific* conditions of successful communication, learning or teaching, the validity claims of knowledge remain unfulfilled. Though at times Kant gives us the impression that he recognizes this fundamental point, the overwhelming idea on the subject he leaves with us is that, unless and until we can *absolutize* the concepts of knowing subject and knowable objects, the shadow of skepticism would continue to linger. Undoubtedly in the chapter on antinomy he refers to the necessity of opening up one's view to the critical examination of others and also parenthetically speaks of "a jury of fallible men." However, it is not at all clear how representative this trend of reasoning is of his basic thought.[28] True, some sympathetic commentators have valiantly tried to show that within the rigid framework of Kant's seemingly dogmatic epistemology enough cues are left to reconstruct a soft epistemology.[29] While I sympathize with this point of view, I find that, *textually* speaking, it is not sustainable. Merely to show that our views must be opened to the *fallible* opponent's criticism is not to admit the fallible character of knowledge itself. On the contrary, Kant's basic arguments are geared up to show (1) how the ascending syntheses in different faculties lead to the constitution of the *universal* object, (2) how the principles of transcendental reason, though operative in and through particular individuals, are in effect universal in their achievements, and (3) how to make the Copernican revolution successful in epistemology. His aim is to prove that this branch of knowledge, like astronomy, needs to be put on some absolutely firm and unquestionable pedestal. I am still inclined to maintain, as I argued earlier, that the proclaimed revolution of Kant has been betrayed by Kant himself.[30]

4. Interpretations of Kant by Modern Neo-Kantians; Husserl's Kantian Insights; Implications of Husserl's Foundationalism

I say this not to minimize in the least the genius of Kant or to rule out the possibility of reinterpreting Kant's epistemology. In fact, as I see the scene, some modern Kantians like Cassirer, Popper, and Sellars have already done it in their own ways. Because my own understanding (or interpretation) of Kant is somewhat close to these thinkers, I feel, it would be appropriate to have a second look at those philosophers who maintain the strong view that Kant's antiskepticism has made unnecessary concessions to the skeptic cause. These philosophers in their bid to attain genuine certainty in knowledge, want to get rid of the tensions found in Kant's *Critique* of reason and return to Cartesian insight and Leibniz's monadological harmony. The strong antiskeptic view that I have in mind may plausibly be attributed to Husserl. I say "attributed" because some Husserlians are not likely to agree with my understanding of Husserl's criticism of Kant. I want to examine his so-called tension-free interpretation of Kant and attempt to purge it of proskeptical empiricist traces because I find his "radical" undertaking, instead of solving the problem, precipitates a "crisis." The transcendental phenomenological search for certitude, though well intended, seems to have ignored the importance of the life world and the human beings environed by it and with whose knowledge we are really concerned.

Following the footsteps of Descartes, Husserl takes us from the natural world to the transcendental world of subjectivity. The method he follows is *constitutive* and phenomenological. His aim is universal constitutive synthesis of all sciences. Philosophy, according to his scheme of thought, is destined to be an *absolutely* certain science of all relatively uncertain sciences. In fact phenomenological inquiries are "constitutional" investigations making gradual advancement. The idea of "universal constitutive synthesis" is not merely *regulative* or heuristic to him. It is not like an "as if" postulate. In a sense it is "*evidently* presupposable" and attainable by "an incessant uncovering of horizons."[31] The proposed transcendental theory of constitution claims to be very powerful and comprehensive and capable of providing us unified pictures not only of nature but also of culture and of humanity as a whole. The process of constitution must not be misconstrued as a priori or arbitrary. It is evidentially warranted expansion of all intentional horizons. The result of constitution is present in intentional anticipation. In a way, one might say, "result" and "anticipation" are "apparent," that is, one is present in the other, simultaneously present, apperceptively available, and embedded in what is said to be "immanent" temporality.[32]

Husserl's concept of horizon performs the job of induction. Unlike the empiricist's concept of induction, this has two dimensions, internal as well as external. Every experience as phenomenon has innumerable anticipative and predelineative cues in it. In a way every phenomenon is situated in a plenum of phenomena, though the plenum itself is not given to us, to our transcendental consciousness, as an object or a unitary whole. This whole or synthesis is more in the nature of a promise. Experiential induction is not merely anticipative but also constitutive (of *objective* cognition). It is marked by an aiming-beyond capacity. This opens up the given to an *external* horizon of cogivens (possible objects) to be constituted.

All our judgments are fulfilments of experiences. Every categorical (form of) judgment articulates the initially undisclosed intentionality of experience. Even abstract ideal objects, mathematical ones for example, are judgmental affirmations of what we get in experience. Through intentional anticipation, by broadening horizons, we get all objects, actual as well as possible, constituted. To put it differently, our constituted unities of objects are said to be rooted in the life world. The "abstract" objects of the logician and the mathematician are no exception to this universal principle or grounding.[33]

The ego constitutes the objects of all sorts for itself but not entirely by itself. The role of the given cannot be ruled out altogether. Where Husserl departs from both Descartes and Kant is to be noted carefully. His constitutive ego is not a thinking substance, which confronts the extended substance. The problem created by the external and extended substance (of the world) does not figure in the Husserlian scheme. The concept of intentionality resolves the problem of defining the relation between the internal horizon and the external horizon of transcendental consciousness. Neither the knowing subject, *ego cogito,* nor the *cogitationes* are to be construed substantively. They are constituted, form a continuum, and are situated in a plenum. The ego is monadic in the Leibnizian sense. Its individuality is not antithetical to its (transcendental-intentional) universality. It can have the givens "of the world" in different modes—as apperceptively cogivens, as items of *empirical* description, as items of eidetic description, and so forth. But the givens are never passively received but always actively constituted.

In this respect his difference from Kant's thesis on the faculty of *passive* sensibility is very significant. Because of its active or constitutive character eidetic description takes the form more of reduction than that of description (as ordinarily understood). Eidos are also products of higher-level constitution. In the constitution of *eidos* de facto ego reaches the height of transcendental or absolute ego. From that point of view transcendental description ceases to be "empirical" as we understand it. What becomes thin in eidetic reduction becomes thinner in transcendental reduction. From that point of view transcendental synthesis constitutes not only the actual givens, actualities, but also nonactualities. Husserl draws our attention to the fact that every sense given, when attended to, brings to our notice certain possible (not actual) and adjoining things that in fact are not being perceived by us. However, those unperceived things, being of a thinner kind, are not excluded from the realm of eidetic or transcendental reduction. In this way, by downgrading the role of the factual, the actual, the individual items of sense-experience, particular images, and so on, Husserl tries to move to the realm of "essential universality."[34] In the process Kant's "empirical realism" is robbed much of its empirical contents as well as realistic character. "Transcendental idealism," though it escapes excision, is given a very new character, that is, primarily constitutive. The presuppositional and the merely regulative characters of ideas are given a good-bye.

Husserl's main problem consists in trying to answer clearly how within "the immanency of conscious" life and on the basis of proclaimed "compelling evidences" knowledge could be credited with objective significance and certainty. Descartes in his search of certainty could ascribe to *ego-cogito* only an *epoche* status. And, therefore, later on, he had to resort to a transcendental deductive argument, deducing cogitating ego from the veracious God. To Husserl, this line of argument is not open. For, in his scheme of thought ego is not isolable from the spatio-temporal world around it. On the contrary, it is a part of it. It also knows itself as a part of it. According to him, the true transcendental phenomenological theory of knowledge is not concerned either with thing-in-itself or being-in-itself; that is, self-enclosed within itself and persistently proving itself elusive. Husserl tries to move, relying mainly on the concept of intentionality, from the Cartesian enclosed worlds, "thinking substance" and "extended substance," to a unified and expanding intermonadic world. In this unified field there is no sharp distinction between ego and nonego, between the empirical and the transcendental. It is distinguished from psychological idealism because it rejects the concept of the empirical given as an external gift of the extended physical world. It is to be distinguished also from the Kantian idealism; for, it dispenses with Kant's concept of the world of things-in-themselves. But this view is also idealistic because it comes out of "self-explication." It is transcendental because this explication knows no boundary. Its horizons, phenomenologically explored, keep on receding. It is like an expanding and open universe. It does not need any external confirmation or supporting evidence. Its main proof is unfolding itself phenomenologically.

Having emphasized the transcendental idealist character of his theory of knowledge, intended to be a universal rigorous science,

Husserl addresses himself to the question whether this self-explicating and self-constituting knowledge would not be doubted by the *other*. Will it not be criticized as solipsistic? Unless he can show that the self is not shut up or enclosed within itself and that, positively speaking, its being consists in self-transcendence itself, the charge of solipsism or at least that of subjective idealism cannot be easily dismissed. It is apparently for this reason that he draws heavily on the monadological structure of the Leibnizian argument. In other words, the ego is never alien. The knowing self is never a lonely self, pitted against the world of objects and populated by other subjects. In the transcendental theory of the objective world both known things and beings are cogiven to and coconstituted by the knowing self. Rightly understood, self-constitution is all-embracing. My ego as a monad is not an isolated self marked by its exclusive ownness. What I constitute, because of being constituted by myself, is not peculiarly my own. For, my own self-constitution is possible by assimilation of others in me. Others are pregiven in my self-constitution. Self-constitution and other-constitution are "appresented" in the fundamental form of consciousness, time. The "Other" though constituted by me has its own constituted sense and in a way it presents myself to me.

Ego in the transcendental attitude first breaks its seemingly solipsistic circle by constituting nature. This constituted *nature,* paradoxically speaking, includes myself and is not alien to me. At this point the shadow of Kant on Husserl is clear. The Kantian nature, like the Husserlian one, is created by human understanding and therefore not quite external or alien to it. In addition, like Kant's nature, it is also available to others. At the second level of constitution the transcendental ego moves up to the level of *animate nature,* containing living beings, my own body in it. Husserl avoids the Kantian body-soul dualism in two complementary ways. The body is constituted by the soul. And the body as a part of animate nature, without ceasing to be the soul's own, points to the soul and gives it a special sense. The body-soul unity constitutes the world and is also contained in the constitution of the world. I, as constituted by the world outside me, am not, strictly speaking, transcendent, falling beyond the ken of the world itself. I am in it and of it. In a way my transcendental ego is a "parenthesized" outcome of the entire objective world. From another standpoint, the world is a "mundanized" form of my own self-apperception. In brief, there is no gap between the self and the other, between what constitutes and what is constituted.

One might pertinently feel at this stage that Husserl's account of world constitution by the transcendental ego is uncritical and that his program in the process of its execution leaves no room for verification or correction. But the author of the proposed rigorous science is not prepared to accept this criticism. According to him, the self and the other, the constituting capacity and the constituted content, are appresent in our experience in such a manner that this appresentation itself could be called verificatory. The anticipations of the self, when fulfilled, turn out to be always harmonious. When the Other is said to be a phenomenological modification of myself, it seems, Husserl leaves no room for any disharmony between myself and what is constituted or modified by myself. The very relation between the self and the Other, the totality of other things and beings, is so defined (in terms of intentionality) by Husserl that the possibility of incompatibility between them is totally ruled out. These sort of "experiential verification" and internal "modification" he speaks of are bound to remind us Leibniz's doctrine of preestablished harmony. If that doctrine sounds "metaphysical" to the Husserlian, one cannot fail to note that it has been given a phenomenological interpretation by Husserl himself.[35] To say that Leibniz's intermonadic harmony and Kant's transcendental subjectivity are objectionably metaphysical and presuppositional, whereas Husserl's transcendental constitution is absolutely presupposition free is to overlook or underplay the phenomenologist's own historical affiliation to the metaphysical tradition authored by Leibniz and Kant, among others. This is also contrary to his thesis that the historical past as a part of the Other is appresent in his own phenomenological "now." If in the name of "modification" we are allowed to re-create the past as we like, then phenomenological "disclosure" of the unity or harmony of the past and the present has to tolerate "everything," even "distortion" or (historical) "non-sense, within it."

Husserl claims that his transcendental phenomenological investigation is not only self-explicative but also self-critical and yet apodictically certain. It is a constant critique of whatever is transcendentally constituted. This constituted history is not naturalistic of the Humean sort; that is, not based on external evidence like documents. It is not a causal series of events subject to external verification and modification. History, truly speaking, is self-explicative, the self's own. Yet it is intermonadically or intersubjectively available and valid. It is "a critical understanding of the total unity of history—*our* history."[36]

The Husserlian conception of critical history takes its cues from Descartes who was always suspicious of the claim of history as a science. History, to him, was "full of sophistry and illusions." If the serial-causal character of prereflective history is accepted, both Descartes and Husserl argue, history cannot be claimed to be a coherent and unified whole and its constituent events cannot be reenacted or repeatedly reflected upon. Obviously, this concept of unitary history is basically anti-Humean and antinaturalistic. It is rejected because, given this conception of history, newly discovered facts may prove inconsistent with the already established ones, necessitating the modification of the latter. But Husserl finds no special reason why the newly and immediately given facts must be deemed to have a stronger say in the constitution of "the self-critical whole of history" that, according to him, by its very nature is bound to be reflective and unitary. History, though constituted as a transcendental unity, is in effect fulfilment of our "truth-seeking will." In this sense one might say its total unity represents a "responsible critique." Like Kant's notion of critique, Husserl's one is also justificatory. The "consummate clarity" that is ascribed to it is due to its self-explicative nature and apodictic method.

To be doubly sure of the reliable character of transcendental history Husserl takes the pains of reminding us that the "self-interpretation" to which it owes its origin and validation is neither isolated nor arbitrary. In its ultimacy "the hidden unity of history" is rooted in intentional inwardness of *our* consciousness. It is clear that in this case Husserl takes me, "I-the-man," as an integral part of *our* living together.[37] Thus he views me and others like myself, the actual authors of history, as immersed in a transcendental life of togetherness, wherein I, in spite of my "superficial" ownness, get lost in unitary consciousness of *all*.

This notion of history seems to be holistic in a questionable sense. It is difficult to see how this ultimate historical sense could be characterized as truly human. How in the whole of history could both primordial nature and individual human beings get lost? Can self-explicative thought be endowed with a peculiar transcendental capacity that may be described as a sort of trick? The transcendental phenomenological view of history as presented by Husserl is performing a sort of vanishing trick on us. The "thick" *empirical* is transformed into a "thin" *ideal* by a mysterious chemistry (called *phenomenological constitution*). In the Kant's scheme of things what is real from the empirical point of view turns out to be ideal from the transcendental one. Kant's universal history is more in the nature of an idea than a systematic account of facts. Somewhat similarly, Husserl's transcendental idealization of history takes away the empirical flesh and blood of the world of facts. As if to console us it is said that what is being taken away, the empirical, was destined to be lost for the fulfilment of our own "truth-seeking will." When reflection is systematically given precedence over perception, the transcendental over the empirical, there arises the distinct possibility of unilateral speculative construction. Notwithstanding his unsatisfactory notion of history, Kant at least warns us against the danger into which empirically unchecked speculation may put us.[38]

The notion of transcendental freedom that has nothing to do with empirical determinism is bound to prove problematic or, to use Kant's own expression, antinomous. Unfortunately, Husserl, unlike Kant, moves on very bravely, relying on the essential or eidetic human competence and ignoring its fallible aspect. That even in the internal horizonal self-explication one may go the wrong way is not at all recognized by Husserl. He thinks that human beings are so rightly equipped with what he calls the universal historical (a priori) sense that they can get all wrongs righted by intentional constitution of their transcendental consciousness. This sort of uncompromising Husserlian apriorism tends to destroy historism. Apriorism and historism hardly go together.

It is not at all surprising that Husserl anticipated this criticism and that he dismissed it. To clarify the grounds of dismissal of this criticism he draws a distinction between the empirically conceived transcendency and the transcendentally constituted transcendency. The former appears to him infected by positivist naivete. When one conceives oneself as one among others, as environed and nourished by nature, and thus suggesting that "others" and "nature" are *alien* and owe one nothing, one is clearly mistaken. If at the naturalistic level the self constituted or shaped by the Other appears to be illusion, Husserl tells us how ultimately it vanishes. One might say, he shows us how the vanishing trick operates rightly and from the right end. *"[E]very thing I,* qua transcendental ego, *know as existing in consequence of myself* and explicate as *constituted in myself* must *belong to me as part of my own essence."*[39] This is a "Copernican shift," shifting from naturalism to transcendentalism, yet retaining nature and others within it. What is more, the self-explicating horizon proves neverending. Self-investigation and self-explication keep on widening and deepening our consciousness. Consequently, in the changing horizons nature and others assume new sense and acquire new cognitive dignity; that is, more and more certainty.

It is instructive to note that at the end of a search for certainty Husserl goes again back to the question of criticism. To him, criticism is self-criticism. The ultimate problems of phenomenology are said to be concerned with self-criticism. Kant's *Critique,* we recall, was basically concerned with the extent and limits of the human cognitive capacity. Similarly phenomenological self-criticism aims at determining the range, limits, and modes of apodicticity. "All transcendental-philosophical theory of knowledge, as *"criticism of knowledge," leads back ultimately to criticism of transcendental phenomenological knowledge."*[40] To start with, criticism is of experience and in the judgmental form. It is essentially self-reflective character of experience itself. Second, this judgmental form is also a subject of a higher-level criticism. But, Husserl hastens to add, this does not mean "endless regress"; however, it is an "endless program."

To him "philosophy is an all-embracing science grounded on an absolute foundation." This foundation, as mentioned earlier, can be deepened and widened through phenomenological explication without entailing endless regress. This foundational science is not answerable to any practical experience that is branded as naive. Justification of philosophy as science does not consist in empirical verification but in uncovering its intentional horizons. Radical criticism cannot come from without; it arises from within. The best criticism is self-criticism and not other-criticism—not the criticism initiated or voiced or put up by other things, beings, or findings. Husserl tries to point out the limitation of the Cartesian axiom, *ego cogito,* pitting ego against the extended external world. Husserl feels that he has shown how phenomenological "skepticism" being essentially self-critical in character, explores new horizons and constitutes an all-embracing self-investigation. Nothing external, spatial or temporal, physical or historical, is allowed to play a critical and corrective role in his theory of knowledge.

Rejection of historism amounts to refusal to learn from experience. To Husserl the "historical" because of its alleged empirical character is an anathema. Yet, because of his commitment to phenomenology, to explore the transcendental in and through the experiential, he can hardly ignore it totally. The only task open to him is to redefine its character or meaning. His main interest in every sphere of his inquiry, science, technology, culture, and so on, is to remove all traces of skepticism and relativism. These evils, enemies of epistemic certitude of philosophy as a rigorous all-embracing science, must be removed. His phenomenological self-investigation convinces him that the roots of evils can be traced to the naturalist mode of construing objects of the external world. Therefore he is totally opposed to what he calls positivist naivete. Because he feels that the historical facts and findings are being construed in a positivist manner, that is, externally, discretely, and causally, he dismisses them summarily. For similar reasons he finds no *foundational* value in the basic empirical statements: verifiers, falsifiers, and so forth. None of these proclaimed evidential statements appears to him really convincing or decisive. The reasons for his dismissive attitude toward the positivist theories of evidence statement is basically identical with those for his rejection of empirical history.

It is "easy" to dismiss historical findings and to criticize naturalism for its alleged naivete but one has to take serious note of the imperative of criticism. If the natural world, the world of culture around us, and everything external are found to be uncritical of the *critical* task, Husserl realizes, he has to look for the fulfilment of the task elsewhere. If the empirical method fails, he is obliged to rely on the transcendental method. From the Other, he feels, he has to return to the self. The search for certitude is essentially an internal search, an inward journey.

Out of the endless wealth of internal consciousness the Husserlian "rigorous science" of certainty is "constituted." The realistic world, which is thoroughly a suspect in transcendental consciousness, was at least indirectly recognized by Husserl in his earlier works like *Logical Investigations*. There he was concerned with the problems of meaning marked by a pronounced platonic orientation. And the theory of transcendental *constitution* was yet to emerge. The shift toward transcendental idealism also was not there. The transition from realism to idealism somehow maintains a continuity in terms of an unending concern with the problems of meaning. The meanings that at one stage were matters of "discovery" gradually turned out to be matters of "constitution." The theory of constitution gives rise to two important problems: (1) antihistorism; and (2) indifference to linguistic theory of meaning. Phenomenologically understood, both the problems are rooted in Husserl's antipresuppositionalism. First, he is not prepared to learn from historical facts and figures as obtained from without. Second, he wants to get *directly* back to things themselves, dispensing with linguistic classification or slicing of the world of objects.

The direct method is intended to do away with *relativism* associated with different ways (linguistic ways) of classifying the objects and different ways of interpreting the facts of history. Once the "naturalist" ways of classification and interpretation are accepted, we are landed into conflicts and contradictions. Only the direct method of intentional-transcendental synthesis can save us from the predicament. For example, the method of eidetic reduction discloses to us "natural kinds" of things, the things as they are. Yet the things are not given to us from without, psychologically or epistemologically. The transcendental subjectivity constitutes them. I constitute them for *us*. It is an antirealist and proidealist internal journey. Embarked upon this journey, *I*, the transcendental subjectivity in me, am not prepared to learn anything from history. Not only "linguistic classification" but also "historical information" are criticized as unreliable guides, and therefore their services are peremptorily dispensed with.[41]

All these innovations and exercises have one basic aim in view and that is *certitude*. The search is direct and without presuppositions. At least that is the claim. The claim is sought to be substantiated by "showing" that there is a direct passage from absolute immediacy of experience to what they really are, their intended meanings or objects. The world is presented or constituted as a world of meanings. Even the meanings of existence and also of *non*-existence are constituted, not given. Acts and their contents are directly bridged by intentionality. What exists is constituted by what may be called a verificatory act and what does not by a nullificatory one. Negatively speaking, reality *simpliciter*, either in the form of positive fact or in that of negative fact, is not given to us. In this way Husserl tries to do away with the dualism between the givens and the conceptual apparatus that converts them into objects of knowledge. Objects are sedimentations, not creations, of our consciousness. Consciousness is absolutely self-founded. All its forms are its own, internal to itself. And yet those forms, being essentially transcendental and intentional in character, are available to each one of us. Husserl's reasoning, as we can see, is relentlessly moving toward an all-comprehensive and absolutely certain science (which is named philosophy).

One might very pertinently raise the question, What does Husserl gain by offering us this so-called doubt-free and absolutely certain form of knowledge? One obvious answer is that this is what it really is. This is not a matter of theorization, nor is it a practical discovery. It is the gradually and horizonally discovered truth of knowledge. Strictly speaking, it is explication of *self*-knowledge. In the Indian idiom, one could express it by saying that it is nothing else than "getting the got." In other words, it is claimed that knowledge by its very nature is doubt free, free from the blemishes of skepticism and relativism. The pall of doubt that we "see" hanging over enclosed knowledge can be opened up by radical *self*-criticism, relentlessly interrogating our own *selves*. In parenthesis, however, it is conceded by Husserl that doubt, which disturbs us so much, obstructs our indubitable self-vision, has its own sense, and is derived from our own constitutive consciousness. To put it differently, Husserl's transcendental idealism holds our self-consciousness *responsible* both for raising doubt and removing it. If we got lost within the charmed circle of doubt and could not somehow get out of it or transcend it, our cancelled, that is, reconstituted, doubt sense could not be identified and sublated as it has been.

Husserl presents a theory of knowledge in which the knowing subject is construed as absolutely free—free from physical, biological, and historical determinants—and therefore capable of grasping what is *certain* knowledge having no blemishes or imperfection in it. This image of knowledge is too good to be accepted and seems to be at variance with our all historically available concepts of knowledge. True, Husserl does not present a close system of knowledge to escape the criticism of skepticism. His system, at least in principle, is open ended, topless, and bottomless. For, he repeatedly reminds us that knowledge in the form of self-explication knows no boundary whatsoever. Yet, he affirms in the same breath, true knowledge is absolutely self-founded. It is a peculiar sort of foundationalism, inwardly expanding without recognizing any outward constraints. Knowledge, as construed by Husserl, is never falsifiable or verifiable from without. It never encounters an unanticipated counterexample. All examples, positive and negative, relevant to knowledge are constituted by the knower (in his or her transcendental capacity). All is smooth and harmonious in it.

Naturally the question is bound to arise, Is this image of knowledge of philosophy as science conducive to the establishment of a rational society and critical culture? Must we say that truth has nothing to do with the rationality of society or the critical character of culture? To say that knowledge's only concern is truth is to miss the point that knowledge is in and born out of our rational critical consciousness. Can this consciousness, in its social and cultural forms, remain impervious to the view of knowledge we accept or reject?

The question just raised does not pertain only to our *knowledge*. It has an equal, if not more important, bearing on our *practical* life. The validity claim of our knowledge can hardly be considered and decided rationally, artificially separating it from the efficacy claim of our praxis. I say *artificially* because our knowledge is not a secular enterprise impervious to the practical influence of our social lives and living. In different contexts Husserl draws our attention to how knowledge is *will* inspired. We *seek* knowledge not only for the sake of knowledge. Additionally, it may be pointed out, following Husserl himself, our cognitive enterprise is sustained by the life world. Denied this sustenance, knowledge becomes abstract and unrelated to the world of praxis. When this unfortunate contingency arises, that is, knowledge is divorced from the life world, we are destined to encounter a crisis. In this case the crisis is not to be taken in a contingent normative sense: it is not that our concern with the life world in the context of seeking truth of knowledge is merely optional. Husserl himself keeps on reminding us that the life world, containing other things and beings and their different properties, is *pregiven* in our consciousness that seeks knowledge.

Rightly understood, it seems to me that the life world is related to our cognitive enterprise in two different ways. First, it is pregiven. But, to be consistent with the basic position of Husserl, one has to qualify this mode of pregivenness. The life world is not *passively* received by us; its mode of being given has an element of *active* constitution in it. Otherwise it is difficult to make good sense of Husserl's assertion that the self constitutes the Other. *I* constitute the world (with "us" in it). Second, if the issue of constitution is interpreted unilaterally, it raises some serious problems. If self-explication is said to be self-constitution, constitution of all, of every thing and being, then the role of the pregiven life world as the necessary sustenance for truth-seeking knowledge is belittled and marginalized, if not denied altogether. Husserl radically differs from Hume and also Kant on the issue of the role and significance of the given. Notwithstanding their difference, Hume and Kant are more or less in favor of taking the given as passive or, what Husserl says, in the naturalist manner. The latter's insistence on taking the given as actively constituted differentiates his position from the former on a very crucial point.

If the life world is allowed to have a relatively independent say of its own that may prove critical at times of what is constituted by the knowing self in the course of its self-explication, the concept of criticism (both as self-criticism and other-criticism), which seems to be so close to Husserl's heart, receives some concrete substance. If the world can never say No to what we provisionally, as *epoche,* suggest as knowledge, then our journey becomes smooth but it remains surprisingly unquestioned, uninterrupted, and unrewarding. The Leibnizian principle of intermonadic harmony and the Kantian one of transcendental intersubjectivity, I feel, have been overused and excessively relied on by Husserl. Consequently, his concern with and claim of self-criticism ultimately turn out to be empty and vacuous. If every nook and corner of the life world is deemed to be constituted by the self, then self-explication as knowledge and its expansion can never be challenged, questioned, and corrected by the former. The self or transcendental subject that makes knowledge possible also make its verification unnecessary and its falsification impossible. Thus it appears as a god surrogate in the life world and performs all the epistemic miracles that could be ascribed only to an all-knowing god.

Yes, this program of Husserl, theoretically speaking, removes the traces of skepticism and relativism from the road leading to philosophy as rigorous science. But what do we gain at the "end" of this journey? First, we are told by Husserl himself that this journey or the program is endless and also in need of continuous self-criticism. If rigorous science is an achievement, much more than an endeavor, it seems to be spurious. If it is claimed to be a victory against skepticism, it seems to be bloodless. If the said victory is so easily achieved, then why does Husserl continuously speak of the *need* for criticism, for self-criticism? If this criticism is intended to be serious, that is, a very critical input for rigorous science, Husserl has to admit that something is *not* there in knowledge, an element of *lack* haunts our cognitive enterprise, unless we go to the extreme absolutist view that what the self lacks is already in the self. There the distinction between "is" and "is not" gets hopelessly blurred. Here the discerning ears may hear the echoes of the Indian doctrine of *līlā,* playfulness, of the Absolute with itself or that of the Cunning of Reason, *List der Vernunft.* I am sure this is not what exactly Husserl had in view. But, of late, I have started hearing that Husserl was informed of some of the basic concepts of Indian philosophy.[42]

Must we be afraid of skepticism, fallibility, and the corrigibility of our knowledge? *Normatively* speaking, most of us appreciate and commend the questioning spirit and the critical attitude. Positively speaking, rarely are we sure of the correctness of the course of action we follow. Nor are we sure of the outcome of our action even when we know and follow the rules or means appropriate to the proposed action. Husserl himself commends the imperative of self-criticism.

But when he spells out his notion of self-criticism, it is, unfortunately, found to be a sustained effort in self-justification. How can I justify my *knowledge* and *action* unless I allow others to criticize them? Other *things* may be critical of my claims, theoretical as well as practical, by asserting their own nature, proving themselves inconsistent with, and as counterexamples to, my claims. Other *beings* may be critical of my enterprises, both theoretical and practical, by propounding and then testing their own modes of understanding and action that are inconsistent with mine. Every theoretical proposition and practical proposal, says Popper, are in a sense *prohibitory;* that is, rule out certain states of affairs and courses of action.

But once I accept the Husserlian doctrine of a transcendental constitution, I find that no thing or being can be inconsistent with my propositions, epistemic or valuational. In the name of criticizing naturalism or positivism Husserl takes away the independent or alien nature of everything, of every being. If all my theories, born of my self-explication, constituted by my transcendental subjectivity, leave nothing external that might prove hostile to it as a sort of counterexample, my truth seeking may turn out to be not only monotonous but also uninformative. If the polytonous or multimorphous character of the world is denied right from the beginning, it is not surprising at all that my efforts to have a unitary picture of the world, an all-embracing and constituted unity of it, does not come across any disturbing or negative finding in it. If my constituted macro structures of the world endlessly replicate and nest in it micro structures, I can*not* be surprised or disturbed by any finding whatsoever. In that case my knowledge can be declared a priori and apodictically certain. The life world(s) with all their empirical, that is, "superficial," variety are destined to conform to my constituted macro-structural unity.

If this is the destiny of phenomenology as a method, then it invites the critical designation "monster-barring" as used by Lakatos in a comparable context.[43] When a theory is threatened by some criticism or counterexample and the author of the theory is determined to save or justify it, that writer is methodologically called upon to "bar" the latter; that is, to make the theory more complex, adding more auxiliary concepts and hypotheses to it, so that it no longer remains open to criticism. In this way, the character of the theory becomes closed, enfolded, and impervious to "alien" influences.

The differences between Husserl's method and the said "monster-barring" method jointly authored by Popper and Lakatos are numerous, but that is not my main concern here. The basic point that engages my attention is their similarity on an unfortunate but important count. Husserl's research program itself has been conceived in such a manner that in the process of its execution no opposition is encountered, still less tolerated—this is precisely being attacked by the Popperian—but in Husserl's case methodology "anticipates," almost dictates and engulfs ontology. *I* become the author of *all* others.

This drawback of phenomenology as a method at least partly accounts for the analytic philosopher's relative indifference to Husserl's research program for a rigorous science. It also accounts for the existential turn of transcendental phenomenology. First, the analytic philosopher or the typical philosopher of science of the Anglo-Saxon world in a characteristic, that is, naturalist, vein conceives of himself or herself as one among many, engaged in a professional work in a particular branch of learning. He or she has no pretension that one's personal way of philosophizing, one's method of science building or analyzing scientific theories authored by others, is unique, enjoys a privileged position and an overarching status. One's approach and attitude are sober. Even at the risk of being dubbed naturalist or positivist one takes one's place as one among others, recognizes others' works and achievements seriously and tries to relate one's own position and work with theirs in a critical manner. The dream of a doubt-free unified science does not blind one to others' works, existence, and their world. This explains the modest character of the philosopher's inquiry and its outcome. Sometimes, like the phenomenologist, one admits that one's philosophy is nothing but a tool or a method of inquiry. It may be a linguistic tool as it was, for example, with the later Wittgenstein, Ryle, Austin, and Grice. Theirs is a sort of phenomenology of language. Sometimes, philosophy has been characterized as a method of philosophy or philosophical reflection on how the scientist works and arrives at conclusions. Like Husserl, many philosophers of science would be prepared to say that their investigations and conclusions are primarily methodological.

At one stage, in 1930s, as we know, some philosophers of science seriously entertained the ambitious research program of *demonstrating* the unity of science in terms of scientific language and method. For both logical and methodological reasons this program has been abandoned. But in one very important respect neither the analytic philosopher nor the philosopher of science thought of discovering a metaphysical or ontological realm that is demonstratively or self-evidently provable and leaves no room for skepticism. By implication, they are prepared to live in a world of cultural plurality, tolerating, if not promoting, the different ways of searching for truth and seeking values. In other words, their works presuppose that we can be, perhaps in fact we are, scientific without dogmatically committing ourselves to a program of unity. Science without unity is not an impossibility.[44]

Second, this open-ended approach to science finds a more radical expression in the shift of phenomenology from its transcendental phase to the existential one. In different ways Heidegger and Sartre try to show that our existence, self-existence, must take precedence over our speculative commitment to the existence of essences or eidos. Influenced by Husserl, they also favor beginning with immediate *self*-experience. But this *self* does not claim to have within it the potentiality of constituting an all-embracing unity of rigorous science. The principle of intentionality is not excessively exploited. The concept of self-explication is not overused. The self is not deemed to be an endlessly growing god with an endless research program of constituting a unified science that is self-shining and absolutely doubt free. The existential turn of phenomenology is marked by the importance it attaches to the recognition of the existence of other selves, other cultures, and the instructive gap between them and the self. The existentialist avoids the errors of the solipsist and the transcendental idealist and does not arrogate the responsibility and capacity of constituting other things and other beings, the world of the Other. In spite of the fact that the existentialist's starting point of philosophizing is self-living, self-existence, he or she finds no comparable doubt-free knowledge of the Other. And that induces, almost compels, the philosopher to enter into discourse or dialogue with others. To enlarge his or her world the existentialist does not fall back on the speculatively presupposed, internal, and the (receding) horizonal resources. Therefore he or she is in need of *communicating* with others and *learning* from them. The most the existentialist knows about self-existence is that one's *solitude* is informed of innumerable possible *relationships* (with others, with other things and beings). The existential phenomenologist's world is open ended and expanding but, fortunately, not enclosed or unified. This philosopher is content with continuity and not speculatively tantalized by the unsupportable dream of unity.

5. The Positive Aspect of Doubt

The price of a doubt-free world is indeed very high, almost impossible to be paid by the mortal and fallible human beings. Unless God is allowed to usurp the position of the human being as the knower, human knowledge cannot achieve the certitude peculiar to God's self-knowledge. The basic point to be borne in mind is this: whether this all-knowing and all-being God is himself self-existent or just one more, though *very* significant, creation (or constitution) of the human mind itself is a widely open question. This is precisely the question that neo-Hegelians, especially Marx, raised against Hegel's theory of knowledge and reality.

Our examination of the views and arguments of such widely different thinkers as the Buddhist, the Naiyāyika, the Vedāntin, Hume, Kant, and Husserl make at least one thing abundantly clear: whatever might be one's *theory* of knowledge, particularly in relation to transcendental things and beings, one's *practical* life, including cognitive attitude to the world(s) of sense and science, shows an unmistakable mixed disposition to acceptance and questioning, questioned acceptance. What seems very important to me is that every *theorist* feels *practically* obliged, in some form or other, to spell out his or her mixed disposition to acceptance and questioning.

12

Skepticism as the Critique of Search: An Epilogue

1. Different Forms of Inference and Learning: The Complex Character of the Ways of Knowing

We search something that we do not have. At times we *re*search into what we already have in an "inadequate" manner. The spirit of search becomes active in us due to a sense of epistemic need or lack. What we call *epistemic need* must not be construed narrowly. It is partly biological and partly physical. Physical causes, biologically screened or interpreted, make us aware of what we need but do not have. The sense of need tells us what we need and which sort of experience may provide it. We learn from experience. The process is endless; however it is not uniform or linear. While some experiences are drawing or inviting in nature, others are disappointing or frustrating. But all experiences are more or less inductive, informative, and help us in knowing what the things around us are, how they are, and what we ourselves are as related to them. Our environment is a field of information; our own bodies are both informed and informative. But the field in which we are placed is not continuously, stably, and uniformly available to us. An element of irremovable *randomness* is in it. The same may be said of our own bodies, of course in a lesser way.

The process of learning from experience is generally known as inductive. Induction is not possible without any base. If past experiences are not somehow or other retained in our memory in a more or less orderly way, inductive learning is not possible. In their zeal to highlight the role of deduction anti-inductivists often underplay the importance of background of learning, organized past experience that makes learning possible. In its maturer form induction yields deduction. The relation between the two is far from antithetical.

Induction and deduction are the two main forms of inference. There are other forms of inference as well; for example, probabilistic or statistical and inverse deductive inference. To say that some forms of inference are demonstrative and definitive and others are not is somewhat misleading if it is not suitably qualified. In fact deduction and induction cannot be strictly isolated. Induction, through repetition or reiteration, gives rise to a deductive propensity. The propensity is obviously cognitive. But repetition or reiteration as such, that is, without some deductive or quasi-deductive presupposition, makes no sense. Unless previous experiences on the basis of their selectively ordered properties are retained in memory, inductive learning from experience proves impossible. The unisolatable or interdependent character of induction and deduction implies that neither mode nor their combination can possibly yield any certain knowledge. This uncertainty is not paralyzing; however, it calls for justification, not only of induction but also of deduction.

Partly because of this reason some writers do not favor the idea of drawing any sharp line of distinction between deductive logic, inductive logic, and probability logic. Analysis of the five-membered structure of inference (*pancvayavī Nyāya*), given earlier, makes the point abundantly clear. According to the Naiyāyikas and Carnap, for instance, logic is logic, neither more nor less. The strength of logic need not be overplayed or underplayed. Contrary to the received opinion, deduction has no superior claim over induction in respect to certainty.

This point may be further clarified and reinforced by pointing out that not only inferential but also perceptual forms of knowledge are not free from uncertainty. One way of putting the point is this. Every perception ends up in perceptual judgment; in perceptual judgment the content of perception is bound more or less to lose its own intrinsic character (*svalakṣaṇa*). What is more, predicative judgment makes general (*sāmānyalakṣaṇa*) and intersubjectively sharable what is in reality peculiar or singular (in terms of space-time identity). This analysis is not peculiar to the psychological and logical empiricists like Hume, Mach, and James, but is traceable also to the Buddhist logicians like Dignāga and Dharmakīrti. Given this phenomenalistic account of knowledge and reality, we are led to certain basic proskeptic theses. First, the future cannot be shown to be determined by the past. Second, it cannot be proved that every event has a sufficiently determinate cause. Third, it is not possible to establish that knowledge is absolutely well-founded. Fourth, nor is it possible to show that scientific knowledge is or even can be complete in its nature. Finally, there cannot be any unified methodology of scientific knowledge that could be regarded either as inductive, deductive, or a combination of both, say, statistical deductive.

2. Induction, Deduction, and Certainty; Logic, Knowledge, and Postulation

That induction does not yield *certain* conclusions is readily conceded by inductivists, but this should not be held against induction as a method of learning. Its utility and positive aspects are numerous. Without this method one cannot move from the given to what is not given in experience. The transition from the known to the unknown is both a practical and a theoretical imperative. On the relatively narrow informative base induction enables us to construct broad cognitive structure. But this movement from narrow base to broad structure can never be absolutely complete or totally comprehensive. We stop where we feel the attained level of comprehensiveness and adequacy, practically speaking, is good enough for our purpose, cognitive or otherwise. By their very nature inductive inference and its variants are perpetually open ended.

In contrast, the structure of deductive inference gives one the initial impression that it can give us certain knowledge. Epistemic certainty is not primarily a matter of structure. The cognitive content that is organized in this or that structure has to be carefully analyzed in order to ascertain its certainty claim, if any. The main reason why in deduction we often believe that the conclusion is certain is our preoccupation with the process of deduction itself. If the rules of inference are followed, the conclusion is taken to be alright, regardless of whether it is true or false. But this is a preanalytic view of deduction. It has more than one purpose to fulfill. To deduce the conclusion from the premise or premises is obviously the most elementary and avowed purpose. Deduction is used also for the purpose of testing, directly or indirectly, theory or law. Through prediction, anticipation, or chance calculation (all presented in some deductive form or other) the cognitive truth claim of a theory, law, and so on is tested. If the deductive consequence is found to be in accord with the concerned theory (used as premise), the latter is taken to be confirmed or favorably tested. This test may be viewed otherwise as well. The deductive conclusion need not be referred to its putative parent theory. Instead we may check whether it is consilient with other observable events and actions. The acceptability of a deductive conclusion is not necessarily confined to its logical deducibility from the premise. Additionally, it is expected to be consilient with other observational items of the concerned domain. The second requirement is specially important in the field of empirical investigation. Whether a particular observational item is consilient with other items of the same domain is not merely a logical question, but also has its epistemological aspects. Even while an observational item does not offend the law of contradiction, its epistemic worth may be in doubt. One may always question whether one is experiencing it rightly or wrongly. Without a satisfactory answer to this question the decision regarding its consistency or consilience with other empirical items is not of much cognitive significance. Therefore the deductive structure or look of an inference by itself cannot guarantee the worth of what is yielded by it.

When some logicians insist on separating logic, deductive or inductive, from epistemology, one must be very careful. I have nothing very particular to say against the autonomy claim of logic. The main question to be borne in mind is the purpose for which the help of logic is being sought by *us*. Logic has no aim of its own: the aim is of the enquirer. And, as I have indicated earlier, logic may be pressed into different types of service—prediction, calculation, testing, and so forth.

The different purposes for which we use logic, broadly speaking, may be viewed in two different ways, subjectively and objectively. A theory or a belief may be accepted by us simply because we like it or it appeals to our emotion or cherished value. For example, I believe that the universe is law governed, although I cannot prove it. To refer to our liking or cherished value is in a weak way citing "reason" or evidence in support of the concerned belief. In this case we may not care to check whether our emotion or value, if questioned, can be defended; we take it as given. So the proposition about that emotion or value, though questionable in principle, is not only accepted but also cited as reason in support of the belief. In this way the acceptability or reliability of the latter is relativized to or made dependent on some reasons or evidence. It is easy for the logician to formulate the logical relations obtained between belief and its evidence, reason or justification. To formulate logically the relation between a set of propositions does not mean much. To show how rationality or probability is conferred on a belief by some emotion, value, or belief is not in itself a very difficult or interesting exercise. This helps us only in understanding why someone accepts a particular belief rather than another in a given set of circumstances (comprising diverse items of like value, belief, and so on).

In cases like these with the help of logic we are able to construct someone's account of accepting a particular belief. In a way this is how concerned persons justifies their beliefs themselves. But their way of justifying, if spelled out, becomes understandable to others. However, this is not to suggest that what becomes thus commonly understandable or intersubjectively sharable is on that very ground endorsed or accepted by others. Therefore, the subjective rationality of belief, in spite of its objective communicability and understandability, is not necessarily acceptable to others. In other words, we may always say that one's belief, though justified in a way, is not justified in the sense that it could command general acceptance.

To strengthen one's case for universal acceptance reference may be made not only to one's own beliefs, values, and so forth, but also to those of others. To repeat my example, in order to make my belief (or theory) of cosmic orderliness acceptable to others I may refer not only to my own emotional satisfaction and partial realization of a cherished value but also refer to similar views expressed by others like the Vedic poets, Spinoza, Kant, Einstein, and many others. Does it make my belief more justified and rationally more acceptable? Obviously additional confirming evidence makes my theory or belief somewhat firmer, but that is not strong enough to substantiate the claim of its universal acceptability.

Besides logical or objective formulability of my belief and its supporting justification, we demand something more from it for the purpose. It is not enough to say that a *coherence* of beliefs with other (evidential or justificatory) beliefs yields or indicates *corresponding* reality. Correspondence cannot be extracted from coherence as such. Pragmatic or practical satisfaction as such does not provide the realistic foundation, if any, of a belief. At times it has been pertinently pointed out by some thinkers of the Buddhist and Cartesian persuasions that our will to be happy or satisfied drives us away from the path of truth and leads to that of ignorance. Therefore, following the cues provided by a cherished value and emotional satisfaction, we may drift away from what is true, from what is warranted by reality.

Coherentism has a strong subjectivist underpinning. Therefore, many epistemologists and probability theorists try to show that their so-called subjective probability is based on some ontological presupposition. Notwithstanding their proempiricist epistemology, they think that the possibility and reliability of the probability estimate of a theory or belief, on ultimate analysis, is traceable to some structural properties of reality. Even when probability is construed in an autonomous way and yet claimed to be self-corrective, the underlying assumption, though very veiled and indirect, is ontological. Subjective probability and its objective or ontological basis are not totally separable. Epistemological uncertainty and ontological indeterminism are two aspects of the same human situation; that is, rooted in the human cybernatic situation in the world, the true nature of which is not exactly available to the concerned person. However, this does not prevent us from searching and researching into the recesses of the world. On the contrary, the *limits* of the knowing mind and the difficult-to-map *secrets* of nature jointly make our task of knowing "the truths" so intriguing and engaging. It is indeed a very curious situation. We know the impediments on our way to knowing the world and our own selves and yet we strongly feel that we are overcoming them maybe very slowly and falteringly. But, while engaged in overcoming those impediments, we are still haunted by a feeling, an informed feeling, that our ways of knowing the world and our place in it are extremely fragmentary, inadequate and therefore questionable.

We strongly want to get out of this "subjective" or "epistemic" ambiguity of the situation. In the process some of us resort to the the method of phenomenology and others resort to some ontological postulates like uniformity of nature, causal necessity, and the limited variety of natural kinds. It is not that the phenomenalist or the phenomenological method cannot be related to the postulates, but when we use the term *postulates* we use it bearing mainly the empiricist in view. Empiricists first resort to postulation and then try to vindicate or justify it in terms of sense-experience (of the objective world). Phenomenologists, on the other hand, try to keep themselves off the postulational strategy. To them the structures and the substructures of the world are primarily a matter of exploration and constitution, not of postulation.

3. Fluxist Ontology and Skeptic Epistemology: Coherentism and Correspondence, Ontological Realism and Epistemological Skepticism, Nature and the Problems of Theorization, the Induction and Biology of Learning

Certain presuppositions or postulations, epistemological as well as ontological, seem to be necessary both in theory and practice. For example, there is literally nothing to be known, the concept of knowledge proves empty. Nothing is left to which the word *knowledge* could be correctly applied. The situation does not appear to be that desperate. Our life and living, thought and action, strongly suggest that there are many objects of knowledge, including knowledge itself. After all knowledge is knowable and in a way nameable. This is not to deny that knowledge of knowledge and the name of knowledge may be imperfect. The object of knowledge may be temporary or durable. Even if it is conceded, for the sake of argument, that nothing is eternal, it has to be admitted that every object endures to that minimum extent necessary for one to know it. The radical fluxists like the Mādhyamik Buddhist or Heracleitus may continue to assert, "nothing is, everything flows." Even to predicate the flowing character of *something* knowable an ontological status has to be accorded it. This is a logical requirement. This is also an epistemological requirement. The possible object of knowledge, whether a flowing river, solid stone, or a mental event, must somehow be cognitively available. It must not be a thing-in-itself. Otherwise how do we make the distinction between what is solid from what is flowing? How is a physical event distinguished from a mental one? The distinction may not be absolute. From a (reductionist) metaphysical point of view one may say, "everything is *basically* matter"; another may say, "everything is *essentially* consciousness." The reductionist position, materialism, spiritualism, or vitalism, must not be superficially understood and lightly dismissed. For, as a metaphysician the reductionist may posit matter or life as *ultimate* reality but can *never* shirk the responsibility to explain the distinction drawn at premetaphysical levels, commonsensical or scientific, between, say, what is physical and what is mental.

I raise this point to rebut the preanalytic contention that the skeptic has no right to speak about knowledge of reality. For, the critic argues, to speak meaningfully about reality the skeptic has to satisfy at least two conditions. First, the skeptic has to believe that there is something real that is knowable and that something can be meaningfully stated. Second, to know what is real and to state it meaningfully one has to further admit the *knowable* existence of the object, knowable in terms of *durability* and *predicability.* In other words, the object of knowledge must be of such nature that one could identify and reidentify it and predicate truly something, property or relation, of it. To explain the cognitive availability of the object it has to be recognized (at least in principle) as an element or item of a kind or class. Without knowing the latter the former cannot be claimed to be *truly* knowable. The object is like an instance or example of what is truly predicable of it.

There can certainly be a dispute regarding the ontological priority of what predicables stand for and what "referring" (subject) expressions stand for. In fact it is there. The minimal realist tries to defend the priority of some austere ontological posits like atoms, point-instants, or sensibilia. Out of the minimal posits all other objects, big and small, concrete and abstract, are sought to be constructed. For the purpose of construction some minimum epistemological presuppositions are called for. For example, without perception or inference as means (*pramāṇa*) of knowing what is there the possibility of construction hardly makes any sense. Perhaps an added epistemic presupposition is absolutely called for: memory. If there is nothing like memory, it is difficult to explain how the objects of perception could possibly be retained in mind and used for construction. Construction is a sort of inference, gluing together cogivens or configurable objects of perception. Instead of epistemological presuppositions some minimal realists prefer some such expressions as *primitive logical techniques*. The idea is that all objects of the world are to be constructed from the remembered minimal posits and with the help of these logical techniques. The usefulness of memory cannot be so easily denied or dispensed with. At this stage I am not going into the possible intervention of the radical and negative skeptic who denies even the very efficacy of memory.

I do not propose to deny the position of the maximal realist who maintains that reality is self-existent, a sort of thing-in-itself, and has nothing to do with its being known. To be known or knownness is not a necessary condition of its existence. "Unknown existence" is not a contradiction in terms. Here for the purpose of my study I need not examine the position in detail.

But the position of the modest realists rightly deserves our attention. On the one hand, they maintain that we, human beings, have the capacity of knowing the different classes of the objects of the world, like tables and chairs, universals and particulars, gods and demons. On the other hand, they make the additional claim that it is possible for us to distinguish between what is mythical, what is commonsensical, what is scientific, and what is metaphysical (in the pejorative sense), and what is philosophical (in the defensible sense). The distinctions drawn between different classes of objects *philosophically* recognized by the modest realists of course presuppose a conceptual framework. One may or may not accept this or that particular framework, but all thinking persons do have their own (but not private) framework. It is largely borrowed from the community—cultural, professional, or linguistic. Our theoretical or professional frameworks and the practical or preanalytic frameworks are not ordinarily coextensive.

The substantial overlap of one's framework with others' frameworks shows two main things. First, how and why our ways of understanding the world are substantially sharable and communicable. Second, why certain objects that make sense in one framework do not do so in another. For example, I may not believe in the gods and demons that my tribal neighbors do. But, at the same time, I can understand their logic in support of their beliefs. To take another and less controversial example, although I accept and follow the allopathic system of a medicines, I know some people who accept and follow the Tibetan (indigeneous) system of medicines. In the matter of treatment of some diseases I find that my preferred system gives better results but not without some bad side effects. I also could see that the Tibetan system preferred by my younger brother gives good results in the treatment of a few other diseases with no (or fewer) bad side effects. What I mean to say is this. Intercultural communication and evaluation of objects and experiences, though possible, are neither complete nor coextensive, even in respect of intracultural communication, though perhaps to a lesser extent.

From this position of the modest realist at least one point of interest clearly comes out. While I can take cognizance of certain exotic things from the point of my framework, my tribal neighbors' framework does not have any place for some of them. On the other hand, the objects that make perfectly good sense within their framework are just not available within my framework. This sort of overlap or fuzziness, partial inclusiveness and partial exclusivity of contents, is a matter of common experience. The more instructive aspect of the problem seems to be this. Some *intelligible* items of one framework figure only in a *bizarre* way in another framework. What (angel or demon, for example) appears clearly intelligible to one may appear "intelligible" to another only in a derivative sense; that is, as objects of somebody else's beliefs. In brief, this suggests, among other things, that skepticism and realism not only can, but also in fact do, go together.

The objects surveyed and systematized in a theory owe their positions in it to having certain properties in which the theorist is interested. Theory construction itself is outcome of an inductive process, because theorization involves generalization. The objects covered by the theory have certain resemblances between them. If the objects of the universe have been unique in the strict sense, they could not be brought under any theory. Theoretical generalization ranging over the objects becomes possible on the grounds that the objects, however numerous they may be, have similarities among themselves at least in respect of some properties.

Negatively speaking, generalization cannot be possible if the objects had an infinite number of independent properties. The properties in respect to which the objects are concommitantly or covariantly relatable must be finite in number. This assertion is based not only on my own experience but also endorsed by others' experiences. This lends an added strength to analoginal induction.

The logical foundation for analogy is the belief that the number of varieties found in nature is limited. It is limited in such a manner that no object is so complex in nature that its properties may be viewed under an infinite number of independent classes.

This assumption of the limited variety in nature is an improved formulation of what is known as uniformity of nature. Compared to the former, the latter seems to lack refinement of form, if not being very naive. It is not enough to say that spatial and temporal differences of objects do not take away the analogies between them in many respects, which is rather a banal assertion. The main question is, In how many respects are they analogical and what are the points of their disanalogy? Further, to make generalization reliable or credible, one has to tell us something about the weight to be given to the points of analogy as well as those of disanalogy. The fact that we are not bewildered by the complexity of the kinds of objects found in nature and give up our investigation into them is believed to be due to this. Some kinds are *basic* and some kinds are not so basic or *derivative*. The former are designated as *generating natures* and the latter *generated natures*. In relation to the former, the latter are bound to be limited. On the basis of our experience of the generating natures and the properties found in them we can generalize and idealize. In the process certain properties are bound to be ignored or left out.

Another assumption that sustains inductive generalization is the denial of an *endless* plurality of causes. If called upon to explain a property or a set of properties, we fail to identify a finite number of conditions or a particular case, our causal explanation is fated to be futile. What is recognized as a cause today may prove, on review and research, to be insufficient or inadequate. In that contingency we are obliged to redefine the scope of the cause or, alternatively, introduce the new concept of collocation of causes. Whether the causes are redefined or we resort to the collocation of causes for generalizing or explaining, at one place or another we have to put a stop to this process of retreat. If to explain the same event its assumed cause is indefinitely revised or conjoined with other similar, adjacent, and deemed causes, the resulting causal explanation loses its force and focus. At the same time, it would be hasty to rule out a priori the possibility of hidden causes, which, if discovered, may provide a better explanation of the event in question. Understandably, the skeptic banks more upon this contingency and draws our attention to it. It is one thing to deny the plurality of causes for a *heuristic* purpose and a different thing to deny the very ontological possibility of it. Inductive methodology is intended to discover the hidden causes or generating natures and not to legislate them. The distinction drawn between "cause of an event" and "cause of cause of the same (event)," though *practically* very important, cannot be theoretically clinched. In the universe itself there is no causal geographical lines or even ligatures.

By this formulation of the issue, how induction works and can be defended to a certain extent, what I try to show is that ontological realism and epistemological skepticism are quite compatible. The point may be substantiated in another way. Every structure of nature has some substructures. But it cannot be said that the substructures are merely the replications of the parent structure. The available analyses of molecules, atoms, and subatomic particles convincingly illustrate this point. Illustrations may be obtained also from the fields of biology, cultural anthropology, and linguistics. In brief, nesting structures and nested substructures are not necessarily homologous. The lesson to be derived from the point is clear. The heterologous relation between the two, in spite of its limited scope, is a word of caution against hasty or sweeping generalizations.

It is true that the *uniform nature* makes induction *easily* possible and that the *limited variety in nature* facilitates it in a *restrictive* way. But neither of these two assumptions is more than of methodological or heuristic value. What leaves this approach unassailed is more important and, therefore, merits more attention; viz., the structure of reality harbors some imponderables that do not easily lend themselves to comprehensive or even restrictive theorization and generalization. An element of randomness or *chance* is irremovable from what may be called incompletely *determined* nature. In other words, I find place both for determinism and chance in the universe as a whole. In physical nature perhaps the elements of determinism are dominant, and therefore, the scope of generalization seems to be broader. At the level of the life sciences the generalizations tend to be relatively restrictive in nature. When we reach the level of human culture, we are advised to be extremely cautious in the matters of generalization, idealization, and prediction. Because the human phenomena are most complex, our generalizations must be extremely circumscribed. The rationality of the point may be persuasively illustrated from economics and art history, for example. When I say this, I must add, I do not even remotely suggest the possibility of doing away with theorization or generalization. One cannot live and move without theory or its practical analogue. The randomness that is in the universe is not to be taken as an advice against theorization itself but only as a caution against hasty and sweeping generalizations.

One recalls at this stage two extreme views regarding the structure of inference. According to the robust realist, the structure of inference is more or less reflective of the structure of reality as represented in the propositions of the inference. The question of inferential validity deemed is to be close to the *truth* of the propositions. Attempts have been made to show that the question of validity cannot be decided purely on logical considerations; epistemological and ontological considerations are also to be taken into account. In some form or other validity hinges on truth of the concerned proposition, and truth in this case is to be construed in terms of correspondence.

Some minimal realists and logical empiricists have tried to unlink the question of inferential validity from that of ontology as mentioned earlier. In their attempt to justify induction in terms of probability they have highlighted the so-called *self-correcting* nature of induction. In the distribution of frequency of alternatives in a probabilistic series they seem to see a sort of limiting tendency or what may be called a *convergence propensity.* Although, under question, they concede that this limit is not surveyable in an infinite series, they take refuge under a weak argument: from the cognitive unavailability of a limit we need not think that there is no limit. On the basis of this weak argument and making reiterative use of the rule of convergence they try to show that the lack of convergence in a lower order may be corrected through its apprisal in a relatively higher order. The correcting posits of the second order are themselves generalizations about the proportions of the alternatives as obtained in the lower order. It has been persuasively argued that through an infinite hierarchy of posits, apprisals, and corrections the point of convergence of relative frequencies that we may statistically attain is not of much epistemological value. Its epistemological value is nil or negligible because of its ontological rootlessness. When I say this, certainly I do not propose to deny the usefulness of a particular order of statistical generalization for checking or testing a generalization of a relatively higher order. In abstract discourses of physics, demography, economics, and so forth it is absolutely necessary. What I am trying to emphasize is that at some stage or other probability calculations, marked by more or less convergence or divergence, have to be made answerable to the ground-level experience. For purely practical considerations this requirement may be deferred or delayed but the relevance of ontology to probability calculation cannot be philosophically ignored. If probabilistic posits are totally unlinked from ontological posits, our inductive ascent would never take us to truth.

Unless the empiricists try seriously to bring scientific methodology closer to the issues of ontology, their use of logic and mathematics will not be of much help to them. In fact it has rightly been pointed out by some philosophers like Popper that an element of normativism must be in the scientific method. Because science is seeking testable truths about the universe, its method must be subservient to this basic aim. The point becomes clear when we see these days that the philosophers, who until recently were concerned mainly with the physical sciences, increasingly are paying serious attention to the problems of the life sciences. That noninductive rules could be of much help in scientific research is evident from the writings of such distinguished scientists as Helmholtz, Poincaré, and Einstein. When some such expressions as *lucky guess, clairvoyance,* or *paranormal perception,* (or *alaukika pratyakṣa*) come out of the pens of eminent scientists, one has to take them seriously. Helmholtz tells us that the solution of many problems was due to "a series of fortunate conjectures."[1] Poincaré observes, "Logic . . . remains barren unless it is fertilised by intuition."[2] Einstein's anti-inductivism is well known. Like Piaget, he thinks that there is a parallel between the progress made in the logical and rational organization of knowledge and the corresponding formative psychological process. In this case genetic epistemology is akin to biology. Therefore one can easily speak in terms of biological organization and cognitive exploration. Bearing this line of thought in mind Einstein observes that scientific thought is the development of prescientific thought. Einstein puts the concept of *Anschauung,* in its special sense taken over from Kant, to a very creative use by highlighting its sense-image-concept link. His theory of light, for example, makes it very clear. Basically the term *Anschauung* is used by the German-speaking people to suggest a strong image-perception link. The Einsteinian notion of thought-experiment is sustained by *Anschauung.* It suggests that some conceptual ramifications are *not* directly relatable to perceptual data. This noninductive and visual mode of thinking works as a memory picture organizing objects for perception. On the one hand, it has a thin psychosomatic base; on the other, it has its extended speculative reach. The point has been carefully looked into by some cognitive psychologists.[3] However, understandably they are not unanimous regarding the nature of the link between the sense content, image, or schema and its freer conceptual form. Is it causal or otherwise? To me it does not appear to be causal.

To understand the self-correcting nature of cognition one need not necessarily follow the inductive lines suggested by, among others, Peirce, Reichenbach, and Salmon; another promising approach is to be found in evolutionary epistemology. The latter has no pretention to be ontologically neutral in either its inspiration or its execution. The evolution of human beings from the apes shows how much seemingly unrelated information could be gradually related in a meaningful manner. Scattered and intermittently received items of information are steadily internalized and organized by the concerned organism. The same point has been worked out more analytically in genetic epistemology. The information we receive from our environment helps us not only in organizing the same but also in inferring something new out of it. On the basis of thin information we build up thick and tall cognitive structure. The process starts right from our childhood. In this connection Popper reminds us the case history of Hellen Keller, which shows how with limited perceptual access to the outer world one could know so much about it. A similar point is made by Quine when he shows how massive cognitive output can be traced to meager sensory input.

This biological view may be interpreted to buttress a modest form of apriorism. Also it may be interpreted in support of quasi-causal relation between what sense-experiences gives us and what we, human beings, build out of them. But the primary purpose for which I refer to the biological aspect of our cognitive process is to highlight the parallel between our epistemic truth seeking and biological education based on sensory information. It is for the same reason that the biology-minded epistemologists speak of the positive analogy between truth value and survival value. When we know the world truly, we can live in it more successfully and less painfully. Analogously, when different sorts of subhuman organisms are more informed of their environment, they are more successful in adapting themselves to it, ensuring their survival and minimizing the threat from the unknown and alien forces. The human effort to know the unknown imponderables is not quite different from the animals' trial-and-error method of mapping the hitherto unmapped environment. In both cases the reach of knowledge or information is more or less limited, leaving much to anticipation, uncertainty, and apprehension. It is mainly on this unmapped terrain of knowledge that the seeds of skepticism lie and, if undetected, sprout.

4. The Human Mind and the Certainty of Knowledge: Evidence, Skepticism, and Foundationalism

We seek evidence both to support and test the truth claim of a hypothesis. The very fact that we seek evidence, confirming, infirming, or both, indicates at least one thing: we are not certain about the admissibility of its truth claim.

Why do we seek evidence at all? What is self-evident does not need any evidence either to confirm or modify it. For example, the Cartesian *cogito* or the Advaita concept of *ātman* is so self-shining or self-evident that the question of supporting evidence makes no sense within the Cartesian system or the monistic system of Śaṁkara. The need for evidence is indicative of some or another, qualitative or quantitative, epistemic lack.

The interesting point to be noted is this. Neither by marshalling supporting evidence can one elevate one's hypothesis to the level of certainty, nor in terms of falsifying evidence can it be totally rejected. Despite the weakness of its formulation, the Duhemian argument against the decisive claim of crucial experiment has to be taken seriously. The consideration on which the noncruciality of so-called crucial experiments rests, when elucidated, brings out some further disturbing implications. Every good hypothesis is rightly believed to be entrenched in several other well-established theories that together form a sort of system. Falsifying evidence or crucial experiments can and obviously does change the hypothesis and, to a certain extent, modifies its parent theoretical system, but it cannot overthrow either one of the two. If neither positive nor negative evidence-experiments can conclusively establish or disestablish a hypothesis, the importance of an evidence is bound to be taken as merely marginal. Positive evidence gathered from a variety of circumstances is undoubtedly of some help in making a hypothesis a little firmer than it had been. Admittedly the weight of the negative evidence or crucial experiments is more than that of the ordinary positive evidence. But in neither case can one, merely on the basis of evidence, be certain about the correctness of the hypothesis. Under the circumstances, the basic contention of the skeptic remains: evidence, whatever may be its quality, quantity, or weight, can never make scientific knowledge certain.

Besides the inconclusive character of falsifying evidence and noncrucial character of crucial experiments, there are several other problematic features of the theory of evidence. The paradox of confirmation is one of them. Not only black ravens but also white handkerchiefs, logically speaking, confirm the hypothesis "All ravens are black." "Raven implies black" is contrapositively equivalent to "Nonblack implies nonravens." An instance of nonblack nonraven is a white handkerchief. One might refer to soft pudding and green cheese as other confirming instances of this kind.

Puzzlingly enough, all these instances (white handkerchief, soft pudding, and green cheese) confirm the raven hypothesis. To what is this puzzle or paradox due? Is it because of the purely syntactical formulation of it as originally offered by Hempel? Can we get out of it by offering a semantic account as parenthetically suggested, for example, by Max Black?[4] Or, is it the contrapositive equivalence condition the main villain in the paradoxical drama? I think that purely logical or formal equivalence fails to take due note of the epistemological significance of the so-called paradox, because our knowledge is basically available in natural or ordinary language, which is extremely rich in its suggestions and nuance and these cannot be captured in pure logical language. *Logically* speaking, there is nothing wrong in taking white handkerchief or green cheese as examples of the generalization "All nonblack (things) are nonraven." Although terms like *raven* and *black* stand for certain (more or less) definitely identifiable objects and properties, the same cannot be said of terms like *nonblack* and *nonraven*. Besides, although the former are open-ended terms, the same cannot be said of the latter, unless semantically spelled out. Another problem is associated with the use of such terms as *raven* and *black* in the generalization "All ravens are black." Neither one nor many black ravens can confirm (in any strong or credible sense) this generalization if its scope is taken to be infinite or empirically unsurveyable.

If we go into the details of the theory of confirmation as ordinarily understood, various other anomalies are bound to disturb us. For example, an evidence, *e*, that confirms two different hypotheses, h_1 or h_2, *separately* cannot confirm the hypothesis resulting from their conjunction, h_1 and h_2. Second, to develop a satisfactory theory of confirmation one must have estimates of quantitative values of degree of confirmation as well as degree of relevance. However, this is not to suggest that the qualitative concept of confirmation based upon a quantitative concept of degree of confirmation necessarily proves very useful. Third, quantifying the degree of relevance, though necessary, is extremely difficult. The same may be said of the degree of confirmation or corroboration. Weighting evidence, though often spoken of, is difficult to indicate in clear terms.[5]

Having briefly looked into the needs and limits of evidence I continue to feel that in terms of evidence we cannot get rid of underdeterminacy of theory. In a way, our theory always exceeds the bounds of evidence. By their very nature evidence, pro and con, enumerative and eliminative, cannot lift us to the level of certainty. Therefore, our truth-seeking theories are always haunted by the specter of skepticism. This is not to be taken necessarily as an admission of the defect of skepticism. On the contrary, its positive application is that it provides the motive force for finding further evidence to make theories more credible and comprehensive. Self-evidentialism may be self-certifying and self-satisfying but not conducive to scientific research and cognitive adventure beyond the immediacy of the given. When we are close to the given we may *feel* certain, but then the scope of our knowledge is extremely narrow. On scrutiny, even that feeling of certainty is found to be baseless. When by conjecture or inference we go well beyond the given, our scope of knowledge enlarges and, at the same time, our question and doubt increase. This brings me to the basic issue of my study. Why do I favor a circumspect form of skepticism or fallibilism?

To talk of skepticism in an omnibus way appears pointless. Almost every philosopher refuses to defend skepticism and *predictably* rejects it. Unless one tell us what one is rejecting and why, one's rejection or acceptance turns out to be equally uninteresting. Therefore, to start with, I briefly mention certain basic forms of skepticism and then argue in support of skepticism as found in philosophy and science. To my mind, for any philosopher of science it is difficult, almost unethical, to reject this rational approach.

Let me say first a word or two on psychological skepticism. Psychological skepticism is as deceptive as psychological certitude, deceptive both positively and negatively. The Cartesian or the Advaitin (nondualist) says that one cannot be wrong about the existence of the self because even the distinction between "the wrong sense" and "the right sense" owes its rationale to the self in question. But there is nothing logical to prevent one from questioning the correctness of this psychological sense of certitude. One can always say, as the Buddhist, Hume, or in a way even Kant do, that the sense-giving self itself is systematically elusive. Therefore, at times instead of the term *psychological sense of certitude* such other terms as *spiritual sense of certitude, transcendental sense of certitude,* and *self-evidencing evidence* are used. As the issue is not merely linguistic but substantive, I do not like to go into the details at this stage. In this connection I have already referred to the views and arguments of the Buddhist, Śaṁkara, Descartes, Hume, Kant, Hegel, Husserl, Heidggger, and Wittgenstein. I think that one cannot encounter one's own self in a static and certain manner. Therefore, it is easy neither to formulate nor to refute the psychological form(s) of skepticism.

To say that the Self (with a capital *S*) is the ultimate subject or knower, the witnessing self (*sākṣī puruṣa*), or the ground of all knowing, cannot itself be an object of knowledge is to re-echo Śaṁkara and Kant, but the ground of knowledge is left unknown. When Dharmakīrti, Hume, or Wittgenstein speaks of the cognitive inavailability of the self, we do not know who (or what) is to be regarded as the knower. Or are we advised to dispense altogether with the notion of the knower? It is true that Descartes and Hegel affirm the cognitive availability of the knowing (human) self. But even they try to show, though in different ways, that what the self knows is not absolutely certain on its own account. Descartes speaks of the necessity of assimilating human knowledge under the veracity of God. Hegel tells us why the knowledge of human beings like us is fragmentary, abstract, and needs dialectical elevation and concretization. In Heidggger's view knowledge is basically in the nature of seeking and not to be taken as a sort of enduring achievement. Even foundationalists like Husserl, in spite of their repeated commitment to the certainty of knowledge, keep simultaneously reminding us of the continuous expanding or growing character of knowledge. Why this expansion? What would be a plausible explanation of the growth of the all-comprehensive structure of knowledge? If certainty can be achieved, why is it not achieved? If something always remains left out to be achieved later, would it be right to assert that knowledge is certain? Or should we not then straightaway admit that *human* knowledge by its very nature is *incomplete,* incomplete forever? In that case "unity of knowledge" or "truth approximation" is nothing more than a heuristic maxim at best or the soothsayer's comforting word at worst.

On this and other related issues that are largely psychological I am of the view that one cannot possibly be more certain about one's knowledge of other things and beings than about one's knowledge of oneself. Whatever we may mean by *self,* whatever ontology of self, thick or thin, Cartesian or Humean, we may subscribe to, it is difficult to deny that *somehow* we know ourselves. In *practical* life we never deny it. Only at the level of *theory* do we air our differences and disputes in a rather contrived vein. If we cannot be sure about our *self*-knowledge, how can we be sure that other forms of knowledge are reliable and credible? I find nothing wrong in confessing that what we often take or believe to be true may turn out to be actually false or insufficiently justified. This is partly because of the complex nature of our own knowing and partly because of the complex nature of what we know. In the final analysis psychological complexity and ontological complexity are found to be responsible for our epistemic uncertainty.

Even the certainty we often ascribe to the forms of logico-mathematical knowledge is found, on scrutiny, to be untenable. Right from the early days philosophers have been enquiring into "the" philosophical foundations of logic and mathematics. While engaged in discovering the foundation of mathematics, they found out not one but several foundations. The versions of foundationalism—logicism, formalism, intuitionism, empiricism, and so on—are not only different but on several counts incompatible. When the mathematical foundationalists themselves confess their inability to be unanimous or sure about their own findings, we need not attach special importance to the certainty claim of the mathematicians: "loss of certainty" seems to be inherent in mathematics. Russell's pessimistic confession on the point quoted earlier seems to be extremely significant.

Similar considerations apply to the field of logic as well. No unique system of logic is acceptable to all logicians. Logic is what logicians as logicians do. "Alternative logics" is not an empty pronouncement or mere rhetoric. If logic and mathematics are assimilated under empiricist or phenomenologist epistemology, it is difficult to draw a tenable line of distinction between the mathematical mode of knowledge and the factual one. If everything could be constructed or constituted out of our experience, why do we not construct them according to our individual likes and dislikes, convenience or inconvenience? Do we have some epistemic constraints and incapacity in it? Or is the constitution of the real world such that we cannot even live in it without *appropriate* forms of logic and mathematics?

What I suggest is this. Even if we do away with the analytic-synthetic distinction, the distinction between the factual discourses and the logico-mathematical ones, at one stage or other, our knowledge, whatever may be the mode of having or constructing it, has to face the world and establish its correctness or legitimacy. Denial of a sharp demarcation between "formal knowledge" and "material knowledge" does not entail epistemological anarchy. It means only the questionability of all forms of knowledge. Questioning itself makes no sense unless there is some possible *answer* to it. There must be a way of finding out the distinction between the (more or less) correct answer and the (more or less) incorrect one. The decision regarding the correctness is neither personal nor unilateral.

It is true that in some cases, particularly in ethics and aesthetics, some of our decisions are found to be highly personal. Even in those cases we try to offer *considerations* in support of our decisions; and the underlying idea is that those decisions are based on *judgments* and that some *principles* underlie those judgments. Without this minimal obligation our axiological judgments could not be meaningfully communicated, shared, and assessed. Though we often differ, for example, in the realm of arts, rarely we refuse to discuss the points of our difference and agreement. By this line of argument again what I modestly submit is this. Relativism, though akin to constructive skepticism, is not incompatible with the claim of genuine (not simulated) knowledge in the realm of values. If in this area of discourse we find our differences more pronounced than in the area of facts or science, it is mainly because the sort of objects with which we are concerned here is of a different kind, providing wider scope for the play of our creative freedom. Not that our freedom is totally absent in science and mathematics. In all areas of our knowledge we are not equally constrained or influenced by the nature of our concerned aims and concerned objects. The worldly objects governed by causal laws or statistical generalizations are not equally compulsive or regular in their influence on the knowing mind and its creative capacity.

When the freedom of the mind is broad and the objective compulsion, causal or otherwise, is less, our judgment of the same reality, same objects, tends to differ. Naturally, it leads to a sort of relativism. It may take different forms, personal, social, or cultural. Negatively speaking, no human judgment, mathematical, scientific, or valuational, can be *universal* in the strict sense. The lack of universability, the presence of relativism or skepticism, or both often leads some philosophers to resort to some form of *transcendental* argument. For example, Kant has tried to show that in his chosen three fields of human knowledge, science, ethics, and aesthetics, we can do away with the defects of relativism and skepticism and attain universal and certain knowledge. In each case he makes liberal use of a set of transcendental arguments. The main thrust of the arguments is to show that certain principles are native to human nature that, if properly applied, give some universally acceptable judgments. In other words, in all branches of knowledge we *can* find out, rather constitute, certain laws or forms of laws that are intersubjective or universal in their scope.

Though I refer to Kant in this connection, several other thinkers have offered comparable transcendental arguments to justify the universality claim of knowledge in different fields. This is a sort of *foundationalism*. It is what may be called the *top-down form of foundationalism.*[6] In terms of unitary principles at the top the diverse details at the lower levels are sought to be organized and, if possible, neatly unified. Top-down foundationalists are determined to remove the danger of skepticism and vindicate universalism. They are bewildered by, almost scared of, the persistent difference in the areas of formal sciences like mathematics, hard sciences like physics, and soft disciplines like anthropology and axiology.

To top-down foundationalists the most disturbing factors are human bodies, their peculiar or local mental make-ups, their diversity of languages (formal and natural), and historical change. By a battery of arguments the transcendentalists try to show how in the "depth" of the mind, notwithstanding their bodily differences, human beings are convergent in their perception, conception, and particularly, apperception. Second, attempts are made to show that the difference in the mental make-ups is purely contingent and superficial. The *basic* structural features of all human minds are alike, if not bound by a sort of preestablished harmony. Third, linguistic difference is said to be surfacial or superficial. The conceptual deep structure underlying different languages is identical. We are strongly advised to take seriously the basic lessons of Bhartṛhari's linguistic monism, Catesian linguistics, Leibniz's program of universal *mathesis,* and the *linguistic universals* of the modern linguists. Finally, historical change is a shadowy movement. The light of reason can enable us to see the real and ahistorical universe in and through the shadows. Reality itself knows no history. *Cultural universals,* like *linguistic universals,* break through the barriers of epochs and cultures. The battery of these arguments is made to converge on the point of rejecting skepticism and vindicating universalism and certainty of knowledge.

The weakness of these antiskeptical arguments are numerous. One can pertinently point out that the person whose mind needs to be freed from the so-called defects of skepticism is certainly not disembodied. The transcendentalists often ascribe the seeds of skepticism mainly to the body, body-based information. The disembodied mind or the mind that is minimally under the influence of somatic irritation and information is expected to be relatively free from the supposed defects of skepticism. What is the point in trying to free the mind absolutely from somatic influences? Can a knowing mind enjoys or suffers no bodily existence and the resulting limitations? In the world of disembodied spirits, I suppose, there is no problem of skepticism requiring its refutation. The obvious purpose of my formulation of the probody argument in a reductio manner is to indicate the extreme weakness of the contrived antibody argument. In brief, skepticism or relativism is not based on the bodily differences of human beings, nor is it due to cultural or linguistic differences. Neither body nor culture is exclusively separative, both have important unifying roles. My body is not peculiarly mine; in a way I owe it to others. Because of its explicit and public nature it is open to and sharable by others. Similarly, our languages and cultures, though distinct in their identity, do not separate us and condemn us to live in isolated islands. In and through language, uttered, written, or bodily, we exceed ourselves, transcend the bounds of our bodies, get in touch with others, and communicate with them.

Extending these lines of argument from language to culture, it may be plausibly argued that in and through the process of acculturation, even as infants, we are put into the web of interhuman relationships. Right from its beginning the world of culture is a world of transcendence. Some so-called alien cultures are silently present in every culture. Strictly speaking, between cultures and languages there is no physical barrier. Different dialects of the same language shed into one another; similarly, the overlap between different cultures, rightly understood, is both intensive and extensive. In the meeting of human minds we encounter no irremovable physical-causal obstacles. The world of culture, like that of knowledge, by its very nature expands and enlarges and, unless deliberately interfered with, does not shrink or degenerate.

Another form of foundationalism that deserves our consideration may be described as bottom up. Though antiskeptical in its aim and inspiration, this form of foundationalism is *not* transcendental. But, like the transcendental approach, it too is unitary. Its main thesis is that we have some basic psychological or ontological elements at our disposal on the basis of which we can build a solid, strong, and unshakable structure. Some empiricists like Russell and Wittgenstein, who are basically opposed to the psychological approach, favor the program of logical empiricism. One of their assumptions is that the uncertainty of knowledge is due to psychologism and the imprecise classical logic traceable to Aristotle and Bacon. They thin down their ontological commitments to a bare minimum of what they call *simple objects*. It is true that both their ontology and logic have undergone significant changes over the decades. The epistemological confidence of the early Russell is not to be found in the later Russell. The language game-theoretic relativism of the later Wittgenstein is a far cry from the Tractatarian rigor.

Neither ontological austerity nor logical rigor can ensure the so-called solidity of the bottom-up foundationalism of the logical empiricists. At one stage most of them used to believe on the basis of basic statements that it would be possible to get rid of metaphysics and to build an epistemological structure taking physics as the paradigm of knowledge. In the verifiability criterion of meaning and the falsifiability criterion of demarcation they found the right way to distinguish between what is scientifically meaningful and what, like myths and metaphysics, though informative in the culture sense, is gibberish from the verificationist and the falsificationist point of view. In their enthusiasm, as we all know, the scientific philosophers draw up an ambitious program of the unity of sciences. They sincerely believed that the discoveries of modern physics, its language, and the new logical techniques would provide them the means necessary to execute the program. Only a few like Popper and Gödel realized the inherent difficulty of executing this program successfully. The fortune of the aspired unity of all sciences did not rest either on physics or logic because, summarily speaking, both physics and logic have been (still are) undergoing ceaseless historical change. No paradigm is perfect; within the paradigms and the interludes between the paradigms we find many imperfections and imprecisions, problems and puzzles, anomalies and paradoxes. These are not necessarily to be construed as purely negative in their implications. On the contrary, correctly understood, they disclose the complex structure of the world we want to capture in science, the endless precisifiability of what we can possibly know. Although no scientist is fond of paradoxes it is to be remembered that the starting point of every scientific hypothesis is some problem or other. What is surprising is not the discovery of this problem or that anomaly but the very fact, as rightly pointed out by Einstein, that our mind can comprehend at all this extremely complex world. To him, human "comprehensibility" of the world appears "awesome." In a way this is a confession that even the best of the scientists, equipped with the best available logico-mathematical techniques, instruments and informational resources, cannot find their way to what the foundationalists and the transcendentalists claim to be certain knowledge. This is not a plea for resigning to skepticism, only a reminder of the finitude of the knower, the complexity of the world to be comprehended, and the resulting fallibility of our comprehension.

5. Constructive Skepticism Defended

The type of modest skepticism or fallibilism that I am defending in this study is nothing new. It is another look at and conceptual reorganization of the available results of history of science, philosophy, and culture. The two extremes I am consistently trying to avoid are (1) the audacious claim that the philosopher can construct a theory of knowledge in which the endlessly complex world could be shown as the *perfect* product of one's mind, and (2) the commonsense view that the world is there exclusively on its own right, no matter how it is viewed by different human beings. The second alternative fails to take into account the cybernetic fact that human beings themselves as knowers are a necessary input to knowing as well as what is known. The knower-known relationship is viewed in a naive and oversimplified manner by (2) and, therefore, is unable to view the cybernetic or circular causal process in which the knowing organism and the known-knowable world are dynamically related. Interestingly enough, the first alternative also fails in a different way to see the effects of the world on the human being as knower and the resulting constraints under which the knowing process is put. In addition, in an anxiety to have *certain* knowledge the philosopher seems to forget that not only the knower but also what is known by their very nature are fated to be fallible and, therefore, questionable.

To appreciate the constructive spirit of skepticism one should not turn toward the universal skeptic. In any rational and serious discussion we should, I think, take the strongest, not the weakest, of the opponents for fruitful exchange. The case of the universal skeptics, taken literally, is negative, because skeptics tell us that they have no thesis of their own to defend and no logic to use for the purpose. Their only aim is to demonstrate the untenability both of the thesis and the logic of the opponent. It is indeed a very astounding claim; namely, that all theses and their supporting logic can be rebutted. Why do the universal skeptics undertake this responsibility at all? Their answer sounds very disarming. On their own they do not initiate any dispute or debate. They are painfully aware that their own position, if any, is as vulnerable as that of their opponents. Left to themselves, they are not interested in philosophical argumentation, which appears to them mere logomachy (*vitaṇḍā*). They know that their own reasoning, if any, is as weak as that of their critics. It is only because of pressure or persuasion that the universal skeptics may get drawn into it. Otherwise, they prefer to remain silent (*muni*; that is, who maintains steadfast silence). The position of universal skepticism is self-referring or, one may say, self-purgative. The only positive construction that one can possibly put on is this. Their preference for silence is based on the realization of the supreme reality, whatever that may be, and also of the futility of discursive reasoning, and not born of any cynicism.

It is instructive to note that even those who are totally opposed to the nonradical forms of skepticism do not fail to recognize, though in a limited way, the importance and relevance of skepticism in the context of truth searching or epistemic certainty. Some sound skeptical at the very *initial* stage of their inquiry. This approach is discernible in different ways in the writings of Descartes, Hegel, and Husserl, for example. The concept of initial skepticism and that of bracketing (or *epoche*) illustrate this methodological skepticism. In the substantive matters of knowledge, commonsensical, scientific, and mathematical, they are not at all skeptic. But having explored a long way through step-by-step arguments the initial skeptics see for themselves and show others that what is seemingly dubitable is really certain and reasons thereof. The possibility of attaining certainty is claimed to rest on the discovery of the real foundation of knowledge.

A distinction is drawn between seeming certainty and real or founded certainty. What ultimately enables the initial skeptics to reach certainty is in fact available to them right from the beginning. The self or *ātman,* which provides the foundation, is there in them all the time. Then why, to start with, do they miss it and how ultimately do they get to it? Is it not what the Vedānta philosophy describes as "getting the got"? Given this formulation, the "got" is always there and yet the process of "getting" is not chimerical. The process has been differently characterized: *via negativa,* dialectical, *neti mārga,* sublative, and unbracketing. The common point of these different approaches, roughly speaking, is that the knowing mind has somehow in it the capacity to realize that what it knows perceptually, conceptually, inferentially, paranormally, or in other ways is more or less inadequate. This inadequacy is partly due to the mind's own limited capacity and partly due to the nature of reality that is not comprehensible in one go. It is not like God's way of knowing itself or the whole of reality as a *totum simul*. Reality is so complex, so vast, that it has to be known gradually, step by step. The knower has the capacity to foresee, more or less inarticulately, the steps beyond the ones already scaled. Metaphorically speaking, the mountain of knowledge has different features in it; viz., inclined slopes, difficult terrains, vast plateaus, steep ridges, and blinding snow storms at high altitudes. Some of these areas are easily climbed or negotiated, others prove extremely difficult to scale. The area of doubt has often been likened to a plateau. It leaves below and behind the plains of ignorance and has before it the promise of the higher peaks of knowledge. Strictly speaking, skepticism does not mean total lack of knowledge, but a form of informed ignorance. The knower knows something warranting him or her to assert that we cannot be sure about what we have known. Perhaps we have in us something more to claim or hope—more than what we have already known; that is, beyond our *immediate* cognitive reach.

It is clear that human knowledge is not to be narrowly defined by the bounds of senses alone. As we are painfully aware, the reach of our senses is very limited. Discovery of highly powerful microscopes, telescopes, and other instruments has now made us further aware of this. And we have good reason to believe that future discoveries along the line would show how little we can know by our bare senses compared to what remains yet to be known.

Apart from the quantitative aspect of this question, some skeptics, the Mādhyamika Buddhists, for example, have pertinently raised the qualitative question on the matter. Whatever we know is particular, discrete, and momentary. It appears just for a moment and then disappears for ever (*kṣaṇabhangura*). It is only because of the traces left behind the momentary things that we are put under the psychological illusory impression of their continuity. We see successive and different drops of water and oil as stream. The antiskeptic may raise the question, "Who are this 'we'?" Are "we", the knowing selves, as momentary as the momentary things and objects of knowledge? Who knows what and how many moments are involved in the moment of knowing?

Not that these points have not been gone into by the Buddhists or other skeptics of the comparable persuasions. But the point with which I am concerned here is that the extreme narrowness of the reach of the senses cannot provide even the minimal foundation of our practical action, not to speak of higher-level theories. The attempt to highlight the limitations of the empirical base of our knowledge is more a theoretical exercise than a practical one. This approach has at times been described as "distortion from below" by writers like Merleau-Ponty, for example. If the empiricist starts from discrete sensibilia or momentary experiences (of things), it is not easy to go up. From isolated sense experiences inductive ascent is impossible. To overcome the psychological handicap posed by this sort of problem, one recalls what logical techniques the acute logicians like Dharmakīrti, Carnap, Goodman, and Quine have resorted to. This is not the place to go into those details. Once the *initial* psychological handicap is taken seriously, as final and irremovable, the constructionist strategy to overcome them does not appear very promising; at least this is the thrust of argument of phenomenologists like Merleau-Ponty. Unless it is justifiably assumed that there are some cues, intentionality (Husserl) or external meaning (Royce), within the so-called discrete sense impressions, the attempts to relate them together as meaningful is bound to appear as alien imposition. In the absence of this imposition, what is there in them to prevent their pulverization or falling apart as unrelated fragments?

If the "distortion from below" puts one type of block on the road to higher-level theory construction, another type of roadblock is encountered by what has been called the "distortion from above," again a term coined by Merleau-Ponty. If sensationalism is one extreme, conceptualism is another. If to map the puzzling mosaic of the empirical world, the philosopher is allowed boundless freedom to *presuppose* or *postulate* extremely abstract concepts or *eidos* of the highest level, we are landed in another kind of problem. Whether we call this position conceptual idealism or transcendental realism, the problem at the ground level proves equally intractable. If the postulated entities, epistemic or ontic, Ideas or Ideas of Reason, need not be answerable to the world of experience in which we live, an element of incredibility is bound to grip our mind. If "corresponding" to presupposed realities and idealities nothing empirical is available, the philosopher is literally allowed to create a ivory tower suspended by some miracle and from which no passage or ladder enables one to descend to the worlds of common sense and science. Skepticism is not peculiar to the radical empiricists whose "logical explanations" of the smallest and the biggest things of the universe appear more or less incredible to us. The presuppositionalist or the postulationist who ask us to believe in the reality of certain things and beings that are not even mediately or remotely relatable to our world of sense-experience does not enlighten us in the least but, on the contrary, evokes in us a sense of disbelief. If the concepts are not required to be applicable to the percepts, empirical realism makes little sense. Kant's treatment of this fundamental problem in the chapter on schematism in the *First Critique* seems to me very insightful. How at times do concepts, mediated or unmediated by schema, result in *mis*application? Kant's apparent failure to satisfactorily answer this question makes his case of empirical realism weak and uncritical, if not indefensible. These considerations lead me to believe that we have to explore the areas of modest skepticism and minimal realism to understand the true implications of constructive skepticism.

The maximal realists have a way of combating skepticism. First, they propound a theory of inflexible reference. The relation between names, proper or general, and their nominata are claimed to be unchangeable. The name-nominata relation, once fixed, is unchangeable and valid in all possible worlds or situations. Second, by insulating referential relation from the context (or form of life) of its use, the maximal realists try to show the irrelevance of the knowing subjects in the determination of meaning. In a veiled way this theory of reference is not only antipsychologist in its inspiration but also indifferent to other objects or items of reality, other than the nominatum or objects in question. Finally, in its stronger form this theory is antihistorical. The name-nominata relation, it is argued, cannot be subject to historical mutation.

Notwithstanding its contemporary influence in some quarters, I am not impressed by it. The grounds of my reservation are several. First, what is special about those persons who *initially* fix the relation between referring expressions and their referents? How can some people, rather than others, be credited with the task of fixing this relation in an inflexible and immutable way? Whatever is fixed by *some* human beings cannot be deemed acceptable to or binding on *all* human beings. Second, referential meanings cannot be dehistorized and decontextualized. Those who believe in this possibility ignore the *human* roots of language and the relation obtained between the world, on the one hand, and different languages, on the other. I do not dispute the relative referential *stability,* but this does not mean that stability is *permanence*. Humankind has no a priori epistemic or meaning-bestowing capacity by which it can explain the so-called inflexible relation between names and nominata. Had this theory been correct, perhaps we would not have encountered the problem of translational indeterminacy or evidential underdeterminacy. Finally, for sociohistorical creatures like ourselves it is not at all possible to grasp the so-called inflexible referential relation. Strictly speaking, it is neither dehistorizable nor decontextualizable. As I have discussed this issue elsewhere, I do not propose to go into its details here.[7]

Another strategy of the maximal realist to stave off skepticism is to construe the knowing person as detached from and causally uninfluenceable by the world to be known. The knower is not organically related to the world; and this knowledge, though in a sense really the *knower's,* is not truly due to him or her. Personal association of the knower with knowledge is purely contingent: there is nothing objective in it. The assumption under which the maximal realist works seems to be this. If the knower is deemed to have a say in the nature or constitution of knowledge, say, of a particular object x, then to a different knower the same object x would appear differently as x_1, x_2, and so on. And this is bound to raise doubt regarding the true identity of x and relativism regarding the nature of knowledge (of x) as available to different knowers, k_1,Ik_2, and so on.

To undo the miseffects of skepticism regarding the nature of the object of knowledge the maximal realist's approach is twofold. First, the realist maintains that the *essence* of the object has nothing to do with how it is perceived, conceived, or judged by the knower. Second, within the knower is a cognitive capacity providing access to the essence of the object. If these two radical claims could be strictly defended, the history of human knowledge would have been quite different. In that case, one might almost say, knowledge would not have undergone so many vicissitudes as evident from the past of the different branches of knowledge. Historism and essentialism hardly go together. Yet when the essentialists like Plato, Kant, or even Hegel speak of history they mean "deformed" and experienceable shadows of essential reality.

The said two essentialist strategies of maximal realist are indefensible and, what is worse, uncalled for. The knower, though distinct from the object, is yet an inseparable part of the world as he or she knows. This inseparability may be understood in different ways, causally, organically, or system theoretically. From whatever point of view we look at the relation, it is difficult to deny that human knowledge, despite its objectivist construals, *a la* Bolzano, Frege, or Popper, are open to both causal and biological influences.[8]

The untenability of maximal realism gives rise to another approach to the whole issue. This is a mixed approach in which one can discern elements of both realism and idealism.

Some objects of experience, to start with, appear to be really real, enduringly real, and not subject to the changing moods of the knower. Objects of illusory experience, for example, do not initially appear to be illusory. The very expression "illusory experience" is *post eventum* and born out of reflection on the experience that appeared to be valid at one stage. The cancellable and corrigible character of experience is not peculiar to the snake-in-rope sort of illusions. These glaring examples are used purely for pedagogic purpose.

Most of our experiences undergo silent but unarrestable change. The roots of change are partly personal and partly impersonal. For example, our experience of time changes as we grow older. The point has been persuasively analyzed by James in his *Psychology*. To the children time moves slowly and to middle-aged people not so slowly; time flows fast to the old people. One easily predictable response to this view is that the nature of the passage of time is entirely a matter of time itself and has nothing to do with the nature of the mode of the percipient's perception. The die-hard realist maintains that the sense of change is due to the *mental* makeup of the persons of different age-groups and that it does *not* disclose the *objective* nature of time. This assertion seems to suggest that there is something to be called time-in-itself, its own inherent metric that has nothing to do with how it is experienced. Certainly this approach exhibits a well-known realist bias but at the same time it ignores the contribution of *experience* to what is experienced; that is, *experience of the object*. The object is not purely external to experience nor the converse; they are distinguishable but inseparable. The rationale of this example may be extended to the case of the experience of space and objects locatable in the space-time framework.

The extent and nature of change of the objects may be explained in different ways, conservatively and radically. In the scheme of conservative explanation objects, in spite of their change, are recognizable and reidentifiable. Their changeability means modification, mutation, accretion, growth, decay, and so forth but not total disappearance or negation.

The objects of perception are not fixed in their nature. Apart from the implicit and interpretative character of judgments underlying perceptions, the objects themselves have different features and facets that are not given to us all at once. Further, objects in nature are not in isolation but in interconnection. It is for our practical investigative purpose and predetermined interest that we view them in one particular way rather than another, at least for the time being. This does not however mean that they could not be reviewed, judged, or interpreted otherwise. What is causally given to our senses need not be taken as something uniquely dictated to our understanding. Our understanding itself has a say, a hermeneutic say, in the matter.[9] To put it differently, the information of the world or its parts as objects seemingly received *passively* are in fact actively and silently processed by our understanding.

Scientific understanding, in spite of its causal structure, is not free from the freedom of different interpretations. Difference of interpretations is quite consistent with their mutual compatibility. So the empirical world that, viewed from one end, seems to be causally and uniquely determined, viewed from another end, is open to different or alternative, but not contradictory, interpretations. The implications of this view are far-reaching. One, human freedom is brought back from its transcendental exile and rehabilitated in the world of *cultural* experience. Our physiological capacities of perception are oriented by cultural *conditions*. Two, freedom is found to be a property not only of the acculturated knowing self but also in a way of the *natural* world of the possible objects of experience itself. Causality and indeterminacy (due to statistical imponderables) are recognized as cohabitable fellow citizens of the same world. In other words, our epistemic freedom has not only its cultural roots but also its ontological warrants.

The recognition of the empirical status of scientific objects does not prevent the transcendentalist from pointing out the negatable character of the former. The causal, statistical, or hermeneutic understanding of the empirical world, viewed transcendentally, may stand not only transmuted beyond recognition but also totally annulled or negated. For example, to Śaṁkara, the world of sense and science, though empirically real and in a sense very concrete, is transcendentally nonexistent. The objects of judgmental understanding are just not available from the highest (*pāramārthika*) point of view.

A comparable line of argument is distinctly discernible in Kant. What is empirically real, *real* to understanding, is transcendentally ideal, *ideal* to reason. However, the transcendentalists do not accept the suggestion that what they call *ideal* has no reality of its own. Simply because of the difference in its nature, unlike the empirical objects, we must not positively assert that the ideal is not real. Instead, we would be advised to recognize the different levels of reality. Being an *Advaitin,* an uncompromising monist (or nondualist) as he is, Śaṁkara will reject the idea of the degrees of reality. He argues at length to refute the dualism of the Sāṁkhya dualism between self, *puruṣa,* and nature, *prakṛti,* and defend causal unity of the Absolute (Brahman) pervading "the empirical-natural" and "the transcendental-spiritual".[10] According to him, reality in the highest sense knows no degree; it is changeable and cancellable only in a nonstrict or lower sense. Abiding and unchanging Reality (inexplicably) sustains the unreal objects of illusory perception and those of empirical judgments that admit of degrees. In this respect, the latter's ontological as well as epistemological positions differ from Kant's. But in a very important sense both may be said to be skeptic *vis-à-vis* the world of sense objects and scientific objects. In this way they try to reconcile realism of one level with idealism of another marked by a mediative or descending skepticism. In the ascending peaks of unitary or nondifferential knowledge the scope of doubt, they claim, is gradually narrowed down. This type of skepticism gives one the impression that it both recognizes and transcends the objects familiar to our practical and scientific life.

At its highest reach the Advaita (literally, nondualistic) form of knowledge is said to be knowledge by identity. At that level knowing *is* being, *not* knowledge as ordinarily understood; that is, knowledge defined in terms of the knower-known distinction. It is stated that whoever knows Brahman becomes Brahman. And therefore the question of doubt, an epistemic state, does not arise at all.

Kant's basic way of doing away with skepticism at the level both of empirically real world and the transcendentally ideal one is to *postulate* the *teleological affinity,* not *causal unity,* between the two. Because this affinity is postulational, it cannot be proven in the way the mathematical objects of pure intuition or physical objects of empirical intuition are. The objects like freedom, the immortal soul, and God are not cognitively available in the strict sense; yet their acceptance is commended and defended by reason in its speculative and practical moments, on the basis of its transcendental assumption that it (reason) cannot be at variance with itself. Faith, hope, and postulation all derive their rationality from the harmonious nature of reason itself. The totality of objects of scientific knowledge and those of hope and faith, the realm of causal necessity and that of moral obligation and aesthetic satisfaction, cannot be incompatible. Because of his noetic dualism Kant admits that the totality of the said two realms is not cognitively available in the stricter sense. Yet, to him, our moral and aesthetic consciousness unmistakably indicate that within the overarching metaphysic of experience mathematics and natural science do serve as *means* for the full and complete development of human reason and values.

Both the *possibilities* and the *uses* of different disciplines, scientific and valuational, have in them an internal censorship that, Kant claims, saves them from both empty speculation and rootless doubt.[11] One may or may not agree with Kant's produalistic metaphysic of experience or the architectonic of reason intended to reconcile empirical realism with transcendental idealism, causality with freedom, keeping both under the surveillance or censorship of reason so that the concerned forms of knowledge do not get stained by skepticism of the Humean origin. Yet one cannot but be impressed by the ingenuity of his argument to relate the two seemingly distinct, if not opposite, realms. Also one cannot fail to note his anxiety to rule out the possible incursion of imponderables or unreason within the causal world. The extension of the transcendental or the strong justificationist structure of argument to the realm of morality and beauty suggests his equal anxiety to expel the possibility of genuine difference in the matters of moral and aesthetic judgment. In the process Kant's top-down structure of transcendental argument becomes enclosed and therefore impervious to "external" or "alien" considerations and criticisms. All his *Critiques* are so framed or formulated that they are not open to any criticism external to them. The sort of self-criticism or internal criticism Kant speaks of is partly internalization of the possible sources of external criticism and partly based on a refusal to take of external criticism that is inconsistent with his overall architectonic of reason.

This transcendental strategy to rebut skepticism ends up in self-justificationism and refusal to listen to others' criticisms. In this respect one must appreciate those thinkers who, like Hegel, recognize *skepsis* as *natural* to human reason engaged in searching the *transcendental*. In terms of *skepsis* they try to define the relation between the empirical and the transcendental, the former being the inarticulate presence of the latter in it and the latter being the outgoing meaning of the former. In plain language, often we fail to see what is visible to others and yet we can mentally visualize what they see. Human reason cannot be so comprehensive as to make futile all our search for further truths or new facts. At no stage of the appropriation of our experience can we feel that what we have known is absolutely certain. This sense of uncertainty is constructive and explorative, not crippling or paralyzing in any way. Rightly understood, *skepsis* is both a driving force and a seeking influence. It drives us from within and goads us to search what is unknown or adds clarity to what is already known. It is a seeking influence in the sense that it is fostered in us by some external or "distant" influence, critical or constructive.

On my part I fail to see how I honestly can possibly claim to have that reason in me which would make my new experience uninformative and searching speculation uncalled for. Always on reflection I find that the information available to me is inadequate and more intriguing than satisfactory or certificatory. Apart from the informative content of experience, it provides a sort of joy. Even the unpleasant and painful experiences have an educative component. In various other ways experience works as a driving force, endlessly driving. Partly this explains my search for what I do not know and research into what I already know but not adequately or satisfactorily enough. In this *skepsis*-haunted disposition of mine I find nothing wrong. On the contrary, I always have a feeling, an informed feeling, that I know much less than what I originally wanted to know in the areas of my interest. It is not a word of humility, nor is it a sin-cleansing confession. It only expresses my sense of discontent with my own present state of knowledge and available (available to me) arguments both for and against it. Simultaneously it expresses also of my will to know more about it and its evidences and examples, both supportive and contrary. This sense of cognitive dissatisfaction and the resulting search and research are what I basically mean by skepticism. I find nothing wrong with it. On the contrary, I think that its absence tends to make us self-righteous and uncritical about our own views and beliefs. What is worse, when we are not genuinely self-critical we often develop a sort of intellectual smugness or arrogance. It stands on the way of our understanding others' views and values. It hinders our rational inquiry. Also it harms our moral nature. When we are smug or develop a vested interest in our own views, we fail to honor and understand others' views and values, and often the defects and disvalues of our own beliefs and practices escape our attention. To prevent such unwelcome contingencies skepticism is our much needed ally in all our critical and constructive inquiry.

Notes

Chapter One. Induction, Probability, and Uncertainty

1. C. G. Hempel, *Aspects of Scientific Explanation,* pp. 53–79. New York: Free Press, 1965.

2. Patrick Suppes, "Concept Formation and Bayesian Decisions," in Jakko Hintikka and Patrick Suppes, eds., *Aspects of Inductive Logic.* Amsterdam: North-Holland Press, 1966. See also his "Probabilistic Inference and the Concept of Total Evidence," in ibid.

3. Hempel, *Aspects,* p. 53.

4. Patrick Suppes, "Probabilistic Inference."

5. Rudolf Carnap and Richard C. Jeffrey, eds. *Studies in Inductive Logic and Probability,* Vol. 1, pp. 13–14. Berkeley: University of California Press, 1971.

6. Ibid.

7. Rudolf Carnap, "Replies and Expositions," in P. A. Schilpp, ed., *The Philosophy of Rudolf Carnap: The Library of Living Philosophers,* p. 972. La Salle, Ill.: Open Court Publishing Co.

8. Imre Lakatos, ed., *The Problem of Inductive Logic,* p. 126. Amsterdam: North-Holland Press, 1968.

9. Ibid., p. 145.

10. H. E. Kyburg, *Probability and the Logic of Rational Belief.* Middletown, Conn.: Wesleyan University Press, 1961. See also, Lakatos, *Problem of Inductive Logic,* and L. J. Cohen and Mary B. Hess, eds. *Application of Inductive Logic,* Oxford: Oxford University Press, 1980.

11. Wesley C. Salmon, in Lakatos, ibid., "The Justification of Induc-

tive Rules of Inference," pp. 24–43, "Salmon's Replies," 74–97, "Who Needs Inductive Acceptance Rules," 139–44.

12. Rudolf Carnap, in Carnap and Jeffrey, *Inductive Logic and Probability.*

13. Patrick Suppes, "Rational Changes of Belief," in Lakatos, *Problem of Inductive Logic,* p. 189.

14. Karl R. Popper, *The Logic of Scientific Discovery,* p. 105. London: Hutchinson, 1959.

15. P. F. Strawson, *Introduction to Logical Theory,* pp. 260–63. London: Methuen, 1952. See also, A. J. Ayer, *The Problem of Knowledge,* pp. 71–75. London: Penguin Books, 1956.

16. A. J. Ayer, *Probability and Evidence,* p. 88. London: Macmillan, 1972.

Chapter Two. Can Induction Be Justified?

1. David Hume, *A Treatise of Human Nature,* ed. L. A. Selby-Bigge. Oxford: Oxford University Press, 1896. David Hume, *Enquiries Concerning Human Understanding,* ed. L. A. Selby-Bigge, 2d ed. Oxford: Oxford University Press, 1902.

2. Donald C. Williams, *The Ground of Induction.* Cambridge, Mass.: Harvard University Press, 1947. For critical review of William's *The Ground of Induction,* see Ernest Nagel in *Journal of Philosophy* 44 (1947): 685–93.

3. Justus Buchler, ed., *Philosophical Writings of Peirce,* p. 202. Mineola, N.Y.: Dover Publications, n.d.

4. William Kneale, *Probability and Induction,* p. 149. Oxford: Oxford University Press, 1949.

5. Ibid., p. 218.

6. J. M. Keynes, *A Treatise on Probability,* p. 51. New York: Harper & Row, 1962.

7. Ibid., pp. 55–56.

8. Roy Harrod, *Foundations of Inductive Logic,* p. 197. London: Macmillan, 1956.

9. Roy Harrod (ibid., pp. 14–15) puts Bayes's theorem in this way. "Let it be possible to enumerate two or more hypotheses which between them

exhaust the possibilities of the case, so that one or other must be true. Let it be possible to assign the definite prior probability to each of the hypotheses. Let it also be possible, assuming in turn that each separate hypothesis is true, to assign in each case a definite probability for the occurrence of an event X in certain defined circumstances. Then if X does occur (or does not occur) in the circumstances in question, this enables us to assign new probabilities to the original hypotheses."

10. Kneale, *Probability and Induction,* p. 205.

11. Harrod, *Foundations,* p. 19.

12. Ibid., p. 253.

13. Ibid., p. 52.

14. Hans Reichenbach, *Experience and Prediction,* p. 382. Chicago: University of Chicago Press, 1938. Reichenbach's comprehensive work on the subject, basically *logical* in character, is *The Theory of Probability.* Berkeley: University of California Press, 1949.

15. Karl Popper, *The Logic of Scientific Discovery,* p. 30. London: Hutchinson, 1959.

16. G. H. von Wright, *A Treatise on Induction and Probability,* p. 21. London: Routledge & Kegan Paul, 1951.

17. William Whewell, *The Philosophy of Inductive Sciences,* Vol. 2, p. 92. London, 1840.

18. S. Barker, *Induction and Hypothesis,* p. 157. Ithaca, N.Y.: Cornell University Press, 1957.

19. J. P. Day, *Inductive Probability,* pp. 274–75. London: Routledge & Kegan Paul, 1961.

20. William S. Jevons, *The Principles of Science,* p. 675. New York: Dover, 1958.

21. Ibid., p. 228.

Chapter Three. Probabalistic Justifications of Induction

1. Rudolf Carnap, *Logical Foundations of Probability,* pp. 207–8. Chicago: University of Chicago Press, 1963.

2. J. M. Keynes, *A Treatise on Probability,* p. 221. New York: Harper & Row, 1962.

3. Ibid., pp. 3–4.

4. Carnap, *Logical Foundations,* p. 44.

5. Karl Popper, *The Logic of Scientific Discovery,* p. 149. London: Hutchinson, 1959.

6. N. R. Hanson's Introduction to Keynes's *Treatise on Probability,* p. ix.

7. Ibid., p. 97.

8. Popper, *Scientific Discovery,* p. 88.

9. Hans Reichenbach, *Experience and Prediction,* p. 333. Chicago: University of Chicago Press, 1938.

10. Ibid., p. 341.

11. Keynes, *Treatise on Probability,* p. 8.

12. Ibid., p. 322.

13. Carnap, *Logical Foundations,* pp. 176–77.

14. Popper, *Scientific Discovery,* pp. 262–65.

15. Keynes, *Treatise on Probability,* pp. 116–17.

16. Ibid., pp. 305–6.

17. Ibid., p. 228.

18. Ibid., p. 228.

19. Ibid., pp. 253–54.

20. R. B. Braithwaite, *Scientific Explanation,* p. 258. New York: Harper & Row, 1953.

21. Keynes, *Treatise on Probability,* p. 249.

22. Ibid., p. 258.

23. Braithwaite, *Scientific Explanation,* p. 275.

24. Keynes, *Treatise on Probability,* p. 264.

25. William Kneale, *Probability and Induction,* p. 211. Oxford: Oxford University Press, 1949.

26. Jean Nicod, *Foundations of Geometry and Induction,* p. 274. London: Routledge & Kegan Paul, 1930.

27. C. G. Hempel, *Aspects of Scientific Explanations,* pp. 10, 21, 25. New York: Free Press, 1970.

28. J. P. Day, *Inductive Probability,* p. 98. London: Routledge & Kegan Paul, 1961.

29. S. Barker, *Induction and Hypothesis,* p. 84. Ithaca, N.Y.: Cornell University Press, 1957.

30. P. F. Strawson, *Introduction to Logical Theory,* p. 261. London: Methuen, 1963.

31. Reichenbach, *Experience and Prediction,* p. 346.

32. Ibid., p. 350.

33. Jerrold Katz, *The Problem of Induction and Its Solution,* pp. 50–52. Chicago: University of Chicago Press, 1962.

34. Reichenbach, *Experience and Prediction,* pp. 346–47.

35. Hans Reichenbach, *Theory of Probability,* p. 474. Berkeley: University of California Press, 1949.

36. Ibid., pp. 447–48.

37. Reichenbach, *Experience and Prediction,* p. 340.

38. Ibid., p. 342.

39. Reichenbach, *Theory of Probability,* p. 475.

40. Reichenbach, *Experience and Prediction,* p. 354.

41. G. H. von Wright, *The Logical Problem of Induction,* Ch. 8. Oxford: Basil Blackwell, 1957.

42. Reichenbach, *Experience and Prediction,* p. 376.

43. Barker, *Induction and Hypothesis,* pp. 65–71.

44. Popper's *Logik der Forschung* was published in 1934, nearly sixteen years earlier than Carnap's main work on the subject, *Logical Foundations of Probability* in 1950. On inductive logic and probability Carnap's notable work, appeared in 1945. See "On Inductive Logic," *Philosophy of Science* 12 (1945): 72–97; and "Two Concepts of Probability," *Journal of Philosophy and Phenomenological Research* 5 (1945): 513–32.

45. Carnap, *Logical Foundations,* pp. 23–36.

46. Y. Bar-Hillel, ed., *Logic, Methodology and Philosophy of Science,* p. 274 ff. Amsterdam: North-Holland Press, 1965.

47. This is a line from Hempel's personal communication to me.

48. Carnap, *Logical Foundations,* p. 572.

49. Popper, *Scientific Discovery,* p. 393.

50. Carnap, *Logical Foundations,* p. v.

51. Ibid., p. 74.

52. Keynes, *Treatise on Probability,* p. 258.

53. Carnap, *Logical Foundations,* pp. 299, 565.

54. Ibid., p. 181.

55. Karl Popper, *Conjectures and Refutations,* p. 291. London: Routledge & Kegan Paul, 1962.

56. J. Hintikka and Patrick Suppes, eds., *Aspects of Inductive Logic.* Amsterdam: North Holland Press, 1966.

57. See for example J. G. Kemeny's contribution, "Carnap's Theory of Probability and Induction," in P. A. Schilpp, ed., *The Philosophy of Rudolf Carnap,* pp. 711–38. La Salle, Ill.: Open Court Publishing Co., 1964.

Chapter Four. Probable Knowledge: Confirmation and Correction

1. W. V. O. Quine, *The Roots of Reference,* p. 20. La Salle, Ill.: Open Court Publishing Co., 1973.

2. Ibid., p. 21.

3. Ibid., pp. 30–32.

4. W. V. O. Quine, "On Popper's Negative Methodology," in P. A. Schilpp, ed., *The Philosophy of Karl Popper,* Vol. 14, Book 1, pp. 218–28. The Library of Living Philosophers. La Salle, Ill.: Open Court Publishing Co., 1974. See, for Popper's response, in Book 2, pp. 989–93. Here both Popper and Quine are concerned mainly with the well-known "paradox of confirmation." Whereas the latter points out the difficulties involved in (1) verifying-falsifying statements of multiple quantifiers, both universal and existential, and (2) exactly defining "evidence," the former tries to show "the complete irrelevance of the problem" based mainly on the failure to indicate the important difference between the notion of confirming *instance* and that of testing (negative) *evidence.*

5. W. V. O. Quine, *Word and Object,* p. 5. Cambridge, Mass.: MIT Press, 1975.

6. Ibid., p. 7.

7. W. V. O. Quine, *Ontological Relativity and Other Essays,* p. 72. New York and London: Columbia University Press, 1969.

8. Karl Popper and John C. Eccles, *The Self and Its Brain,* p. 120. Berlin, New York, and London: Springer International, 1977.

9. Karl Popper, *Conjectures and Refutations: The Growth of Scientific Knowledge,* p. 219. London: Routledge & Kegan Paul, 1962.

10. Ibid., p. 391.

11. W. V. O. Quine, *The Roots of Reference,* p. 34. La Salle, Ill.: Open Court Publishing Co., 1973.

12. Karl R. Popper, *The Logic of Scientific Discovery,* p. 103. London: Hutchinson, 1959.

13. Popper, *Conjectures and Refutations,* p. 388.

14. Quine, *Roots of Reference,* p. 3.

15. W. V. O. Quine, *Philosophy of Logic,* p. 99. Englewood Cliffs, N.J.: Prentice-Hall, 1970.

16. In Popper, *Conjectures and Refutations,* pp. 253–92.

17. P. F. Strawson, *Skepticism and Naturalism: Some Varieties,* pp. 32–50. London: Methuen, 1985.

18. Popper, *Conjectures and Refutations,* p. 209.

19. Ibid., p. 214.

20. Ludwig Wittgenstein, *Tractatus Logico-Philosophicus,* trans. D. F. Pears and B. F. McGuinness, 5.15. London: Routledge & Kegan Paul, 1961. See also 5.151–5.156.

21. Ibid., 2.0201, 3.144.

22. Ibid., 4.023.

23. Ibid., 5.634.

24. Ibid., 6.362.

25. Rudolf Carnap, *Logical Foundations of Probability,* p. 565. Chicago: Chicago University Press, 1950.

26. Ibid., pp. 114–18.

27. See "Induction by Enumeration and Induction by Elimination," in Imre Lakatos, ed., *The Problem of Inductive Logic.* Amsterdam: North-

Holland Publishing Company, 1968. Reference to other works of Hintikka on the subject may also be found in this paper.

28. Ibid., p. 201.

29. J. M. Keynes, *A Treatise on Probability,* pp. 66, 68, 71, 219–21, 258–64. New York: Harper & Row, 1962.

30. Hintikka, "Induction by Enumeration," p. 203.

31. G. H. von Wright, *The Logical Problem of Induction,* 2d ed., p. 119. Oxford: Basil Blackwell, 1957.

32. Imre Lakatos, ed., *The Problem of Inductive Logic,* p. 219. Amsterdam: North-Holland Publishing Company, 1968.

33. Alex C. Michalos, *The Popper-Carnap Controversy.* The Hague: Martinus Nijhoff, 1971.

34. Popper, *Logic of Scientific Discovery,* p. 363.

35. Wright, *Logical Problem,* pp. 183–84.

36. Ibid., pp. 166–67.

37. Rudolf Carnap, "Probability and Content-Measure," in P. K. Feyerabend and Grover Maxwell, eds., *Mind, Matter and Method,* p. 248. Minneapolis: University of Minnesota Press, 1966.

38. Carnap, *Logical Foundations,* pp. 273–79.

39. Ibid., p. 257.

40. C. G. Hempel, "Inductive-Nomological versus Statistical Explanation," in H. Feigl and G. Maxwell, eds., *Minnesota Studies in the Philosophy of Science,* Vol. 3, pp. 149–69. Minneapolis: University of Minnesota Press, 1962.

41. J. Hintikka and J. Pietarinen, "Semantic Information and Inductive Logic," in J. Hintikka and Patrick Suppes, eds., *Aspects of Inductive Logic,* p. 97. Amsterdam: North-Holland Publishing Company, 1966.

42. Isaac Levi, *Hard Choices,* Cambridge: Cambridge University Press, 1986.

43. A. K. Sen, *Choice of the Welfare and Measurement.* Cambridge, Mass.: MIT Press, 1982.

44. H. Putnam, *Mathematics, Matter and Method: Philosophical Papers,* Vol. 1, p. 268. Cambridge: Cambridge University Press, 1975.

45. Ibid., p. 304.

46. Radu J. Bogdan, "Two Turns in Induction," in L. Jonathan Cohen and Mary Hesse, eds., *Applications of Inductive Logic,* p. 413. Oxford: Clarendon Press, 1980.

47. J. W. N. Watkins, *Science and Scepticism,* London: Hutchinson, 1984.

48. Imre Lakatos, *Methodology of Scientific Research Programmes: Philosophical Papers,* Vol. 1. Cambridge: Cambridge University Press, 1978.

49. Gerald Radnitzky and W. W. Bartley, III, eds., *Evolutionary Knowledge, Rationality and the Sociology of Knowledge,* p. 19. La Salle, Ill.: Open Court Publishing Co., 1987.

50. Alfred North Whitehead, *Process and Reality* (corrected edition), ed. D. R. Griffin and D. W. Sherburne, pp. 128–29, 199–207. New York: Free Press, 1978.

51. Alfred North Whitehead, *Adventure of Ideas,* p. 112. New York: Free Press, 1967.

Chapter Five. What Is Wrong With Skepticism?

1. *The Dialectical Method of Nagarjuna: Vigrahavyavartani,* ed. E. H. Johnston and Arnold Kunst, trans. by Kamaleshwar Bhattacharaya, p. 112. Delhi: Motilal Banarsidas, 1986.

2. Here Matilal refers to Ayer's formulation of skepticism as found in his book, *The Problem of Knowledge* (1955). See B. K. Matilal, *Perception: An Essay on Classical Indian Theories of Knowledge,* p. 50. Oxford: Clarendon Press, 1986.

3. *A Letter from a Gentleman to his Friend in Edinburgh,* ed. by Ernst C. Hassner and John V. Price, Edinburgh: Edinburgh University Press, 1967.

4. B. Russell, *Human Knowledge: Its Scope and Limits,* p. 9. London: Allen & Unwin, 1948.

5. David Hume, *Enquiries Concerning the Human Understanding,* ed. L. A. Selby-Bigge, 2d ed., p. 160. Oxford: Oxford University Press, 1902.

6. Russell, *Human Knowledge,* pp. 66–67.

7. Ibid., pp. 143–44.
8. Ibid., p. 516.

9. C. G. Hempel, *Aspects of Scientific Explanation.* New York: Free Press, 1965.

10. Matilal, *Perception,* p. 67.

11. Sextus Empiricus, *Outlines of Pyrrhonism,* 3 vols., trans. R. G. Bury, London & Camb., Mass., 1933–36, Book 2, Sec. 20.

12. Francis Bacon, *Novum Organum,* Book 1, Sec. 33 in J. Spedding, R. L. Ellis, and D. D. Heath, eds., *The Works of Francis Bacon,* London: Longman, 1857–59.

13. See *Outlines of Pyrrhonis,* Book 2, Sec. 188.

14. See ibid., Book 1, Sec. 208.

15. John Locke, *An Essay Concerning Human Understanding,* Book 4, Ch. 14, Sec. 1.

16. David Hume, *An Enquiry Concerning Human Understanding,* Sec. 12, Pt. 2.

17. Blaise Pascal, *Pascal's Pensees*. London: Routledge and Kegan Paul, 1950.

18. George Santayana, *Scepticism and Animal Faith*. New York: Dover, 1955.

19. David Hume, *A Treatise of Human Nature,* Book 1. "Of the Understanding," Book 2. "Of the Passions." Book 3. "Of Morals." L. A. Selby-Bigge, ed., Hume, *Enquiri*. Oxford: Oxford University Press, 1888.

20. Nicholas Rescher, *Scepticism: A Critical Reappraisal*. Oxford: Basil Blackwell, 1980.

21. *The Philosophical Works of Descartes,* trans. by E. S. Haldane and G. R. T. Ross, Vol. 1, p. 148. Cambridge: Cambridge University Press, 1911.

22. Robert E. Alexander, "The Problem of Metaphysical Doubt and Its Removal," in R. J. Butler, ed., *Cartesian Studies*. Oxford: Basil Blackwell, 1972.

23. L. J. Beck, *The Metaphysics of Descartes: A Study of the Meditations*. Oxford: Clarendon Press, 1965.

24. Anthony Kenny, *Descartes: A Study of His Philosophy.* New York: 1968.

25. Peter Unger, *Ignorance: A Case for Skepticism*. Oxford: Clarendon Press, 1975.

26. Edmund Husserl, *Cartesian Meditations: An Introduction to Phenomenology,* trans. Dorion Cairns. The Hague: Martinus Nijhoff, 1973.

27. G. E. Moore, *Philosophical Papers*. London: Allen & Unwin, 1959.

28. Barry Stroud, *The Significance of Philosophical Scepticism,* p. 38. Oxford: Clarendon Press, 1984.

29. Immanuel Kant, *Critique of Pure Reason,* trans. N. Kemp Smith, B.x.1. London: Macmillan, 1973.

30. Ibid., B 275.

31. Ibid., A 760–B 788.

32 Ibid., B 406.

33. Ibid., A 761–B 789.

34. Stroud, *Philosophical Scepticism,* p. 128.

35. Kant, *Pure Reason,* A 376, A386–87.

36. Ibid., B 275.

37. Ibid., B 276.

38. Ibid., A 11–12–B 25–26.

39. Ibid., A 12–B 26.

40. Ibid., A 371–A 378.

41. Both in the *Cartesian Meditations,* and in *The Crisis of European Sciences and Transcendental Phenomenology,* trans. David Carr. Evanston, Ill.: Northwestern University Press, 1970.

Chapter Six. Husserl and Popper on Skepticism: Some Problems

1. Edmund Husserl, *Logical Investigations,* Vol. 1, trans. J. N. Findlay, pp. 135–38. London: Routledge & Kegan Paul, 1970.

2. Edmund Husserl, *Cartesian Meditations: An Introduction to Phenomenology,* trans. Dorion Cairns, pp. 150–51. The Hague: Martinus Nijhoff, 1973.

3. Edmund Husserl, *The Crisis of European Sciences and Transcendental Phenomenology,* trans. David Carr, pp. 12–13. Evanston, Ill.: Northwestern University Press, 1970.

4. J. J. Kockelmans, "Husserl and Kant on the Pure Ego," in Frederick Elliston and Peter MacCormick, eds., *Husserl: Exposition and Appraisals.* Notre Dame: Ind.: University of Notre Dame Press, 1977.

5. Karl Popper, *Conjectures and Refutations: Growth of Scientific Knowledge,* p. 72. London: Routledge & Kegan Paul, 1969.

6. Karl Popper, *The Open Society and Its Enemies,* vol. 2, p. 309. London: Routledge & Kegan Paul, 1974.

7. Karl Popper, *Realism and the Aim of Science,* pp. 139–45. Totowa, N.J.: Rowman & Littlefield, 1983.

8. John Watkins, *Science and Scepticism,* pp. 249–254. London: Hutchinson, 1984.

9. Karl Popper, *The Logic of Scientific Discovery,* Ch. 5. London: Hutchinson, 1959.

10. Ibid., p. 96.

11. Rudolf Carnap, *The Logical Structure of the World and Pseudoproblems in Philosophy,* trans. Rolf A. George. London: Routledge & Kegan Paul, 1967.

12. Popper, *Scientific Discovery,* p. 104.

13. Ibid., p. 103.

14. Watkins, *Science and Scepticism,* p. 253.

15. A. J. Ayer, "Truth Verification and Verisimilitude," in P. A. Schilpp, ed., *The Philosophy of Karl Popper,* Vol. II, pp. 688–89. The Library of Living Philosophers, La Salle, Ill.: Open Court Publishing Co., 1974.

16. Ibid., p. 1114.

17. Watkins, *Science and Scepticism,* p. 260.

18. Ibid., pp. 261–62; see also pp. 276–78.

19. Popper, *Scientific Discovery,* p. 111.

20. Watkins, *Science and Scepticism,* p. 261.

21. Hilary Putnam, *Realism and Reason: Philosophical Papers*. Vol. 3, pp. 84, 86, 225–26. Cambridge: Cambridge University Press, 1986.

22. G. E. Moore, *Philosophical Papers*. London: Allen & Unwin, 1959.

23. W. V. O. Quine, *Word and Object,* p. 243. Cambridge, Mass.: MIT Press, 1960.

24. Ibid., p. 248.

25. Ibid., p. 263.

26. Hilary Putnam, *Mind, Language and Reality: Philosophical Papers,* Vol. 2, pp. 162–64; see also, pp. 177–78. Cambridge: Cambridge University Press, 1975.

27. Popper, *Open Society,* Vol. 2, pp. 13–15, 18, 290, 291.

28. Hilary Putnam, *Mathematics, Matter and Method: Philosophical Papers,* Vol. 1, pp. 254–55. Cambridge: Cambridge University Press, 1975.

29. P. K. Mukhopadhyay, *Indian Realism: A Rigorous Descriptive Metaphysics,* Calcutta: K. P. Bagchi & Co., 1984.

30. Gregory J. Darling, *An Evaluation of the Vedāntic Critique of Buddhism.* Delhi: Motilal Banarsidass, 1987.

31. Edmund Husserl, *Logical Investigations,* Vol. 1, p. 115.

32. Ibid., p. 140.

33. David Woodruff Smith and Ronald McIntyre, *Husserl and Intentionality,* Ch. 4. Dordrecht: D. Reidel, 1982. See also Husserl, *Ideas,* trans. W. R. Boyce Gibson, Secs. 89, 129, 133. New York: Collier Books, 1962.

34. Husserl, *Logical Investigations,* Vol. I, pp. 365–66, Vol. II, 624–25, 634–35.

35. Edmund Husserl, *Experience and Judgement: Investigations in a Genealogy of Logic,* ed. Ludwig Landgrebe, trans. by James S. Churchill and Karl Americs. Evanston, Ill.: Northwestern University Press, 1973.

36. Ibid., p. 92.

37. Quine, *Word and Object,* pp. 270–76.

38. *Vedānta-Śūtras: Commentary by Śaṁkara,* I.2.21, I.4.22, II.1.14, III.2.1–8, IV.1.4–5. See trans. George Thibaut, Indian edition. Delhi: Motilal Banarsidass, 1962. Part I, pp. 135–39, 278–83, 320–30; Part II, pp. 133–47, 340–45. Delhi: Motilal Banarsidass, 1962. See also, Karl H. Potter, *The Encyclopaedia of Indian Philosophy,* Vol. 3. *Advaita Vedanta,* pp. 69–71, 120–21, 367–70, 411–15. Delhi: Motilal Banarsidass, 1981.

39. Pandit Vindhyesvari Prasad Dvivedi, ed., *Brahmasūtra* with a Commentary by Bhaskarāchāry, p. 124. Benaras: Vidya Vilash Press, 1915. See also, Daniel H. Ingalls, "Śaṁkara's Argument against the Buddhist," *Philosophy East and West* 3, no. 4 (1954).

40. Vidhushekhara Bhattacharya, ed. and trans. *Āgamaśāstra of Gauḍapāda,* pp. 58, 164, 179. Calcutta: Calcutta University, 1943.

41. T. K. Balasubrahmanyam, *The Works of Śrī Śaṅkarachārya,* vol. 1. Srirangam: Sri Vani Vilash Press, n.d.: See also Darling, *Vedāntic Critique.*

42. Husserl, *Ideas,* p. 14.

43. Husserl, *The Crisis,* pp. 178–83.

44. Husserl, *Experience and Judgement,* p. 365.

45. Ibid, p. 311.

46. Husserl, *Ideas,* p. 96.

47. Husserl, *The Crisis,* pp. 186–87.

48. Ibid., pp. 180, 196–97.

49. Husserl, *Cartesian Meditations,* pp. 147, 149, 153.

50. Harrison Hall, "Was Husserl a Realist or an Idealist?" in Hubert L. Dreyfus, ed. *Husserl, Intentionality and Cognitive Science.* Cambridge, Mass.: MIT Press, 1982.

51. D. P. Chattopadhyaya, *Anthropology and Historiography of Science.* Athens, Ohio: Ohio University Press, 1990.

52. Husserl, *Cartesian Meditations,* p. 157.

Chapter Seven. For and Against Skepticism: Moore and Wittgenstein

1. "The Nature of Judgment," in G. E. Moore, *Essays in Retrospect,* ed. A. Ambrose and M. Lazerowitz, p. 177. London: Allen & Unwin, 1970.

2. "The Refutation of Idealism," in G. E. Moore, *Philosophical Studies,* pp. 5–12. London: Routledge & Kegan Paul, 1922.

3. "Defence of Commonsense," in G. E. Moore, *Philosophical Papers* p. 42. London and Humanities Press, New York: 1959. p. 45.

4. B. Russell, *An Inquiry into Meaning and Truth,* p. 13. Middlesex: Penguin, 1967.

5. G. E. Moore, "Proof of an External World," *Philosophical Papers,* pp. 143–46. London: Allen & Unwin, 1959. See also *Some Main Problems,* pp. 119–23.

6. G. E. Moore, *Philosophical Papers,* p. 250. Here I have freely drawn upon an admirable exposition of David O'Connor, *The Metaphysics of G. E. Moore.* Dordrecht: D. Reidel, 1982.

7. Hilary Putnam, *Realism and Reason: Philosophical Papers,* vol. 3, pp xviii–ixx, 225–26. Cambridge: Cambridge University Press, 1986.

8. Moore, *Philosophical Papers,* pp. 146–47.

9. S. A. Kripke, *Wittgenstein: On Rules and Private Language,* p. 106. Oxford: Basil Blackwell, 1983.

10. L. Wittgenstein, *On Certainty,* ed. G. E. M. Anscombe and G. H. von Wright, para 114. New York: Harper & Row, 1972.

11. Ibid., para. 108–11.

12. Ibid., Sec. *On Certainty.*

13. Ibid., para. 507.

14. Ibid., para. 509.

15. Ibid., para. 450.

16. Ibid., para. 449.

17. Ibid., para. 457–58.

18. Ibid., para. 501.

19. Anthony Kenny, *Wittgenstein,* p. 217–18. London: Penguin Press, 1973.

20. L. Wittgenstein, *Tractatus Logico-Philosophicus,* trans. D. F. Pears and B. F. McGuiness. London: Routledge & Kegan Paul, 1972.

21. Robert J. Fogelin, "Wittgenstein and Classical Scepticism," in Stuart Shankar, ed., *Ludwig Wittgenstein: Critical Assessments,* Vol. 2, London: Croom Helm, 1986.

22. George Berkeley, *The Principles of Human Knowledge,* Secs. 29–34.

23. Ibid., Sec. 40.

24. David Hume, *A Treatise of Human Nature,* ed. L. A. Selby-Bigge, Book 1, Part 4, Sec. 1. Oxford: Clarendon Press, 1888.

25. Ibid., Book 1, Part 3, Sec. 9, p. 124.

Chapter Eight. Examination of Some Views on Skepticism

1. Morris Kline, *Mathematics: The Loss of Certainty,* p. 218. Oxford: Oxford University Press, 1980.

2. Ibid., p. 291.

3. Bertrand Russell, *Portraits from Memory and Other Essays.* London: George Allen & Unwin, 1956.

4. D. P. Chattopadhyaya, "Models and Metaphors in Arts, Science and Mathematics," in *Knowledge, Freedom and Language,* Delhi: Motilal

Banarsidass, 1989. See also Jacques Hadamard, *The Psychology of Invention in the Mathematical Field.* Princeton, N.J., 1945.

5. *Remarks on the Foundations of Mathematics,* pp. 272–73. Oxford: Basil Blackwell, 1978.

6. Imre Lakatos, *Proofs and Refutations.* Cambridge: Cambridge University Press, 1976. See also William Asprary and Philip Kitcher, eds., *History and Philosophy of Modern Mathematics: Minnesota Studies in the Philosophy of Science,* vol. 11. Minneapolis: University of Minnesota Press, 1988.

7. Kline, *Mathematics,* p. 27.

8. Gaurinath Sastri, *The Philosophy of Word and Meaning: Some Indian Approaches with Special Reference to the Philosophy of Bhartṛhari.* Calcutta: Sanskrit College, 1959. See also Kunjunni Raja, *Indian Theories of Meaning.* Madras: Adyar Library, 1977.

9. George D. Romanos, *Quine and Analytic Philosophy,* pp. 102–4. Cambridge, Mass.: MIT Press, 1983.

10. W. V. O. Quine. *Ontological Relativity and Other Essays,* p. 72. New York: Columbia University Press, 1969.

11. Donald Davidson, "A Coherence Theory of Truth and Knowledge," in Ernest Lepore, ed., *Truth and Interpretations: Perspectives on the Philosophy of Donald Davidson,* p. 307. Oxford: Basil Blackwell, 1986.

12. Ibid., p. 327.

13. Ibid., p. 328.

14. W. V. O. Quine, *The Roots of Reference,* pp. 38–39. La Salle, Ill.: Open Court Publishing Co., 1974.

15. D. P. Chattopadhyaya, *History, Society and Polity.* Delhi: Macmillan, 1976. See also its enlarged 2nd Edn., *Sri Aurobindo* and *Karl Marx: Integral Sociolgy and Dialectical Sociology.* Delhi: Motilal Banarsidass, 1988.

16. Fred Dretske, *Knowledge and the Flow of Information.* Cambridge, Mass.: MIT Press, 1982. See, for criticism, Hilary Putnam's "Probability and the Mental," in D. P. Chattopadhyaya, ed., *Humans, Meanings and Existences. Jadavpur Studies in Philosophy* 5. Delhi: Macmillan, 1983.

17. D. P. Chattopadhyaya, "Bolzano and Frege: A Note on Ontology," in D. P. Chattopadhyaya and P. K. Sen, eds., *Logic, Ontology and Action: Jadavpur Studies in Philosophy* 1. Delhi: Macmillan, 1979.

18. D. P. Chattopadhyaya, *Individuals and Worlds: Essays in Anthropological Rationalism.* Delhi: Oxford University Press, 1976. See also

idem., *Knowledge, Freedom and Language*. Delhi: Motilal Banarsidass, 1989.

19. H. G. Gadamer, *Truth and Method,* p. 311. New York: Crossroad Publishing Co., 1978.

Chapter Nine. Hegel and Heidegger on Skepticism: Some Problems

1. H. G. Gadamer, *Truth and Method,* p. 311. New York: Crossroad Publishing Co., 1978.

2. Karl Popper, *The Logic of Scientific Discovery,* p. 30. London: Hutchinson, 1959.

3. Francis Bacon, *Novum Organum: Aphorisms Concerning the Interpretations of Nature and the Kingdom of Man,* Vol. 30, p. 107. *The Great Books of the Western World.* Chicago: University of Chicago Press, 1952.

4. Gadamer, *Truth and Method,* p. 317.

5. M. Heidegger, *Hegel's Concept of Experience,* trans. K. R. Dove, pp. 141–48. New York: Harper & Row, 1970.

6. Ibid., pp. 66–67.

7. Heidegger, *Hegel's Concept of Experience,* p. 90.

8. Ibid., p. 104.

9. Ibid., p. 129.

10. Ibid., pp. 130–31.

11. Martin Heidegger, *Being and Time,* trans. John Macquarrie and Edward Robinson, pp. 271–72. New York: Harper & Row, 1962.

12. Ibid., p. 269.

13. Ibid., p. 300.

14. H. G. Gadamer, *Hegel's Dialectic,* trans. P. C. Smith, p. 98. New Haven, Conn.: Yale University Press, 1976.

Chapter Ten. The Antiskeptical Ontology of Kant and Hegel and the Proskeptical Thesis of Duhem and Quine

1. D. P. Chattopadhyaya, *Individuals and Worlds: Essays in Anthropological Rationalism.* Delhi: Oxford University Press, 1976.

2. I. Kant, *Critique of Pure Reason,* trans. N. Kemp Smith, A686–87–B714–15. London: Macmillan, 1973.

3. Ibid., A689–B717.

4. G. W. Leibniz, *Discourses on Metaphysics—Correspondence with Arnauld–Monadology,* trans. G. R. Montgomery, pp. 33–34. La Salle, Ill.: Open Court Publishing Co., 1973.

5. G. W. Leibniz, *New Essays on Human Understanding,* trans. and ed. P. Remnant and J. Bennett, Ch. 6, pp. 399–400. Cambridge: Cambridge University Press, 1982.

6. C. D. Broad, *Leibniz: An Introduction,* ed. C. Lewy, pp. 133–34. Cambridge: Cambridge University Press, 1975.

7. G. W. F. Hegel, *The Phenomenology of Mind,* trans. J. B. Baillie, p. 219. London: George Allen & Unwin, 1966.

8. Kant, *Critique of Pure Reason,* A257–58–B313–14.

9. Ibid., A650–51–B678–79.

10. Hegel, *Phenomenology of Mind,* p. 140.

11. Ibid., pp. 246–50.

12. Kant, *Critique of Pure Reason,* A658–68–B686–96. See also his *Critique of Judgment,* trans. J. H. Bernard, pp. 327–39. New York: Hafner Publishing Company, 1972.

13. Pierre Duhem, *The Aim and Structure of Physical Theory,* Ch. 1. New York: Atheneum Publishers, 1962.

14. W. V. O. Quine, *From a Logical Point of View.* Cambridge, Mass.: Harvard University Press, 1953. See also *Word and Object,* p. 67. Cambridge, Mass.: MIT Press, 1975.

15. Duhem, *Aim and Structure,* pp. 18, 335.

16. Kant, *Critique of Pure Reason,* A294–295–B350–351.

17. See, for example, Thomas Reid, *Inquiry into the Human Mind* (1764); and S. A. Grave's *The Scottish Philosophy of Commonsense.* Oxford: 1960.

18. R. B. Braithwaite, *Scientific Explanation,* p. 294. New York: Harper & Row, 1960.

19. J. S. Mill, *Auguste Comte and Positivism,* London, 1965, p. 57. Quoted in Gerd Buchdahl, *Metaphysics and the Philosophy of Science,* p. 27. Oxford: Basil Blackwell, 1969.

20. J. M. Keynes, *A Treatise on Probability,* p. 222. New York: Harper & Row, 1962.

21. Ibid., p. 234.

22. Gilbert Ryle, *The Concept of Mind,* pp. 116–17. Harmondsworth: Penguin Books, 1963.

23. Rudolf Carnap, *Logical Foundations of Probability,* p. 575. Chicago: University of Chicago Press, 1963.

24. Hans Reichenbach, *The Rise of Scientific Philosophy,* p. 239. Berkeley: University of California Press, 1951.

25. Ibid., p. 246.

26. Ibid., p. 266. See also Reichenbach, *The Philosophy of Space and Time,* pp. 287–88. New York: Dover Books, 1958.

27. W. V. O. Quine, *Theories and Things,* pp. 70–71. Cambridge, Mass.: Harvard (Belknap) Press, 1981.

28. Stanley L. Jaki, *Uneasy Genius: The Life and Works of Pierre Duhem,* pp. 368–70. Dordrecht: Martinus Nijhoff, 1987.

29. Duhem, *Aim and Structure,* p. 147.

30. Ibid., p. 137.

31. Ibid., p. 143.

32. Ibid., pp. 158–59.

33. Ibid., p. 216; see also pp. 183, 187, 206–7.

34. W. V. Quine, *Philosophy of Logic,* pp. 95–102. Englewood Cliffs, N.J.: Prentice-Hall, 1970. Hilary Putnam, *Realism and Reason,* pp. 81–86. Cambridge: Cambridge University Press, 1983. Michael Dummett, *Truth and Other Enigmas,* pp. xxix–xxxii, 173–76, 275–78, 358–61. London: Duckworth, 1978.

35. Dummett, *Truth,* p. 289.

36. See, for example, the papers of Donald Davidson, Richard Rorty, Collin McGinn, Peter D. Klein, and Ernest Sosa in Ernest Lepore, ed., *Truth and Interpretation: Perspectives on the Philosophy of Donald Davidson.* Oxford: Basil Blackwell, 1986.

Chapter Eleven. Between Dogmatism and Skepticism

1. The intimate relation between willing and knowing has been emphasized by philosophers of widely different persuasions like Buddhists and

pragmatists. Whereas the former point out the "will" or "desire" element underlying cognitive enterprise, the latter highlight how in determining the truth value of knowledge we in effect are involved in acts of evaluation.

In different ways, C. I. Lewis, Stuart Hampshire, and Jürgen Habermas, among others, have persuasively drawn our attention to the intimate relation obtained between interest, knowledge, and action. See, for example, Lewis, *An Analysis of Knowledge and Valuation,* pp. 366–78. La Salle, Ill.: Open Court Publishing Co., 1971. Hampshire, *Thought and Action,* pp. 270–73. London: Chatto and Windus, 1970. Habermas, *Knowledge and Human Interests,* pp. 191–213. Boston: Beacon Press, 1971.

2. G. W. Leibniz, *New Essays on Human Understanding* (abridged ed.), trans. and ed. P. Remnant and J. Bennett, pp. 361–89. Cambridge: Cambridge University Press, 1982.

3. See, for example, Anthony Kenny, "Descartes on the Will" and E. J. Ashworth, "Descartes' Theory of Clear and Distinct Ideas," in R. J. Butler, ed., *Cartesian Studies*. Oxford: Basil Blackwell, 1972.

4. D. P. Chattopadhyaya, *Individuals and Worlds: Essays in Anthropological Rationalism,* Ch. 2, "Reason: Sovereign, Autonomous and Human." Delhi: Oxford University Press, 1976.

5. The fallibilists I have in view are not only Heracletus and Sextus, Nāgārjuna and Dharmakīrti, but also Hume, Peirce, Popper, and Ricoeur. Needless to say the skeptical issues they are concerned with are not identical, not even at bottom.

6. *The Nyāya Sūtras of Gotama,* trans. Satish Chandra Vidyabhusana, pp. 7–8. New Delhi: Oriental Books Reprint Corporation, 1975.

7. T. R. V. Murti, *The Central Philosophy of Buddhism,* pp. 126–31. London: Unwin Paperbacks, 1980.

8. F. T. Stcherbatsky, *The Conception of Buddhist Nirvana* (with the Sanskrit Text of Mādhyamika-Kārikā), pp. 27–41. Varanasi: Bharatiya Vidya Prakashan, 1975.

9. F. T. Stcherbatsky, *Buddhist Logic,* Vol. 1, p. 512. New York: Dover Publications (no date).

10. See Stcherbatsky's *Buddhist Nirvāṇa,* pp. 52, 40ff, 99ff, 107, 111–13, 141-43, 168. According to Nagarjuna: "the universe viewed as a whole is the Absolute, viewed as a process it is the phenomenal," p. 52.

11. Jonathan Bennett, *Locke, Berkeley and Hume: Central Themes,* pp. 181. Oxford: Clarendon Press, 1971. See also George Berkeley, *Principles of Human Knowledge,* in *Great Books of the Western World,* Vol. 35, pp. 420–22, 442–44. Chicago: University of Chicago Press, 1952. And Bennett's pa-

per, "Berkeley and God," in C. B. Martin and D. H. Armstrong, eds., *Locke and Berkeley.* London: Macmillan, (no date).

12. Quoted by Richard H. Popkin, "David Hume: His Pyrrhonism and his Critique of Pyrrhonism," p. 53, in *Hume,* ed. V. C. Chappell. London: Macmillan, 1968.

13. David Hume, *A Treatise of Hume Nature,* ed. L. A. Selby-Bigge, p. 218. Oxford: Clarendon Press, 1888.

14. D. P. Chattopadhyaya, *Individuals and Societies: A Methodological Inquiry,* 2nd ed., pp. 109ff, 157, 162, 266–67. Calcutta: Scientific Book Agency, 1975.

15. D. P. Chattopadhyaya, *Environment, Evolution and Values: Studies in Man, Society and Science,* pp. 96, 136–37. New Delhi: South Asian Publishers, 1982.

16. Karl Potter, eds., *Encyclopaedia of Indian Philosophies: Nyāya Vaiśeṣika,* pp. 158–59. Delhi: Motilal Banarsidass, 1977. See also, pp. 154–60, 313–14, 456, 506, 543–44, 579–80.

17. I. Kant, *Critique of Pure Reason,* A671–B699.

18. Ibid., A697–703–B724–731.

19. See "Copernicus Betrayed," in D. P. Chattopadhyaya, *Individuals and Worlds: Essays in Anthropological Rationalism.* Delhi: Oxford University Press, 1976.

20. David Hume, *Dialogues Concerning Natural Religion,* ed. Norman Kemp Smith, 2d ed., p. 219, nn. London: 1947.

21. I. Kant, *Philosophical Correspondence,* ed. and trans. A. Zweig, p. 252. Chicago: University of Chicago Press, 1967. Quoted in Ermanno Vencivengo, *Kant's Copernican Revolution,* p. 169. New York and Oxford: Oxford University Press, 1987.

22. Kant, *Critique of Pure Reason,* A506–7–B534–35.

23. Ibid., A vii–viii.

24. Ibid., A216–18–B263–65.

25. Ibid., A408–B434 and A681–B709.

26. Ibid., A277–B333.

27. Ibid., A133–34–B172–73.

28. Ibid., A472(76)–B503–4.

29. Vencivengo, *Kant's Copernican Revolution.*

30. Chattopadhyaya, *Individuals and Worlds.*

31. Edmund Husserl, *Cartesian Meditations,* trans. by Dorion Cairns, p. 54. The Hague: Martinus Nijhoff, 1973.

32. Ibid., p. 63–64.

33. Edmund Husserl, *Experience and Judgement,* ed. Ludwig Landgrebe, trans. James Churchill and Karl Amerika, pp. 32–22, 44–45. Evanston, Ill.: Northwestern University Press, 1973.

34. Husserl, *Cartesian Meditations,* pp. 68–72.

35. Ibid., pp. 115–16, 120–21, 150–51.

36. Edmund Husserl, *The Crisis of European Sciences and Transcendental Phenomenology: An Introduction to Phenomenological Philosophy,* trans. David Carr, p. 71. Evanston, Ill.: Northwestern University Press, 1970.

37. Ibid., p. 108.

38. I have discussed this question elsewhere. See, for example, D. P. Chattopadhyaya, *Environment, Evolution and Values: Studies in Man, Society and Science,* pp. 24–41. New Delhi: South Asian Publishers, 1982.

39. Husserl, *Cartesian Meditations,* p. 149.

40. Ibid., p. 152.

41. Leszek Kolakowski, *Husserl and the Search for Certitude,* pp. 52–57. Chicago: University of Chicago Press, 1987.

42. D. P. Chattopadhyaya, Lester Embree, and J. N. Mohanty, eds. *Phenomenology and Indian Philosophy.* Proceedings of the Joint Conference of the Indian Council of Philosophical Research (ICPR) and the Centre of Advanced Research in Phenomenology (CARP) held at New Delhi in January 1988. See, in particular, Karl Schuhmann, "Husserl and Indian Thought." Delhi: Motilal Banarsidass, 1991.

43. Imre Lakatos, *Proof and Refutations: The Logic of Mathematical Discovery,* ed. John Worrall and Elie Zahar, pp. 14–23, 30–33, 42–43, 83–86. Cambridge: Cambridge University Press, 1976.

44. See, for example, IIya Prigogine, *From Being to Becoming: Time and Complexity in the Physical Sciences.* San Francisco: W. H. Freeman, 1980. Karl Popper, *The Open Universe,* ed. W. W. Bartley, III. Totowa, N.J.: Rowman & Littlefield, 1982. Joseph Margolis, *Science without Unity: Reconciling the Human and Natural Sciences,* Oxford: Basil Blackwell, 1987.

Chapter Twelve. Skepticism as the Critique of Search: An Epilogue

1. L. Koenigsberger, *Hermann von Helmholtz,* trans. F. A. Welby, Oxford: Clarendon Press, 1906. Quoted in W. I. B. Beveridge, *The Art of Scientific Investigation,* p. 81. New York: Vintage Books, n.d.

2. Henry Poincaré, *Science and Method.* New York: Dover Publications, n.d.

3. Arthur I. Miller, *Imagery in Scientific Thought,* pp. 242–48. Cambridge, Mass.: MIT Press, 1987.

4. Max Black, "Notes on the 'Paradoxes of Confirmation'," in Jaakko Hintikka and Patrick Suppes, *Aspects of Inductive Logic,* pp. 182–85. Amsterdam: North Holland, 1966.

5. See, for example, the papers of Wesley C. Salmon and Clark Glymour in *The Concept of Evidence,* ed. Peter Achinstein. Oxford: Oxford University Press, 1983.

6. Joseph Margolis, *Science without Unity.* Oxford: Basil Blackwell, 1987.

7. D. P. Chattopadhyaya, *Knowledge, Freedom and Language: An Interwoven Fabric of Man, Time and World,* Chs. 1, 2, and 11. Delhi: Motilal Banarsidass, 1989.

8. Ibid., Ch. 12, "Bolzano and Frege: A Note on Ontology." See also, the papers of Karl Popper, W. W. Bartley III, and Donald Campbell in *Evolutionary Epistemology: Rationality and the Sociology of Knowledge,* ed. Gerard Radnitzky and W. W. Bartley III. La Salle, Ill.: Open Court Publishing Co., 1987.

9. See, for example, the interesting work, Patrick A. Heelan, *Space-Perception and Philosophy of Science,* pp. 8–13, 147–54, 168–72. Berkeley: University of California Press, 1983.

10. *Vedānta-Sūtras, Śaṁkara's Commentary,* trans. George Thibaut, *Sacred Books of the East,* Vol. 34, I.I, 10 and 11, pp. 60–64. Delhi: Motilal Banarsidass, 1968.

11. I. Kant, *Critique of Pure Reason,* trans. N. Kemp Smith, A848–51–B876–79. London: Macmillan, 1973.

Chapter Twelve: Skepticism as the Crucible of Science: An Epilogue

1. L. Koenigsberger, *Hermann von Helmholtz*, trans. F. A. Welby, Oxford: Clarendon Press, 1906. Quoted in W. I. B. Beveridge, *The Art of Scientific Investigation*, p. 6. New York: Vintage Books, n.d.

2. Henry Poincaré, *Science and Method*, New York: Dover Publications, n.d.

3. Arthur I. Miller, *Imagery in Scientific Thought*, pp. 242–48. Cambridge, Mass.: MIT Press, 1987.

4. Max Black, "Notes on the Paradoxes of Confirmation," in Jaakko Hintikka and Patrick Suppes, *Aspects of Inductive Logic*, pp. 182–85. Amsterdam: North Holland, 1966.

5. See, for example, the papers of Wesley C. Salmon and Clark Glymour in *The Concept of Evidence*, ed. Peter Achinstein, Oxford: Oxford University Press, 1983.

6. Joseph Agassi, *Science in Flux*, [illegible]. Oxford: Basil Blackwell, 1987.

7. D. P. Chattopadhyaya, *Knowledge, Freedom and Language: An Interwoven Fabric of Man, Time and World*, Chs. 1, 2, and 11. Delhi: Motilal Banarsidass, 1989.

8. Ibid., Ch. 12: "Bolzano and Frege: A Note on Ontology". See also the papers of Karl Popper, W. W. Bartley III and Donald Campbell in *Evolutionary Epistemology, Rationality and the Sociology of Knowledge*, ed. Gerard Radnitzky and W. W. Bartley III, La Salle, Ill.: Open Court Publishing Co., 1987.

9. See, for example, the interesting work of Patrick A. Heelan, *Space-Perception and the Philosophy of Science*, pp. 3–15, 147–51, 168–192. Berkeley: University of California Press, 1983.

10. *Vedānta-Sūtras*, *Saṃkara's Commentary*, trans. George Thibaut, *Sacred Books of the East*, Vol. 34, I.[illegible] and II, pp. 60–[illegible]. Delhi: Motilal Banarsidass, 1988.

11. I. Kant, *Critique of Pure Reason*, trans. N. Kemp Smith, A838–[illegible], B866–[illegible]. London: Macmillan, 1973.

Index of Names

Index of Subjects